1 – 19 PROGRESSION

Endorsed for
Pearson Edexcel Qualifications

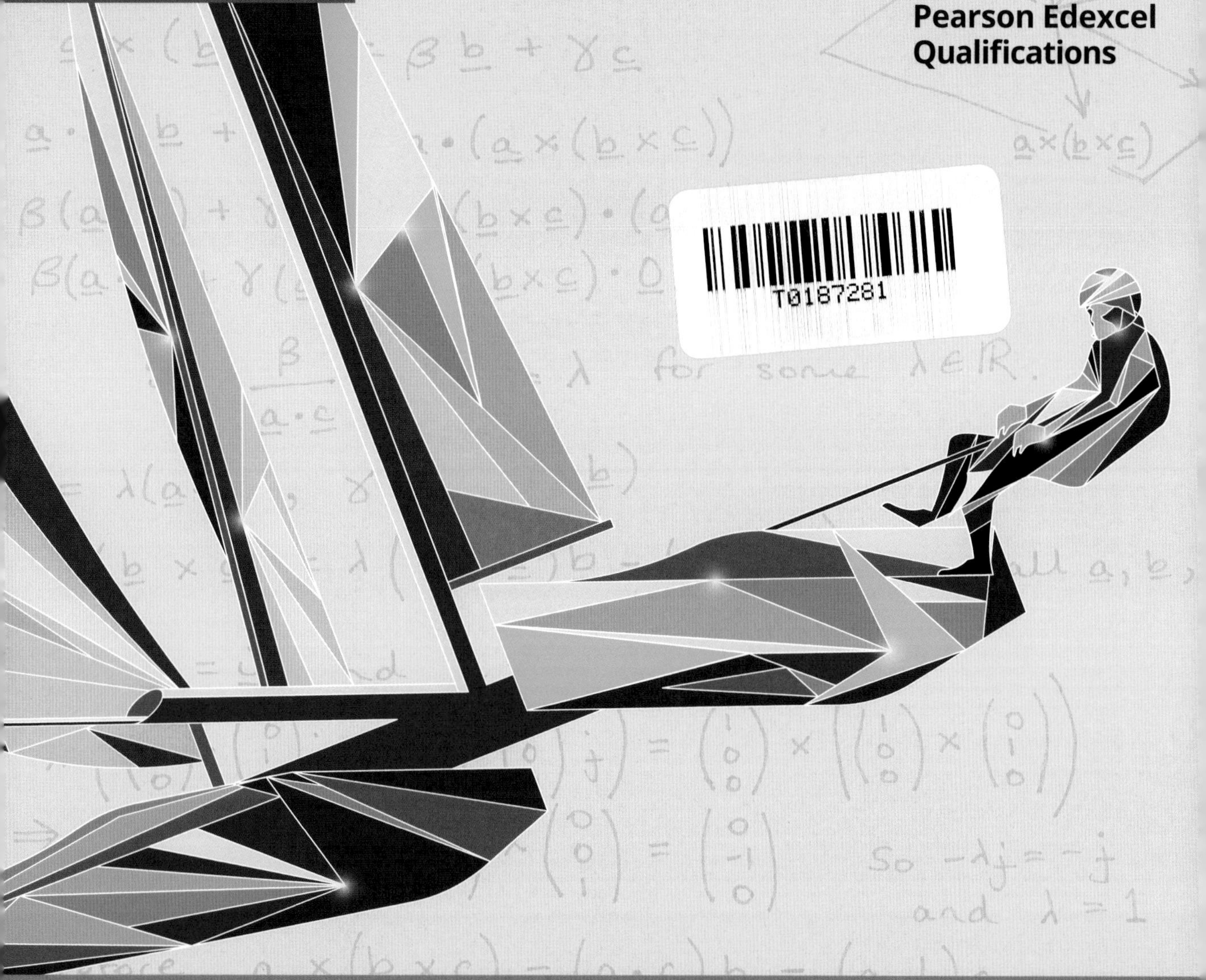

Pearson Edexcel AS and A level Further Mathematics

Further Pure Mathematics 1
FP1

Series Editor: Harry Smith
Authors: Greg Attwood, Ian Bettison, Jack Barraclough, Tom Begley, Lee Cope, Bronwen Moran, Laurence Pateman, Keith Pledger, Harry Smith, Geoff Staley, Dave Wilkins

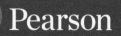

Pearson

Published by Pearson Education Limited, 80 Strand, London WC2R 0RL.

www.pearsonschoolsandfecolleges.co.uk

Copies of official specifications for all Pearson qualifications may be found on the website: qualifications.pearson.com

Edited by Tech-Set Ltd, Gateshead
Typeset by Tech-Set Ltd, Gateshead

Cover illustration Marcus@kja-artists

First published 2018

24
10 9

British Library Cataloguing in Publication Data
A catalogue record for this book is available from the British Library

ISBN 978 1292 18335 0

Printed in the UK by Bell and Bain Ltd, Glasgow

Acknowledgements
The authors and publisher would like to thank the following for their kind permission to reproduce their photographs:

(Key: b-bottom; c-centre; l-left; r-right; t-top)

123RF: Destinacigdem 62, 105cr; **Alamy Stock Photo:** Steve Morgan, 32, 105cl, Kevin Britland 92, 105r; **Getty Images:** Andrzej Wojcicki/Science Photo Library 01, 105l; **Shutterstock:** Chaikom 116, 191l, Jag_cz 131, 191cl, SamJonah 149, 191c, Bart Sadowski 161, 191cr, Sky Antonio, 179, 191r

All other images © Pearson Education

A note from the publisher

In order to ensure that this resource offers high-quality support for the associated Pearson qualification, it has been through a review process by the awarding body. This process confirms that this resource fully covers the teaching and learning content of the specification or part of a specification at which it is aimed. It also confirms that it demonstrates an appropriate balance between the development of subject skills, knowledge and understanding, in addition to preparation for assessment.

Endorsement does not cover any guidance on assessment activities or processes (e.g. practice questions or advice on how to answer assessment questions), included in the resource nor does it prescribe any particular approach to the teaching or delivery of a related course.

While the publishers have made every attempt to ensure that advice on the qualification and its assessment is accurate, the official specification and associated assessment guidance materials are the only authoritative source of information and should always be referred to for definitive guidance.

Pearson examiners have not contributed to any sections in this resource relevant to examination papers for which they have responsibility.

Examiners will not use endorsed resources as a source of material for any assessment set by Pearson.

Endorsement of a resource does not mean that the resource is required to achieve this Pearson qualification, nor does it mean that it is the only suitable material available to support the qualification, and any resource lists produced by the awarding body shall include this and other appropriate resources.

Pearson has robust editorial processes, including answer and fact checks, to ensure the accuracy of the content in this publication, and every effort is made to ensure this publication is free of errors. We are, however, only human, and occasionally errors do occur. Pearson is not liable for any misunderstandings that arise as a result of errors in this publication, but it is our priority to ensure that the content is accurate. If you spot an error, please do contact us at resourcescorrections@pearson.com so we can make sure it is corrected.

Contents

● = A level only

Overarching themes

The following three overarching themes have been fully integrated throughout the Pearson Edexcel AS and A level Mathematics series, so they can be applied alongside your learning and practice.

1. Mathematical argument, language and proof

- Rigorous and consistent approach throughout
- Notation boxes explain key mathematical language and symbols
- Dedicated sections on mathematical proof explain key principles and strategies
- Opportunities to critique arguments and justify methods

2. Mathematical problem solving

- Hundreds of problem-solving questions, fully integrated into the main exercises
- Problem-solving boxes provide tips and strategies
- Structured and unstructured questions to build confidence
- Challenge boxes provide extra stretch

The Mathematical Problem-solving cycle

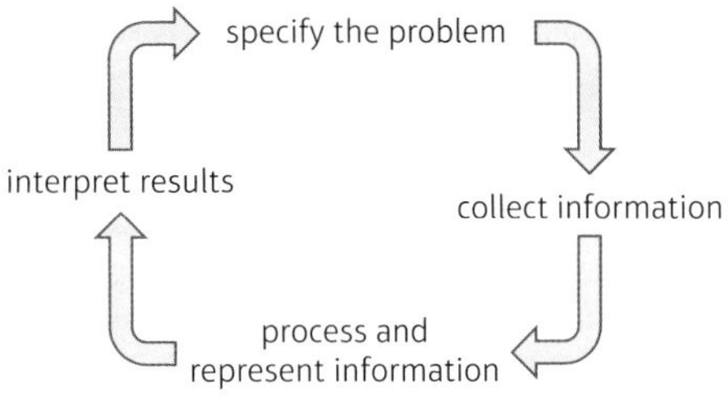

3. Mathematical modelling

- Dedicated modelling sections in relevant topics provide plenty of practice where you need it
- Examples and exercises include qualitative questions that allow you to interpret answers in the context of the model
- Dedicated chapter in Statistics & Mechanics Year 1/AS explains the principles of modelling in mechanics

Finding your way around the book

Access an online digital edition using the code at the front of the book.

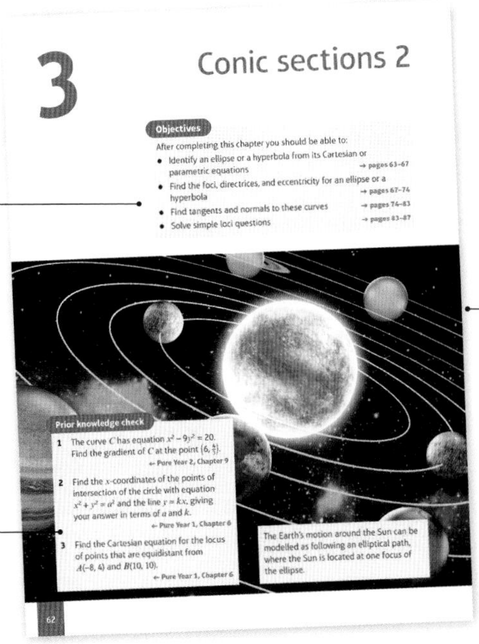

3 Conic sections 2

Objectives

After completing this chapter you should be able to:

- Identify an ellipse or a hyperbola from its Cartesian or parametric equations → pages 63–67
- Find the foci, directrices, and eccentricity for an ellipse or a hyperbola → pages 67–74
- Find tangents and normals to these curves → pages 74–83
- Solve simple loci questions → pages 83–87

Prior knowledge check

1 The curve C has equation $x^2 - 9y^2 = 20$. Find the gradient of C at the point $(6, \frac{4}{3})$. ← Pure Year 2, Chapter 9

2 Find the x-coordinates of the points of intersection of the circle with equation $x^2 + y^2 = a^2$ and the line $y = kx$, giving your answer in terms of a and k. ← Pure Year 1, Chapter 6

3 Find the Cartesian equation for the locus of points that are equidistant from $A(-8, 4)$ and $B(10, 10)$. ← Pure Year 1, Chapter 6

The Earth's motion around the Sun can be modelled as following an elliptical path, where the Sun is located at one focus of the ellipse.

62

Each chapter starts with a list of objectives

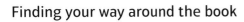

The real world applications of the maths you are about to learn are highlighted at the start of the chapter with links to relevant questions in the chapter

The *Prior knowledge check* helps make sure you are ready to start the chapter

A level content is clearly flagged

Exercise questions are carefully graded so they increase in difficulty and gradually bring you up to exam standard

Exercises are packed with exam-style questions to ensure you are ready for the exams

Challenge boxes give you a chance to tackle some more difficult questions

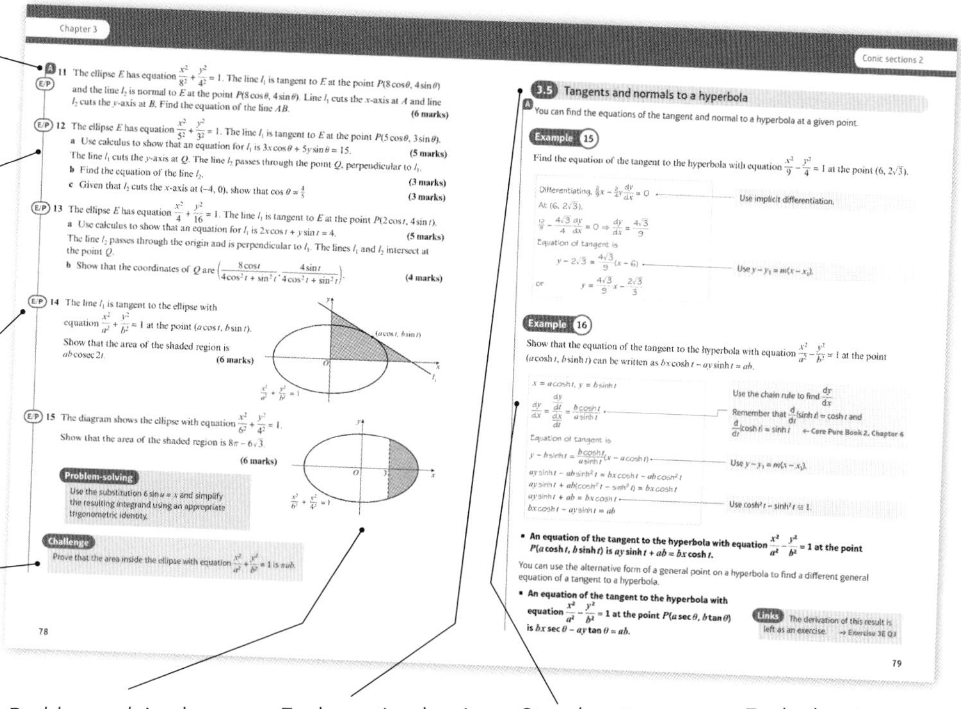

Exam-style questions are flagged with Ⓔ

Problem-solving questions are flagged with Ⓟ

Problem-solving boxes provide hints, tips and strategies, and *Watch out* boxes highlight areas where students often lose marks in their exams

Each section begins with explanation and key learning points

Step-by-step worked examples focus on the key types of questions you'll need to tackle

Each chapter ends with a *Mixed exercise* and a *Summary of key points*

Every few chapters a *Review exercise* helps you consolidate your learning with lots of exam-style questions

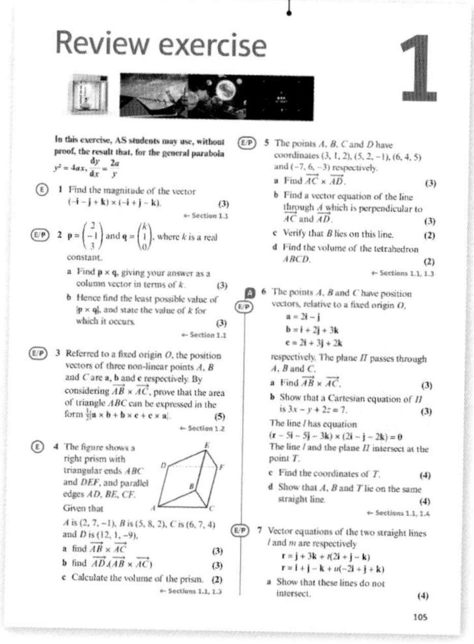

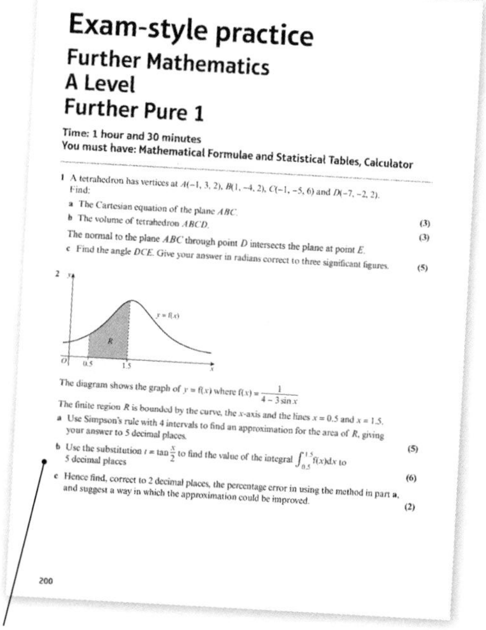

AS and A level practice papers at the back of the book help you prepare for the real thing.

Extra online content

Whenever you see an *Online* box, it means that there is extra online content available to support you.

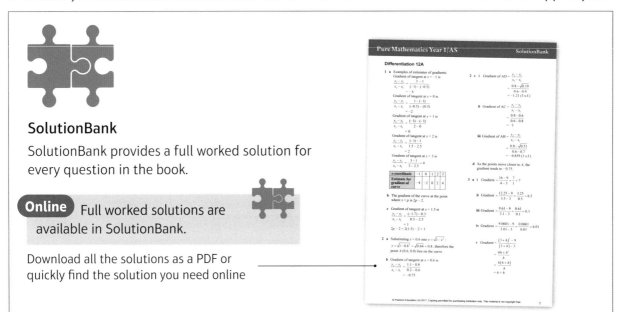

SolutionBank

SolutionBank provides a full worked solution for every question in the book.

Online Full worked solutions are available in SolutionBank.

Download all the solutions as a PDF or quickly find the solution you need online

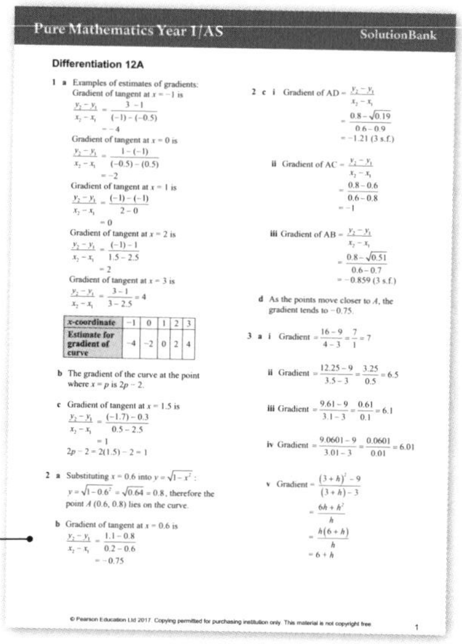

Pure Mathematics Year 1/AS   SolutionBank

Differentiation 12A

1 a Examples of estimates of gradients:
Gradient of tangent at $x = -1$ is
$\frac{y_2 - y_1}{x_2 - x_1} = \frac{3 - 1}{(-1) - (-0.5)} = -4$
Gradient of tangent at $x = 0$ is
$\frac{y_2 - y_1}{x_2 - x_1} = \frac{1 - (-1)}{(-0.5) - (0.5)} = -2$
Gradient of tangent at $x = 1$ is
$\frac{y_2 - y_1}{x_2 - x_1} = \frac{(-1) - (-1)}{2 - 0} = 0$
Gradient of tangent at $x = 2$ is
$\frac{y_2 - y_1}{x_2 - x_1} = \frac{(-1) - 1}{1.5 - 2.5} = 2$
Gradient of tangent at $x = 3$ is
$\frac{y_2 - y_1}{x_2 - x_1} = \frac{3 - 1}{3 - 2.5} = 4$

x-coordinate	−1	0	1	2	3
Estimate for gradient of curve	−4	−2	0	2	4

b The gradient of the curve at the point where $x = p$ is $2p - 2$.

c Gradient of tangent at $x = 1.5$ is
$\frac{y_2 - y_1}{x_2 - x_1} = \frac{(-1.7) - 0.3}{0.5 - 2.5} = 1$
$2p - 2 = 2(1.5) - 2 = 1$

2 a Substituting $x = 0.6$ into $y = \sqrt{1 - x^2}$:
$y = \sqrt{1 - 0.6^2} = \sqrt{0.64} = 0.8$, therefore the point A (0.6, 0.8) lies on the curve.

b Gradient of tangent at $x = 0.6$ is
$\frac{y_2 - y_1}{x_2 - x_1} = \frac{1.1 - 0.8}{0.2 - 0.6} = -0.75$

2 c i Gradient of AD $= \frac{y_2 - y_1}{x_2 - x_1} = \frac{0.8 - \sqrt{0.19}}{0.6 - 0.9} = -1.21$ (3 s.f.)

ii Gradient of AC $= \frac{y_2 - y_1}{x_2 - x_1} = \frac{0.8 - 0.6}{0.6 - 0.8} = -1$

iii Gradient of AB $= \frac{y_2 - y_1}{x_2 - x_1} = \frac{0.8 - \sqrt{0.51}}{0.6 - 0.7} = -0.859$ (3 s.f.)

d As the points move closer to A, the gradient tends to −0.75.

3 a i Gradient $= \frac{16 - 9}{4 - 3} = \frac{7}{1} = 7$

ii Gradient $= \frac{12.25 - 9}{3.5 - 3} = \frac{3.25}{0.5} = 6.5$

iii Gradient $= \frac{9.61 - 9}{3.1 - 3} = \frac{0.61}{0.1} = 6.1$

iv Gradient $= \frac{9.0601 - 9}{3.01 - 3} = \frac{0.0601}{0.01} = 6.01$

v Gradient $= \frac{(3 + h)^2 - 9}{(3 + h) - 3} = \frac{6h + h^2}{h} = \frac{h(6 + h)}{h} = 6 + h$

© Pearson Education Ltd 2017. Copying permitted for purchasing institution only. This material is not copyright free.   1

Use of technology

Explore topics in more detail, visualise problems and consolidate your understanding using pre-made GeoGebra activities.

Online Find the point of intersection graphically using technology.

GeoGebra

GeoGebra-powered interactives

Interact with the maths you are learning using GeoGebra's easy-to-use tools

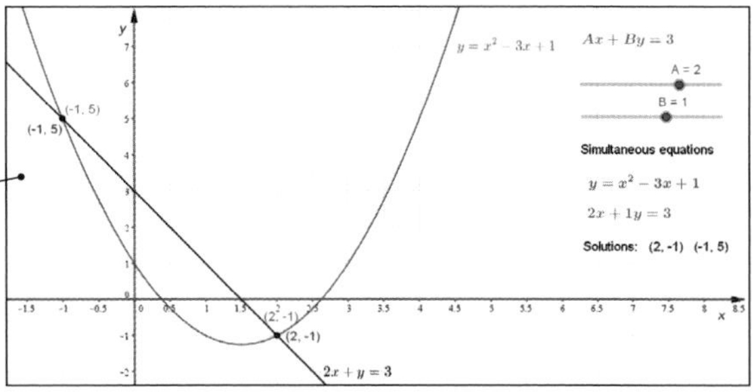

Access all the extra online content for free at:

www.pearsonschools.co.uk/fp1maths

You can also access the extra online content by scanning this QR code:

Vectors

1

Objectives

After completing this chapter you should be able to:

- Find the vector product $\mathbf{a} \times \mathbf{b}$ of two vectors $\mathbf{a}$ and $\mathbf{b}$ → pages 2–6
- Interpret $|\mathbf{a} \times \mathbf{b}|$ as an area → pages 7–11
- Find the scalar triple product $\mathbf{a}.\mathbf{b} \times \mathbf{c}$ of three vectors $\mathbf{a}$, $\mathbf{b}$ and $\mathbf{c}$, and be able to interpret it as a volume → pages 11–16
- Write the vector equation of a line in the form $(\mathbf{r} - \mathbf{a}) \times \mathbf{b} = \mathbf{0}$ → pages 16–20
- Find the direction ratios and direction cosines of a line → pages 17–20
- Use vectors in problems involving points, lines and planes and use the equivalent Cartesian forms for the equations of lines and planes → pages 20–25

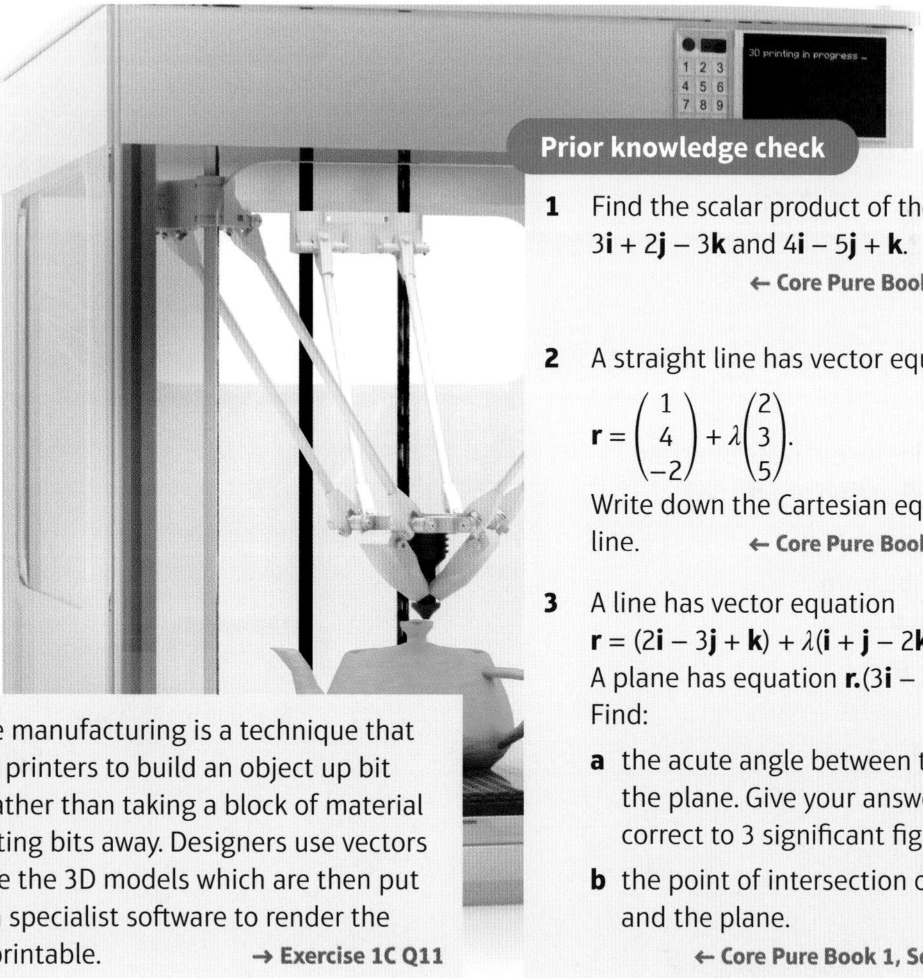

Additive manufacturing is a technique that uses 3D printers to build an object up bit by bit rather than taking a block of material and cutting bits away. Designers use vectors to create the 3D models which are then put through specialist software to render the object printable. → Exercise 1C Q11

Prior knowledge check

1 Find the scalar product of the vectors $3\mathbf{i} + 2\mathbf{j} - 3\mathbf{k}$ and $4\mathbf{i} - 5\mathbf{j} + \mathbf{k}$.
← Core Pure Book 1, Section 9.3

2 A straight line has vector equation

$$\mathbf{r} = \begin{pmatrix} 1 \\ 4 \\ -2 \end{pmatrix} + \lambda\begin{pmatrix} 2 \\ 3 \\ 5 \end{pmatrix}.$$

Write down the Cartesian equation of the line. ← Core Pure Book 1, Section 9.1

3 A line has vector equation $\mathbf{r} = (2\mathbf{i} - 3\mathbf{j} + \mathbf{k}) + \lambda(\mathbf{i} + \mathbf{j} - 2\mathbf{k})$.
A plane has equation $\mathbf{r}.(3\mathbf{i} - 2\mathbf{j} + 2\mathbf{k}) = 2$.
Find:

a the acute angle between the line and the plane. Give your answer in radians correct to 3 significant figures.

b the point of intersection of the line and the plane.
← Core Pure Book 1, Sections 9.4, 9.5

1.1 Vector product

You have already encountered the **scalar** (or **dot**) **product** of two vectors.

The scalar (or dot) product of two vectors **a** and **b** is written as **a.b**, and defined as

$\mathbf{a}.\mathbf{b} = |\mathbf{a}||\mathbf{b}|\cos\theta$,

where θ is the angle between **a** and **b**.

The scalar product produces a number (or scalar) as an answer. It is useful to define a second type of product that gives an answer as a vector.

- **The vector (or cross) product of the vectors a and b is defined as**

  $\mathbf{a} \times \mathbf{b} = |\mathbf{a}||\mathbf{b}|\sin\theta\,\hat{\mathbf{n}}$

  where θ is the angle between a and b.

Links If $\mathbf{a} = \begin{pmatrix} x_1 \\ y_1 \\ z_1 \end{pmatrix}$ and $\mathbf{b} = \begin{pmatrix} x_2 \\ y_2 \\ z_2 \end{pmatrix}$

then $\mathbf{a}.\mathbf{b} = x_1x_2 + y_1y_2 + z_1z_2$.

← **Core Pure Book 1, Chapter 9**

Online Use GeoGebra to explore the cross product of two vectors.

Notation $\hat{\mathbf{n}}$ is the unit vector that is perpendicular to both **a** and **b**.

Since $0 \leqslant \theta \leqslant 180°$, $|\mathbf{a}||\mathbf{b}|\sin\theta$ is a positive scalar quantity. This means that $\mathbf{a} \times \mathbf{b}$ is a vector quantity with magnitude $|\mathbf{a}||\mathbf{b}|\sin\theta$ that acts in the direction of $\hat{\mathbf{n}}$.

The direction of $\hat{\mathbf{n}}$ is that in which a right-handed screw would move when turned from **a** to **b**.

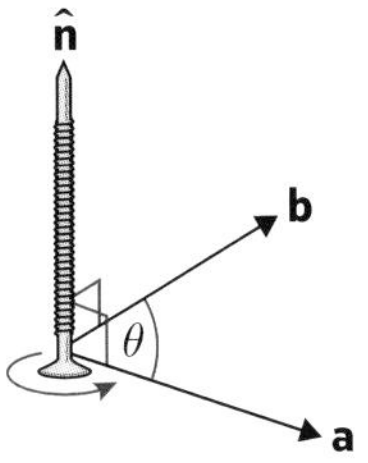

If the turn is in the opposite sense, i.e. from **b** to **a**, then the movement of the screw is in the opposite direction to $\hat{\mathbf{n}}$, i.e. in the direction of $-\hat{\mathbf{n}}$.

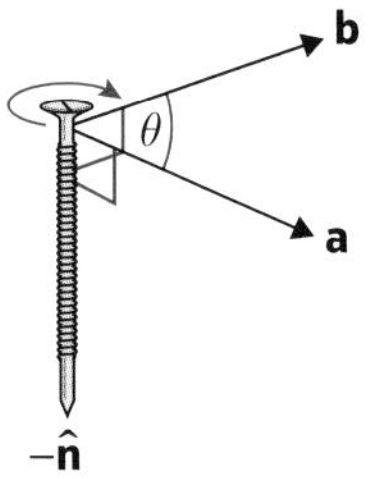

So $\mathbf{b} \times \mathbf{a} = |\mathbf{b}||\mathbf{a}|\sin\theta\,(-\hat{\mathbf{n}})$
$= -|\mathbf{a}||\mathbf{b}|\sin\theta\hat{\mathbf{n}}$
$= -\mathbf{a} \times \mathbf{b}$

- $\mathbf{b} \times \mathbf{a} = -\mathbf{a} \times \mathbf{b}$

Problem-solving

You can also use a 'right-hand rule' to determine the direction of $\hat{\mathbf{n}}$, and hence the direction of $\mathbf{a} \times \mathbf{b}$.
If **a** is your first finger, and **b** is your second finger, then $\mathbf{a} \times \mathbf{b}$ acts in the direction of your thumb:

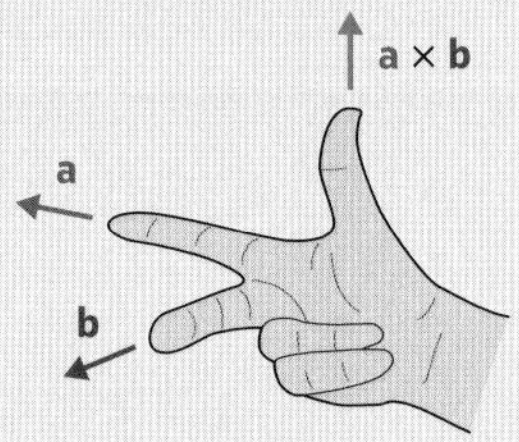

Watch out The vector product is not commutative: the order of multiplication matters.

Example 1

Find the values of:

a $\mathbf{i} \times \mathbf{i}$ **b** $\mathbf{j} \times \mathbf{k}$ **c** $\mathbf{i} \times \mathbf{k}$.

a $\mathbf{i} \times \mathbf{i} = \mathbf{0}$

$\sin\theta = 0$, as the angle between **i** and itself is zero.

b $\mathbf{j} \times \mathbf{k} = 1 \times 1 \times \sin 90°\mathbf{i} = \mathbf{i}$

The angle between **j** and **k** is 90° and, as **j** and **k** are unit vectors, each has magnitude 1 unit.

c $\mathbf{i} \times \mathbf{k} = -\mathbf{k} \times \mathbf{i} = -1 \times 1 \times \sin 90°\mathbf{j} = -\mathbf{j}$

Use the right-hand rule. If **i** is your first finger and **k** is your second finger, your thumb will point **away** from **j**, so $\mathbf{i} \times \mathbf{k} = -\mathbf{j}$.

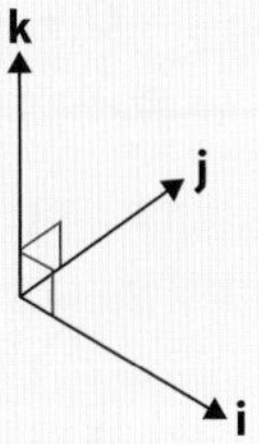

- $\mathbf{i} \times \mathbf{i} = \mathbf{0}$
- $\mathbf{j} \times \mathbf{j} = \mathbf{0}$
- $\mathbf{k} \times \mathbf{k} = \mathbf{0}$
- $\mathbf{i} \times \mathbf{j} = \mathbf{k}$ **and** $\mathbf{j} \times \mathbf{i} = -\mathbf{k}$
- $\mathbf{j} \times \mathbf{k} = \mathbf{i}$ **and** $\mathbf{k} \times \mathbf{j} = -\mathbf{i}$
- $\mathbf{k} \times \mathbf{i} = \mathbf{j}$ **and** $\mathbf{i} \times \mathbf{k} = -\mathbf{j}$

As $\mathbf{a} \times \mathbf{b} = |\mathbf{a}||\mathbf{b}| \sin\theta\,\hat{\mathbf{n}}$, $\mathbf{a} \times \mathbf{b} = \mathbf{0}$ implies that $\mathbf{a} = \mathbf{0}$, $\mathbf{b} = \mathbf{0}$ or $\sin\theta = 0$.

$\sin\theta = 0$ implies that $\theta = 0$ or 180°, so **a** and **b** must be parallel.

- **If $\mathbf{a} \times \mathbf{b} = \mathbf{0}$ then either $\mathbf{a} = \mathbf{0}$, $\mathbf{b} = \mathbf{0}$ or a and b are parallel.**

Example 2

Given that $\mathbf{a} = \begin{pmatrix} a_1 \\ a_2 \\ a_3 \end{pmatrix}$ and $\mathbf{b} = \begin{pmatrix} b_1 \\ b_2 \\ b_3 \end{pmatrix}$ find $\mathbf{a} \times \mathbf{b}$.

$$\begin{aligned}
\mathbf{a} \times \mathbf{b} &= (a_1\mathbf{i} + a_2\mathbf{j} + a_3\mathbf{k}) \times (b_1\mathbf{i} + b_2\mathbf{j} + b_3\mathbf{k}) \\
&= a_1b_1(\mathbf{i} \times \mathbf{i}) + a_1b_2(\mathbf{i} \times \mathbf{j}) + a_1b_3(\mathbf{i} \times \mathbf{k}) \\
&\quad + a_2b_1(\mathbf{j} \times \mathbf{i}) + a_2b_2(\mathbf{j} \times \mathbf{j}) + a_2b_3(\mathbf{j} \times \mathbf{k}) \\
&\quad + a_3b_1(\mathbf{k} \times \mathbf{i}) + a_3b_2(\mathbf{k} \times \mathbf{j}) + a_3b_3(\mathbf{k} \times \mathbf{k}) \\
&= a_1b_2\mathbf{k} + a_1b_3(-\mathbf{j}) + a_2b_1(-\mathbf{k}) + a_2b_3(\mathbf{i}) + a_3b_1(\mathbf{j}) + a_3b_2(-\mathbf{i}) \\
&= (a_2b_3 - a_3b_2)\mathbf{i} + (a_3b_1 - a_1b_3)\mathbf{j} + (a_1b_2 - a_2b_1)\mathbf{k}
\end{aligned}$$

Simplify the cross product and collect like terms.

In determinant form,

$$\mathbf{a} \times \mathbf{b} = \begin{vmatrix} \mathbf{i} & \mathbf{j} & \mathbf{k} \\ a_1 & a_2 & a_3 \\ b_1 & b_2 & b_3 \end{vmatrix} = \mathbf{i}\begin{vmatrix} a_2 & a_3 \\ b_2 & b_3 \end{vmatrix} - \mathbf{j}\begin{vmatrix} a_1 & a_3 \\ b_1 & b_3 \end{vmatrix} + \mathbf{k}\begin{vmatrix} a_1 & a_2 \\ b_1 & b_2 \end{vmatrix}$$

$$= (a_2b_3 - a_3b_2)\mathbf{i} + (a_3b_1 - a_1b_3)\mathbf{j} + (a_1b_2 - a_2b_1)\mathbf{k}$$

You can write each component as the determinant of a 2 × 2 matrix, or the whole vector product as a determinant of a 3 × 3 matrix.

← **Core Pure Book 1, Chapter 6**

Notation You may assume the vector product is **distributive** over vector addition. This means that $\mathbf{a} \times (\mathbf{b} + \mathbf{c}) = (\mathbf{a} \times \mathbf{b}) + (\mathbf{a} \times \mathbf{c})$

- $\mathbf{a} \times \mathbf{b} = (a_2b_3 - a_3b_2)\mathbf{i} + (a_3b_1 - a_1b_3)\mathbf{j} + (a_1b_2 - a_2b_1)\mathbf{k}$
- $\mathbf{a} \times \mathbf{b} = \begin{vmatrix} \mathbf{i} & \mathbf{j} & \mathbf{k} \\ a_1 & a_2 & a_3 \\ b_1 & b_2 & b_3 \end{vmatrix} = \mathbf{i}\begin{vmatrix} a_2 & a_3 \\ b_2 & b_3 \end{vmatrix} - \mathbf{j}\begin{vmatrix} a_1 & a_3 \\ b_1 & b_3 \end{vmatrix} + \mathbf{k}\begin{vmatrix} a_1 & a_2 \\ b_1 & b_2 \end{vmatrix}$

Example 3

Given that $\mathbf{a} = 2\mathbf{i} - 3\mathbf{j}$ and $\mathbf{b} = 4\mathbf{i} + \mathbf{j} - \mathbf{k}$, find $\mathbf{a} \times \mathbf{b}$:

a directly

b by a method involving a determinant.

c Verify that $\mathbf{a} \times \mathbf{b}$ is perpendicular to both **a** and **b**.

a $(2\mathbf{i} - 3\mathbf{j}) \times (4\mathbf{i} + \mathbf{j} - \mathbf{k})$

$= 8(\mathbf{i} \times \mathbf{i}) + 2(\mathbf{i} \times \mathbf{j}) - 2(\mathbf{i} \times \mathbf{k}) - 12(\mathbf{j} \times \mathbf{i}) - 3(\mathbf{j} \times \mathbf{j}) + 3(\mathbf{j} \times \mathbf{k})$

Use the distributive property to multiply out the brackets.

$= \mathbf{0} + 2\mathbf{k} + 2\mathbf{j} + 12\mathbf{k} - \mathbf{0} + 3\mathbf{i}$

Simplify the cross products of unit vectors.

$= 3\mathbf{i} + 2\mathbf{j} + 14\mathbf{k}$

b $\begin{vmatrix} \mathbf{i} & \mathbf{j} & \mathbf{k} \\ 2 & -3 & 0 \\ 4 & 1 & -1 \end{vmatrix} = \mathbf{i}\begin{vmatrix} -3 & 0 \\ 1 & -1 \end{vmatrix} - \mathbf{j}\begin{vmatrix} 2 & 0 \\ 4 & -1 \end{vmatrix} + \mathbf{k}\begin{vmatrix} 2 & -3 \\ 4 & 1 \end{vmatrix}$

$= \mathbf{i}(3 - 0) - \mathbf{j}(-2 - 0) + \mathbf{k}(2 + 12)$

$= 3\mathbf{i} + 2\mathbf{j} + 14\mathbf{k}$

Problem-solving

Using the discriminant is usually a quicker way to evaluate the cross product.

c $(3\mathbf{i} + 2\mathbf{j} + 14\mathbf{k}).(2\mathbf{i} - 3\mathbf{j}) = (3 \times 2) + (2 \times (-3)) + (14 \times 0) = 0$

$(3\mathbf{i} + 2\mathbf{j} + 14\mathbf{k}).(4\mathbf{i} + \mathbf{j} - \mathbf{k}) = (3 \times 4) + (2 \times 1) + (14 \times (-1)) = 0$

Work out $(\mathbf{a} \times \mathbf{b}).\mathbf{a}$ and $(\mathbf{a} \times \mathbf{b}).\mathbf{b}$. If both answers are 0 then $\mathbf{a} \times \mathbf{b}$ is perpendicular to both **a** and **b**.

Example 4

Find a unit vector perpendicular to both $(4\mathbf{i} + 3\mathbf{j} + 2\mathbf{k})$ and $(8\mathbf{i} + 3\mathbf{j} + 3\mathbf{k})$.

The vector product will give a perpendicular vector.

$\begin{vmatrix} \mathbf{i} & \mathbf{j} & \mathbf{k} \\ 4 & 3 & 2 \\ 8 & 3 & 3 \end{vmatrix} = \mathbf{i}\begin{vmatrix} 3 & 2 \\ 3 & 3 \end{vmatrix} - \mathbf{j}\begin{vmatrix} 4 & 2 \\ 8 & 3 \end{vmatrix} + \mathbf{k}\begin{vmatrix} 4 & 3 \\ 8 & 3 \end{vmatrix}$

$= \mathbf{i}(9 - 6) - \mathbf{j}(12 - 16) + \mathbf{k}(12 - 24)$

$= 3\mathbf{i} + 4\mathbf{j} - 12\mathbf{k}$

Watch out You can find vector products using your calculator. But you might encounter a vector with an unknown in it, so it is important that you know how to find the vector product manually.

Since $|3\mathbf{i} + 4\mathbf{j} - 12\mathbf{k}| = \sqrt{3^2 + 4^2 + (-12)^2} = 13$

Find the magnitude of your product vector.

a suitable unit vector is $\frac{1}{13}(3\mathbf{i} + 4\mathbf{j} - 12\mathbf{k})$.

Divide the vector by its magnitude to obtain a unit vector.

Example 5

Find the sine of the acute angle between the vectors $\mathbf{a} = 2\mathbf{i} + \mathbf{j} + 2\mathbf{k}$ and $\mathbf{b} = -3\mathbf{j} + 4\mathbf{k}$.

$\mathbf{a} \times \mathbf{b} = |\mathbf{a}||\mathbf{b}|\sin\theta\,\hat{\mathbf{n}}$

So $\dfrac{|\mathbf{a} \times \mathbf{b}|}{|\mathbf{a}||\mathbf{b}|} = \sin\theta$

Rearrange the formula to make $\sin\theta$ the subject. $|\hat{\mathbf{n}}| = 1$ so $|\mathbf{a} \times \mathbf{b}| = |\mathbf{a}||\mathbf{b}|\sin\theta$.

$$\mathbf{a} \times \mathbf{b} = \begin{vmatrix} \mathbf{i} & \mathbf{j} & \mathbf{k} \\ 2 & 1 & 2 \\ 0 & -3 & 4 \end{vmatrix}$$
$$= \mathbf{i}(4 + 6) - \mathbf{j}(8 - 0) + \mathbf{k}(-6 - 0)$$
$$= 10\mathbf{i} - 8\mathbf{j} - 6\mathbf{k}$$

Calculate the vector product.

and $|10\mathbf{i} - 8\mathbf{j} - 6\mathbf{k}| = \sqrt{100 + 64 + 36}$

Find the magnitude of $\mathbf{a} \times \mathbf{b}$.

So $\sin\theta = \dfrac{\sqrt{200}}{\sqrt{2^2 + 1^2 + 2^2}\sqrt{(-3)^2 + 4^2}}$

Also find the magnitude of **a** and of **b** and substitute the three surds into the formula for $\sin\theta$.

$$= \frac{\sqrt{200}}{\sqrt{9}\sqrt{25}}$$
$$= \frac{10\sqrt{2}}{3 \times 5}$$
$$= \frac{2\sqrt{2}}{3}$$

Simplify your answer.

Watch out In general, to find the angle between two vectors use the scalar product. This gives the cosine of the angle. Immediately we know whether the angle is acute or obtuse. In this example it is not clear whether the angle θ is acute or obtuse. This is similar to the ambiguous case when using the sine rule.

Exercise 1A

1 Simplify:

a $5\mathbf{j} \times \mathbf{k}$ **b** $3\mathbf{i} \times \mathbf{k}$ **c** $\mathbf{k} \times 3\mathbf{i}$

d $3\mathbf{i} \times (9\mathbf{i} - \mathbf{j} + \mathbf{k})$ **e** $2\mathbf{j} \times (3\mathbf{i} + \mathbf{j} - \mathbf{k})$ **f** $(3\mathbf{i} + \mathbf{j} - \mathbf{k}) \times 2\mathbf{j}$

g $\begin{pmatrix} 5 \\ 2 \\ -1 \end{pmatrix} \times \begin{pmatrix} 1 \\ -1 \\ 3 \end{pmatrix}$ **h** $\begin{pmatrix} 2 \\ -1 \\ 6 \end{pmatrix} \times \begin{pmatrix} 1 \\ -2 \\ 3 \end{pmatrix}$ **i** $\begin{pmatrix} 1 \\ 5 \\ -4 \end{pmatrix} \times \begin{pmatrix} 2 \\ -1 \\ -1 \end{pmatrix}$ **j** $\begin{pmatrix} 3 \\ 0 \\ 2 \end{pmatrix} \times \begin{pmatrix} 1 \\ -1 \\ 2 \end{pmatrix}$

2 Find the vector product of the vectors **a** and **b**, leaving your answers in terms of λ in each case.

a $\mathbf{a} = \lambda\mathbf{i} + 2\mathbf{j} + \mathbf{k}$ $\quad \mathbf{b} = \mathbf{i} - 3\mathbf{k}$

b $\mathbf{a} = 2\mathbf{i} - \mathbf{j} + 7\mathbf{k}$ $\quad \mathbf{b} = \mathbf{i} - \lambda\mathbf{j} + 3\mathbf{k}$

3 Find a unit vector that is perpendicular to both $2\mathbf{i} - \mathbf{j}$ and to $4\mathbf{i} + \mathbf{j} + 3\mathbf{k}$.

4 Find a unit vector that is perpendicular to both $4\mathbf{i} + \mathbf{k}$ and $\mathbf{j} - \sqrt{2}\mathbf{k}$.

5 Find a unit vector that is perpendicular to both $\mathbf{i} - \mathbf{j}$ and $3\mathbf{i} + 4\mathbf{j} - 6\mathbf{k}$.

6 Find a unit vector that is perpendicular to both $\begin{pmatrix} 1 \\ 6 \\ 4 \end{pmatrix}$ and to $\begin{pmatrix} 5 \\ 9 \\ 8 \end{pmatrix}$.

7 Find a vector of magnitude 5 which is perpendicular to both $\begin{pmatrix} 4 \\ 0 \\ 1 \end{pmatrix}$ and $\begin{pmatrix} 0 \\ \sqrt{2} \\ 1 \end{pmatrix}$.

8 Find the magnitude of $(\mathbf{i} + \mathbf{j} - \mathbf{k}) \times (\mathbf{i} - \mathbf{j} + \mathbf{k})$.

9 Given that $\mathbf{a} = -\mathbf{i} + 2\mathbf{j} - 5\mathbf{k}$ and $\mathbf{b} = 5\mathbf{i} - 2\mathbf{j} + \mathbf{k}$, find:

a $\mathbf{a}.\mathbf{b}$

b $\mathbf{a} \times \mathbf{b}$

c the unit vector in the direction $\mathbf{a} \times \mathbf{b}$.

10 Find the sine of the angle between each of the following pairs of vectors **a** and **b**. You may leave your answers as surds, in their simplest form.

a $\mathbf{a} = 3\mathbf{i} - 4\mathbf{j}$, $\mathbf{b} = 2\mathbf{i} + 2\mathbf{j} + \mathbf{k}$

b $\mathbf{a} = \mathbf{j} + 2\mathbf{k}$, $\mathbf{b} = 5\mathbf{i} + 4\mathbf{j} - 2\mathbf{k}$

c $\mathbf{a} = 5\mathbf{i} + 2\mathbf{j} + 2\mathbf{k}$, $\mathbf{b} = 4\mathbf{i} + 4\mathbf{j} + \mathbf{k}$

11 The line l_1 has equation $\mathbf{r} = \mathbf{i} - \mathbf{j} + \lambda(\mathbf{i} + 2\mathbf{j} + 3\mathbf{k})$ and the line l_2 has equation $\mathbf{r} = 2\mathbf{i} + \mathbf{j} + \mathbf{k} + \mu(2\mathbf{i} - \mathbf{j} + \mathbf{k})$. Find a vector that is perpendicular to both l_1 and l_2.

(P) **12** It is given that $\mathbf{a} = \begin{pmatrix} 1 \\ 3 \\ -1 \end{pmatrix}$ and $\mathbf{b} = \begin{pmatrix} 2 \\ u \\ v \end{pmatrix}$ and that $\mathbf{a} \times \mathbf{b} = \begin{pmatrix} w \\ -6 \\ -7 \end{pmatrix}$, where u, v and w are scalar constants. Find the values of u, v and w.

(P) **13** Given that $\mathbf{p} = a\mathbf{i} - \mathbf{j} + 4\mathbf{k}$, that $\mathbf{q} = \mathbf{j} - \mathbf{k}$ and that their vector product $\mathbf{q} \times \mathbf{p} = 3\mathbf{i} - \mathbf{j} + b\mathbf{k}$ where a and b are scalar constants,

a find the values of a and b

b find the value of the cosine of the angle between **p** and **q**.

(P) **14** If $\mathbf{a} \times \mathbf{b} = \mathbf{0}$, $\mathbf{a} = 2\mathbf{i} + \mathbf{j} - \mathbf{k}$ and $\mathbf{b} = 3\mathbf{i} + \lambda\mathbf{j} + \mu\mathbf{k}$, where λ and μ are scalar constants, find the values of λ and μ.

(P) **15** If three vectors **a**, **b** and **c** satisfy $\mathbf{a} + \mathbf{b} + \mathbf{c} = \mathbf{0}$, show that

$$\mathbf{a} \times \mathbf{b} = \mathbf{b} \times \mathbf{c} = \mathbf{c} \times \mathbf{a}$$

Challenge

a is a non-zero vector and **b** and **c** are non-parallel vectors.

Given that $\mathbf{a} \times \mathbf{b} = \mathbf{c} \times \mathbf{a}$, show that **a** is parallel to $\mathbf{b} + \mathbf{c}$.

1.2 Finding areas

You can use the vector product to solve problems involving areas of triangles and parallelograms.

Example 6

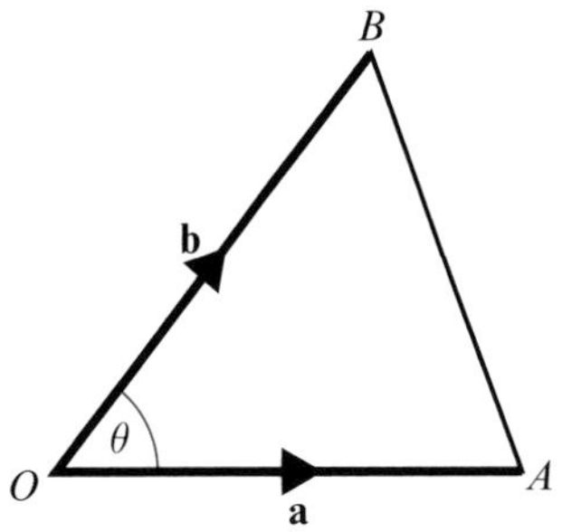

Find the area of triangle OAB, where O is the origin, A is the point with position vector **a** and B is the point with position vector **b**.

Area of triangle $OAB = \frac{1}{2}(OA)(OB)\sin\theta$

$= \frac{1}{2}|\mathbf{a}||\mathbf{b}|\sin\theta$

$= \frac{1}{2}|\mathbf{a} \times \mathbf{b}|$

Use the formula for area of triangle, Area $= \frac{1}{2}ab\sin C$, and let the angle $AOB = \theta$.

Use the definition of vector product to obtain this result.

- **If A and B have position vectors a and b respectively, then**
  Area of triangle $OAB = \frac{1}{2}|\mathbf{a} \times \mathbf{b}|$

Example 7

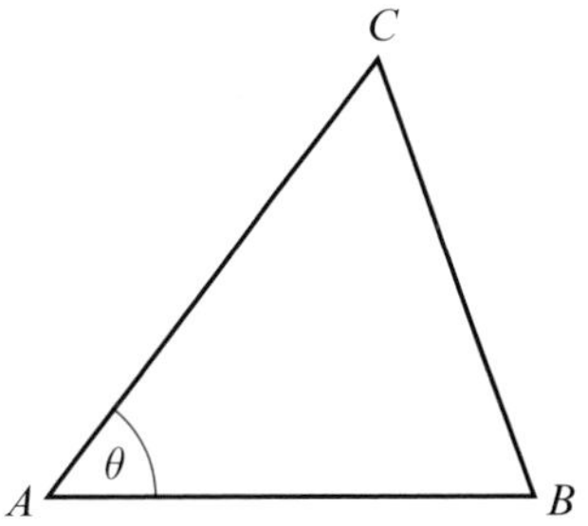

Find the area of triangle ABC, where the position vectors of A, B and C are **a**, **b** and **c** respectively.

Area of triangle $ABC = \frac{1}{2}(AB)(AC)\sin\theta$

$= \frac{1}{2}|\mathbf{b} - \mathbf{a}||\mathbf{c} - \mathbf{a}|\sin\theta$

$= \frac{1}{2}|(\mathbf{b} - \mathbf{a}) \times (\mathbf{c} - \mathbf{a})|$

$= \frac{1}{2}|(\mathbf{b} \times \mathbf{c}) - (\mathbf{b} \times \mathbf{a}) - (\mathbf{a} \times \mathbf{c}) + (\mathbf{a} \times \mathbf{a})|$

$= \frac{1}{2}|(\mathbf{b} \times \mathbf{c}) + (\mathbf{c} \times \mathbf{a}) + (\mathbf{a} \times \mathbf{b})|$

$= \frac{1}{2}|(\mathbf{a} \times \mathbf{b}) + (\mathbf{b} \times \mathbf{c}) + (\mathbf{c} \times \mathbf{a})|$

Let the angle $BAC = \theta$.

Use the definition of the vector product.

Expand using the distributive law.

Use $\mathbf{a} \times \mathbf{a} = \mathbf{0}$, $\mathbf{a} \times \mathbf{b} = -\mathbf{b} \times \mathbf{a}$ and $\mathbf{c} \times \mathbf{a} = -\mathbf{a} \times \mathbf{c}$.

- **If A, B and C have position vectors a, b and c respectively, then**
  Area of triangle $ABC = \frac{1}{2}|\overrightarrow{AB} \times \overrightarrow{AC}|$
  $= \frac{1}{2}|(\mathbf{b} - \mathbf{a}) \times (\mathbf{c} - \mathbf{a})|$
  $= \frac{1}{2}|(\mathbf{a} \times \mathbf{b}) + (\mathbf{b} \times \mathbf{c}) + (\mathbf{c} \times \mathbf{a})|$

Example 8

Find the area of the parallelogram $ABCD$, where the position vectors of A, B and D are **a**, **b** and **d** respectively.

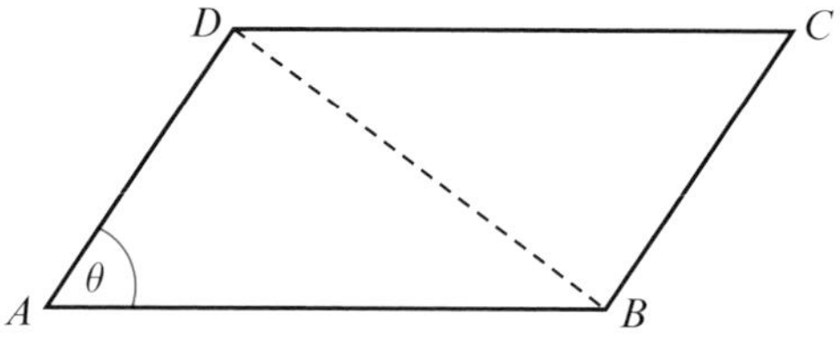

Area of parallelogram $ABCD$

= area of triangle ABD + area of triangle BCD

= 2 × area of triangle ABD

= $(AB)(AD)\sin\theta$

= $|(\mathbf{b} - \mathbf{a}) \times (\mathbf{d} - \mathbf{a})|$

= $|(\mathbf{a} \times \mathbf{b}) + (\mathbf{b} \times \mathbf{d}) + (\mathbf{d} \times \mathbf{a})|$

The two triangles are congruent so have equal area.

θ is the angle BAD.

- **If A and B have position vectors a and b respectively, then**

  Area of parallelogram $OABC = |\mathbf{a} \times \mathbf{b}|$

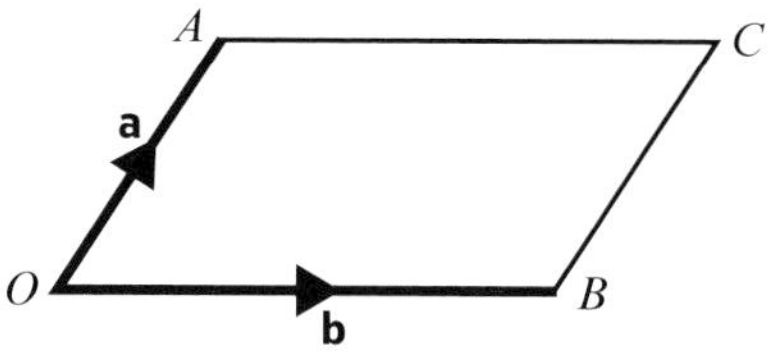

- **If A, B, C and D have position vectors a, b, c and d respectively, then**

  Area of parallelogram $ABCD = |\overrightarrow{AB} \times \overrightarrow{AD}|$

  $= |(\mathbf{b} - \mathbf{a}) \times (\mathbf{d} - \mathbf{a})|$

  $= |(\mathbf{a} \times \mathbf{b}) + (\mathbf{b} \times \mathbf{d}) + (\mathbf{d} \times \mathbf{a})|$

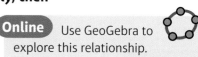

Online Use GeoGebra to explore this relationship.

Example 9

Find the area of triangle OAB, where O is the origin, A is the point with position vector $\mathbf{i} - \mathbf{j}$ and B is the point with position vector $3\mathbf{i} + 4\mathbf{j} - 6\mathbf{k}$.

Area of triangle $OAB = \frac{1}{2}|(\mathbf{i} - \mathbf{j}) \times (3\mathbf{i} + 4\mathbf{j} - 6\mathbf{k})|$

$$(\mathbf{i} - \mathbf{j}) \times (3\mathbf{i} + 4\mathbf{j} - 6\mathbf{k}) = \begin{vmatrix} \mathbf{i} & \mathbf{j} & \mathbf{k} \\ 1 & -1 & 0 \\ 3 & 4 & -6 \end{vmatrix}$$

$$= 6\mathbf{i} + 6\mathbf{j} + 7\mathbf{k}$$

First find the vector product using the determinant method.

So area of triangle $= \frac{1}{2}|6\mathbf{i} + 6\mathbf{j} + 7\mathbf{k}| = \frac{1}{2}\sqrt{6^2 + 6^2 + 7^2}$

$= \frac{\sqrt{121}}{2} = 5.5$

Then use this to find the area of the triangle.

Example 10

Find the area of triangle ABC, where the position vectors of A, B and C are $4\mathbf{i} - 2\mathbf{j} + \mathbf{k}$, $-12\mathbf{i} + 14\mathbf{j} + \mathbf{k}$ and $-4\mathbf{i} - 2\mathbf{j} + \mathbf{k}$ respectively.

$\overrightarrow{AB} = (-12\mathbf{i} + 14\mathbf{j} + \mathbf{k}) - (4\mathbf{i} - 2\mathbf{j} + \mathbf{k}) = -16\mathbf{i} + 16\mathbf{j}$

$\overrightarrow{AC} = (-4\mathbf{i} - 2\mathbf{j} + \mathbf{k}) - (4\mathbf{i} - 2\mathbf{j} + \mathbf{k}) = -8\mathbf{i}$

Find vectors representing two of the sides of the triangle.

$$\overrightarrow{AB} \times \overrightarrow{AC} = \begin{vmatrix} \mathbf{i} & \mathbf{j} & \mathbf{k} \\ -16 & 16 & 0 \\ -8 & 0 & 0 \end{vmatrix} = 128\mathbf{k}$$

So area of triangle $ABC = \frac{1}{2}|128\mathbf{k}| = 64$

Area of triangle $= \frac{1}{2}\left|\overrightarrow{AB} \times \overrightarrow{AC}\right|$. Find $\overrightarrow{AB} \times \overrightarrow{AC}$ using the discriminant method, then find half its modulus. Remember that $|p\mathbf{k}| = p$ for any scalar p.

Example 11

Find the area of the parallelogram $ABCD$, where the position vectors of A, B and D are $2\mathbf{i} + \mathbf{j} - \mathbf{k}$, $6\mathbf{i} + 4\mathbf{j} - 3\mathbf{k}$ and $14\mathbf{i} + 7\mathbf{j} - 6\mathbf{k}$ respectively.

Area of parallelogram $ABCD = \left|\overrightarrow{AB} \times \overrightarrow{AD}\right|$

$\overrightarrow{AB} = (6\mathbf{i} + 4\mathbf{j} - 3\mathbf{k}) - (2\mathbf{i} + \mathbf{j} - \mathbf{k}) = 4\mathbf{i} + 3\mathbf{j} - 2\mathbf{k}$

$\overrightarrow{AD} = (14\mathbf{i} + 7\mathbf{j} - 6\mathbf{k}) - (2\mathbf{i} + \mathbf{j} - \mathbf{k}) = 12\mathbf{i} + 6\mathbf{j} - 5\mathbf{k}$

Find vectors representing two adjacent sides of the parallelogram.

$$\overrightarrow{AB} \times \overrightarrow{AD} = \begin{vmatrix} \mathbf{i} & \mathbf{j} & \mathbf{k} \\ 4 & 3 & -2 \\ 12 & 6 & -5 \end{vmatrix} = -3\mathbf{i} - 4\mathbf{j} - 12\mathbf{k}$$

So area of parallelogram $= |-3\mathbf{i} - 4\mathbf{j} - 12\mathbf{k}| = 13$

Area of parallelogram $= \left|\overrightarrow{AB} \times \overrightarrow{AD}\right|$.

Exercise 1B

1 Find the area of triangle OAB, where O is the origin, A is the point with position vector $\mathbf{a}$ and B is the point with position vector $\mathbf{b}$ in the following cases.

a $\mathbf{a} = \mathbf{i} + \mathbf{j} - 4\mathbf{k}$　　$\mathbf{b} = 2\mathbf{i} - \mathbf{j} - 2\mathbf{k}$

b $\mathbf{a} = 3\mathbf{i} + 4\mathbf{j} - 5\mathbf{k}$　　$\mathbf{b} = 2\mathbf{i} + \mathbf{j} - 2\mathbf{k}$

c $\mathbf{a} = \begin{pmatrix} 2 \\ 3 \\ 0 \end{pmatrix}$　　$\mathbf{b} = \begin{pmatrix} 2 \\ 6 \\ -9 \end{pmatrix}$

2 Find the area of triangle ABC, where the position vectors of A, B and C are $\mathbf{a}$, $\mathbf{b}$ and $\mathbf{c}$ respectively, in the following cases:

a $\mathbf{a} = \mathbf{i} - \mathbf{j} - \mathbf{k}$　　$\mathbf{b} = 4\mathbf{i} + \mathbf{j} + \mathbf{k}$　　$\mathbf{c} = 4\mathbf{i} - 3\mathbf{j} + \mathbf{k}$

b $\mathbf{a} = \begin{pmatrix} 0 \\ 1 \\ 2 \end{pmatrix}$　$\mathbf{b} = \begin{pmatrix} 1 \\ 0 \\ 2 \end{pmatrix}$　$\mathbf{c} = \begin{pmatrix} 2 \\ 0 \\ -10 \end{pmatrix}$

3 Find the area of the triangle with vertices $A(1, 0, 2)$, $B(2, -2, 0)$ and $C(3, -1, 1)$.

4 Find the area of the triangle with vertices $A(-1, 1, 1)$, $B(1, 0, 2)$ and $C(0, 3, 4)$.

5 Find the area of the parallelogram $ABCD$, shown in the diagram, where the position vectors of A, B and D are $\mathbf{i} + \mathbf{j} + \mathbf{k}$, $-3\mathbf{i} + 4\mathbf{j} + \mathbf{k}$ and $2\mathbf{i} - \mathbf{j}$ respectively.

6 Find the area of the parallelogram $ABCD$, shown in the diagram, in which the vertices A, B and D have coordinates $(0, 5, 3)$, $(2, 1, -1)$ and $(1, 6, 6)$ respectively.

7 Find the area of the parallelogram $ABCD$, shown in the diagram, where the position vectors of A, B and D are $\mathbf{j}$, $\mathbf{i} + 4\mathbf{j} + \mathbf{k}$ and $2\mathbf{i} + 6\mathbf{j} + 3\mathbf{k}$ respectively.

(P) **8** Relative to an origin O, the points P and Q have position vectors $\mathbf{p}$ and $\mathbf{q}$ respectively, where $\mathbf{p} = a(\mathbf{i} + \mathbf{j} + 2\mathbf{k})$, $\mathbf{q} = a(2\mathbf{i} + \mathbf{j} + 3\mathbf{k})$ and $a > 0$.

Find the area of triangle OPQ, giving your answer in terms of a.

(P) **9** **a** Prove that the area of the parallelogram $ABCD$ is $|(\mathbf{b} - \mathbf{a}) \times (\mathbf{c} - \mathbf{a})|$

b Show that $(\mathbf{b} - \mathbf{a}) \times (\mathbf{c} - \mathbf{a}) = (\mathbf{b} - \mathbf{a}) \times (\mathbf{d} - \mathbf{a})$ implies that $(\mathbf{b} - \mathbf{a}) \times (\mathbf{c} - \mathbf{d}) = \mathbf{0}$, and explain the geometrical significance of this vector product.

(E) **10** The position vectors of the points A, B and C relative to an origin O are $2\mathbf{i} - \mathbf{j} - \mathbf{k}$, $6\mathbf{i} - 2\mathbf{k}$ and $3\mathbf{i} + 3\mathbf{j}$ respectively.

Find:

a $\overrightarrow{AC} \times \overrightarrow{BC}$ **(3 marks)**

b the exact area of triangle ABC. **(2 marks)**

(E) **11** The sail of a yacht is modelled as a triangle with vertices at $A(-3, 2, -4)$, $B(-2, -3, 1)$ and $C(1, 2, -1)$, where the dimensions are in metres.

a Find $\overrightarrow{AB} \times \overrightarrow{AC}$. **(3 marks)**

b Hence find the area of fabric needed to construct the sail according to this model. **(2 marks)**

c Suggest, with a reason, whether the actual area of fabric needed to construct the sail will be larger or smaller than this value. **(1 mark)**

(E) **12** A jeweller makes gold pendants in the shape of a parallelogram $ABCD$ where sides AB and DC are equal and parallel. She designs the pendants in 3D space and models the pendants as having vertices $A(-1, 2, 0)$, $B(3, -3, -2)$ and $D(-2, 0, 3)$ where each unit represents 1 cm.

a Find the coordinates of point C. **(2 marks)**

Given that gold costs £595 per cm^3, and that the pendants will be 3 mm thick,

b find, correct to the nearest pound, the cost of making one pendant. **(4 marks)**

Challenge

In the diagram below, $ABCD$ and $CDEF$ are parallelograms which lie in the same plane.

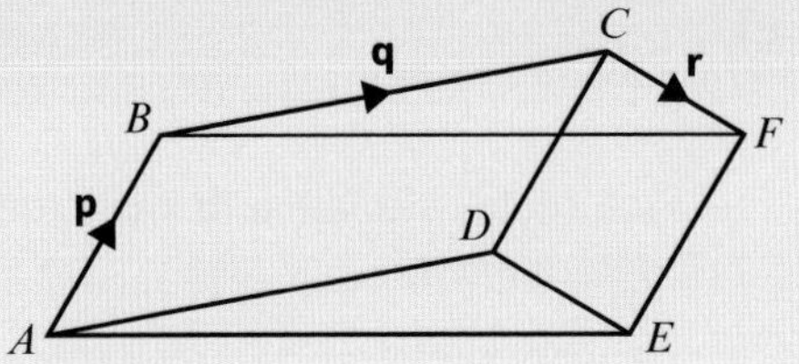

$\overrightarrow{AB} = \mathbf{p}$, $\overrightarrow{BC} = \mathbf{q}$ and $\overrightarrow{CF} = \mathbf{r}$

By considering area, show that $|\mathbf{p} \times (\mathbf{q} + \mathbf{r})| = |\mathbf{p} \times \mathbf{q}| + |\mathbf{p} \times \mathbf{r}|$.

1.3 Scalar triple product

You can find the **scalar triple product** of three vectors **a**, **b** and **c**, and use it to find the volume of a parallelepiped and of a tetrahedron.

Online Use GeoGebra to explore the scalar triple product.

Notation A **parallelepiped** is a three-dimensional solid with six parallelogram-shaped faces.

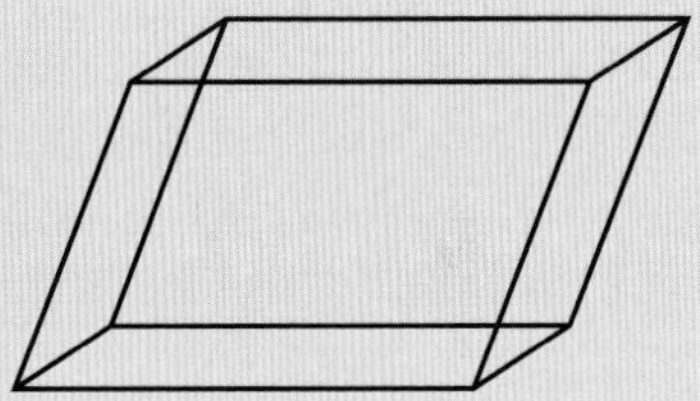

You know that $\mathbf{b} \times \mathbf{c} = (b_2c_3 - b_3c_2)\mathbf{i} + (b_3c_1 - b_1c_3)\mathbf{j} + (b_1c_2 - b_2c_1)\mathbf{k}$, where $\mathbf{b} = b_1\mathbf{i} + b_2\mathbf{j} + b_3\mathbf{k}$ and $\mathbf{c} = c_1\mathbf{i} + c_2\mathbf{j} + c_3\mathbf{k}$.

So if $\mathbf{a} = a_1\mathbf{i} + a_2\mathbf{j} + a_3\mathbf{k}$, then

- $\mathbf{a}.(\mathbf{b} \times \mathbf{c}) = a_1(b_2c_3 - b_3c_2) + a_2(b_3c_1 - b_1c_3) + a_3(b_1c_2 - b_2c_1)$

This can also be written as

- $\mathbf{a}.(\mathbf{b} \times \mathbf{c}) = \begin{vmatrix} a_1 & a_2 & a_3 \\ b_1 & b_2 & b_3 \\ c_1 & c_2 & c_3 \end{vmatrix}$**, and a.(b × c) is known as the scalar triple product.**

Example 12

Given that $\mathbf{a} = 3\mathbf{i} - \mathbf{j} + 4\mathbf{k}$, $\mathbf{b} = \mathbf{i} + \mathbf{j} - \mathbf{k}$ and $\mathbf{c} = 2\mathbf{i} + 3\mathbf{j} + 5\mathbf{k}$, find

a $\mathbf{a}.(\mathbf{b} \times \mathbf{c})$ **b** $\mathbf{b}.(\mathbf{c} \times \mathbf{a})$ **c** $\mathbf{a}.(\mathbf{a} \times \mathbf{c})$

a $\mathbf{b} \times \mathbf{c} = \begin{vmatrix} \mathbf{i} & \mathbf{j} & \mathbf{k} \\ 1 & 1 & -1 \\ 2 & 3 & 5 \end{vmatrix} = 8\mathbf{i} - 7\mathbf{j} + \mathbf{k}$

So $\mathbf{a}.(\mathbf{b} \times \mathbf{c}) = (3\mathbf{i} - \mathbf{j} + 4\mathbf{k}).(8\mathbf{i} - 7\mathbf{j} + \mathbf{k})$
$= 24 + 7 + 4$
$= 35$

You could calculate $\mathbf{a}.(\mathbf{b} \times \mathbf{c})$ directly as a determinant:

$$\begin{vmatrix} 3 & -1 & 4 \\ 1 & 1 & -1 \\ 2 & 3 & 5 \end{vmatrix} = 3\begin{vmatrix} 1 & -1 \\ 3 & 5 \end{vmatrix} - (-1)\begin{vmatrix} 1 & -1 \\ 2 & 5 \end{vmatrix} + 4\begin{vmatrix} 1 & 1 \\ 2 & 3 \end{vmatrix}$$
$$= 24 + 7 + 4 = 35$$

b $\mathbf{c} \times \mathbf{a} = \begin{vmatrix} \mathbf{i} & \mathbf{j} & \mathbf{k} \\ 2 & 3 & 5 \\ 3 & -1 & 4 \end{vmatrix} = 17\mathbf{i} + 7\mathbf{j} - 11\mathbf{k}$

So $\mathbf{b}.(\mathbf{c} \times \mathbf{a}) = (\mathbf{i} + \mathbf{j} - \mathbf{k}).(17\mathbf{i} + 7\mathbf{j} - 11\mathbf{k})$
$= 17 + 7 + 11$
$= 35$

Notice that $\mathbf{a}.(\mathbf{b} \times \mathbf{c}) = \mathbf{b}.(\mathbf{c} \times \mathbf{a})$

c $\mathbf{a} \times \mathbf{c} = -\mathbf{c} \times \mathbf{a} = -17\mathbf{i} - 7\mathbf{j} + 11\mathbf{k}$

Use the result that $\mathbf{a} \times \mathbf{c} = -\mathbf{c} \times \mathbf{a}$

So $\mathbf{a}.(\mathbf{a} \times \mathbf{c}) = (3\mathbf{i} - \mathbf{j} + 4\mathbf{k}).(-17\mathbf{i} - 7\mathbf{j} + 11\mathbf{k})$
$= -51 + 7 + 44$
$= 0$

This scalar product is zero since $\mathbf{a} \times \mathbf{c}$ is perpendicular to $\mathbf{a}$.

The above worked example illustrates two important points.

- **The scalar triple product is cyclic:**
  $\mathbf{a}.(\mathbf{b} \times \mathbf{c}) = \mathbf{b}.(\mathbf{c} \times \mathbf{a}) = \mathbf{c}.(\mathbf{a} \times \mathbf{b})$
- **If a vector is repeated then the scalar triple product is equal to zero:**
  $\mathbf{a}.(\mathbf{a} \times \mathbf{p}) = \mathbf{a}.(\mathbf{p} \times \mathbf{a}) = 0$ **for any vector p.**

Hint You can use the first of these to prove the second:
$\mathbf{a}.(\mathbf{a} \times \mathbf{p}) = \mathbf{p}.(\mathbf{a} \times \mathbf{a}) = \mathbf{p}.\mathbf{0} = 0$

Example 13

Find the volume of the parallelepiped shown in the figure, given that O is the origin and A, B and C have position vectors **a**, **b** and **c** respectively. The angle between **b** and **c** is θ and the angle between the perpendicular height and **a** is ϕ.

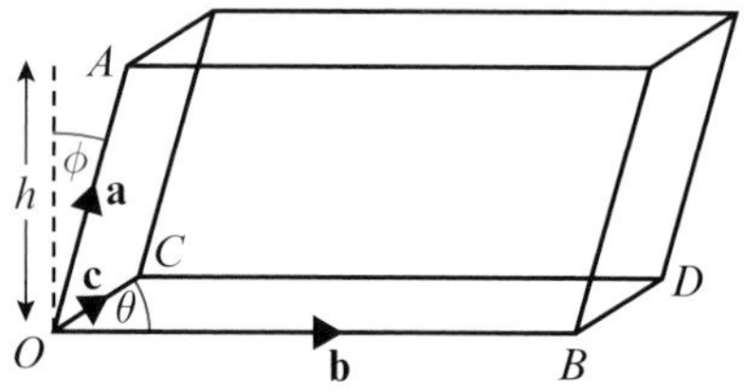

The volume of the parallelepiped is given by (area of base) × h where h is the perpendicular distance between the base and the top face.

The base, $OBDC$ is a parallelogram and its area is $|\mathbf{b} \times \mathbf{c}|$.

So the volume of the parallelepiped is $|\mathbf{b} \times \mathbf{c}|h$

But $h = OA\cos\phi$ — As $\cos\phi = \frac{h}{OA}$

So volume is $|\mathbf{b} \times \mathbf{c}|OA\cos\phi$

$= |\mathbf{b} \times \mathbf{c}||\mathbf{a}|\cos\phi$ — Since $\mathbf{b} \times \mathbf{c}$ is in the direction of the perpendicular height, ϕ is the angle between vector **a** and vector $\mathbf{b} \times \mathbf{c}$.

$= \mathbf{a}.(\mathbf{b} \times \mathbf{c})$ — From the definition of scalar product.

- **If three sides of a parallelepiped are given by vectors a, b and c as shown in the diagram, then the volume of the parallelepiped is given by $|\mathbf{a}.(\mathbf{b} \times \mathbf{c})|$.**

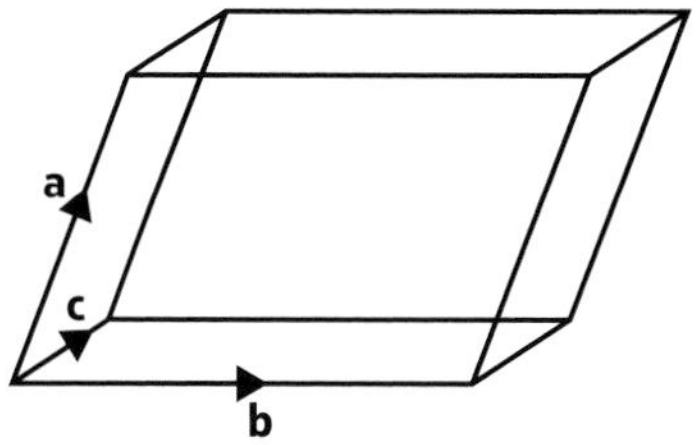

Note **a**, **b** and **c** can be any three non-parallel sides of the parallelepiped.

Example 14

Find the volume of the tetrahedron shown in the figure, given that O is the origin and A, B and C have position vectors **a**, **b** and **c** respectively. The angle between **b** and **c** is θ and the angle between the perpendicular height and **a** is ϕ.

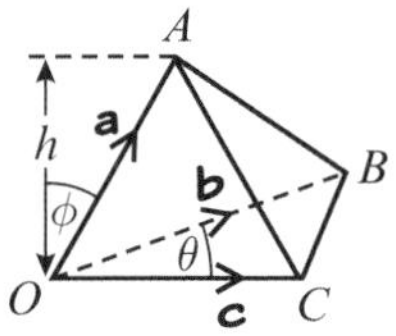

The volume of the tetrahedron is given by the formula $\frac{1}{3}$(area of base) × h where h is the perpendicular height. — The volume of a pyramid is $\frac{1}{3}$(area of base) × h.

The triangular base, OBC has area $\frac{1}{2}|\mathbf{b} \times \mathbf{c}|$

And $h = OA\cos\phi = |\mathbf{a}|\cos\phi$

So volume of tetrahedron is $\frac{1}{3} \times \frac{1}{2}|\mathbf{b} \times \mathbf{c}||\mathbf{a}|\cos\phi$

$= \frac{1}{6}\mathbf{a}.(\mathbf{b} \times \mathbf{c})$ — As in Example 13, $\mathbf{b} \times \mathbf{c}$ is in the direction of the perpendicular height, so ϕ is the angle between vector **a** and vector $\mathbf{b} \times \mathbf{c}$.

- **If three sides of a tetrahedron are given by vectors a, b and c as shown in the diagram, then the volume of the tetrahedron is given by $\frac{1}{6}|\mathbf{a}.(\mathbf{b} \times \mathbf{c})|$.**

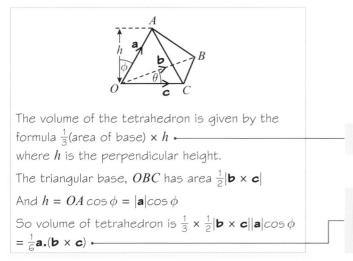

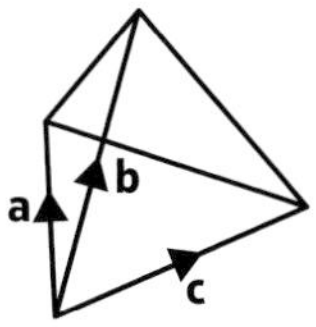

Note **a**, **b** and **c** can be any three non-coplanar sides of the tetrahedron.

Example 15

Find the volume of a tetrahedron which has vertices at (1, 1, −1), (2, 4, −1), (3, 0, −2) and (0, 4, 5).

If the vertices are labelled A, B, C and D in the order given above and have position vectors **a**, **b**, **c** and **d** respectively, then:

$\overrightarrow{AB} = \mathbf{b} - \mathbf{a} = \mathbf{i} + 3\mathbf{j}$

$\overrightarrow{AC} = \mathbf{c} - \mathbf{a} = 2\mathbf{i} - \mathbf{j} - \mathbf{k}$

$\overrightarrow{AD} = \mathbf{d} - \mathbf{a} = -\mathbf{i} + 3\mathbf{j} + 6\mathbf{k}$

Find expressions for the vectors describing the displacement from one of the vertices to the other three.

Volume of tetrahedron $= \frac{1}{6}|\overrightarrow{AB}.(\overrightarrow{AC} \times \overrightarrow{AD})|$

$$\overrightarrow{AB}.(\overrightarrow{AC} \times \overrightarrow{AD}) = \begin{vmatrix} 1 & 3 & 0 \\ 2 & -1 & -1 \\ -1 & 3 & 6 \end{vmatrix} = -36$$

Use the scalar triple product to find the volume.

So the volume is $\frac{1}{6}|-36| = 6$.

Problem-solving

$\overrightarrow{AB}.(\overrightarrow{AC} \times \overrightarrow{AD})$ is negative. If you swapped any pair of vectors in this scalar triple product the answer would be 6 instead of −6.

For example, $\overrightarrow{AC}.(\overrightarrow{AB} \times \overrightarrow{AD}) = 6$.

Exercise 1C

1 Given that $\mathbf{a} = 5\mathbf{i} + 2\mathbf{j} - \mathbf{k}$, $\mathbf{b} = \mathbf{i} + \mathbf{j} + \mathbf{k}$ and $\mathbf{c} = 3\mathbf{i} + 4\mathbf{k}$, find:

a $\mathbf{a}.(\mathbf{b} \times \mathbf{c})$ **b** $\mathbf{b}.(\mathbf{c} \times \mathbf{a})$ **c** $\mathbf{c}.(\mathbf{a} \times \mathbf{b})$

(P) 2 Given that $\mathbf{a} = \mathbf{i} - \mathbf{j} - 2\mathbf{k}$, $\mathbf{b} = 2\mathbf{i} + \mathbf{j} - \mathbf{k}$ and $\mathbf{c} = 2\mathbf{i} - 3\mathbf{j} - 5\mathbf{k}$, find $\mathbf{a}.(\mathbf{b} \times \mathbf{c})$. What can you deduce about the vectors **a**, **b** and **c**?

3 Find the volume of the parallelepiped $ABCDEFGH$ where the vertices A, B, D and E have coordinates (0, 0, 0), (3, 0, 1), (1, 2, 0) and (1, 1, 3) respectively.

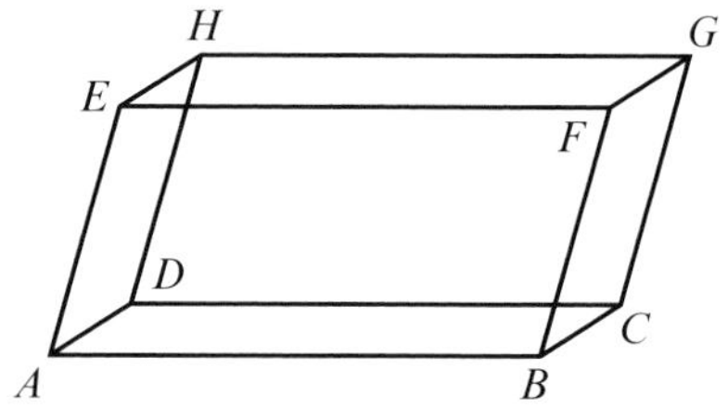

4 Find the volume of the parallelepiped $ABCDEFGH$ where the vertices A, B, D and E have coordinates (−1, 0, 1), (3, 0, −1), (2, 2, 0) and (2, 1, 2) respectively.

5 A tetrahedron has vertices at $A(1, 2, 3)$, $B(4, 3, 4)$, $C(1, 3, 1)$ and $D(3, 1, 4)$.
Find the volume of the tetrahedron.

6 A tetrahedron has vertices at $A(2, 2, 1)$, $B(3, -1, 2)$, $C(1, 1, 3)$ and $D(3, 1, 4)$.

a Find the area of face BCD.

b Find a unit vector normal to the face BCD.

c Find the volume of the tetrahedron.

7 A tetrahedron has vertices at $A(0, 0, 0)$, $B(2, 0, 0)$, $C(1, \sqrt{3}, 0)$ and $D\left(1, \frac{\sqrt{3}}{3}, \frac{2\sqrt{6}}{3}\right)$.

a Show that the tetrahedron is regular.

b Find the volume of the tetrahedron.

8 A tetrahedron $OABC$ has its vertices at the points $O(0, 0, 0)$, $A(1, 2, -1)$, $B(-1, 1, 2)$ and $C(2, -1, 1)$.

a Write down expressions for $\overrightarrow{AB}$ and $\overrightarrow{AC}$ in terms of $\mathbf{i}$, $\mathbf{j}$ and $\mathbf{k}$ and find $\overrightarrow{AB} \times \overrightarrow{AC}$. **(3 marks)**

b Deduce the area of triangle ABC. **(2 marks)**

c Find the volume of the tetrahedron. **(3 marks)**

9 The points A, B, C and D have position vectors $\mathbf{a}$, $\mathbf{b}$, $\mathbf{c}$ and $\mathbf{d}$ respectively, where

$\mathbf{a} = 2\mathbf{i} + \mathbf{j}$ $\quad$ $\mathbf{b} = 3\mathbf{i} - \mathbf{j} + \mathbf{k}$ $\quad$ $\mathbf{c} = -2\mathbf{j} - \mathbf{k}$ $\quad$ $\mathbf{d} = 2\mathbf{i} - \mathbf{j} + 3\mathbf{k}$

a Find $\overrightarrow{AB} \times \overrightarrow{BC}$ and $\overrightarrow{BD} \times \overrightarrow{DC}$. **(4 marks)**

b Hence find:

i the area of triangle ABC **(2 marks)**

ii the volume of the tetrahedron $ABCD$. **(3 marks)**

10 The edges OP, OQ and OR of a tetrahedron $OPQR$ are the vectors $\mathbf{a}$, $\mathbf{b}$ and $\mathbf{c}$ respectively, where

$\mathbf{a} = 2\mathbf{i} + 4\mathbf{j}$ $\quad$ $\mathbf{b} = 2\mathbf{i} - \mathbf{j} + 3\mathbf{k}$ $\quad$ $\mathbf{c} = 4\mathbf{i} - 2\mathbf{j} + 5\mathbf{k}$

a Evaluate $\mathbf{b} \times \mathbf{c}$ and deduce that OP is perpendicular to the plane OQR. **(4 marks)**

b Write down the length of OP and the area of triangle OQR and hence the volume of the tetrahedron. **(3 marks)**

c Verify your result by evaluating $\mathbf{a}.(\mathbf{b} \times \mathbf{c})$. **(2 marks)**

11 An architect is designing landscaping sculptures in the shape of tetrahedra. She designs them in 3D software with the origin as her starting point. The position vectors of vertices A, B and C from the origin are $3\mathbf{i} + 2\mathbf{j} + \mathbf{k}$, $2\mathbf{i} - \mathbf{j} - 4\mathbf{k}$ and $-2\mathbf{i} + 4\mathbf{j} - 2\mathbf{k}$.

a Find $\overrightarrow{OB} \times \overrightarrow{OC}$. **(3 marks)**

She prints solid prototype models using a 3D printer and a scale of 1 unit in her design representing 2 cm on the model. The density of the plastic used by the printer is 1.13 g/cm^3.

b Find, to the nearest gram, the mass of one prototype model. **(5 marks)**

12 A scientist is studying the crystal structure of a mineral. The crystal forms a lattice with parallelepipedal unit cells. He models one cell as having vertices with coordinates (0, 0, 0), (0.6, 0.6, 0), (0.9, −0.9 ,0), (−0.4, −0.4, −1.3), (0.2, 0.2, −1.3), (1.1, −0.7, −1.3), (0.5, −1.3, −1.3) and (1.5, −0.3, 0).

Crystallographers measure distances in angstroms, where 10 angstroms is equal to one nanometre (10^{-9} metres).

Find the volume of the unit cell of the crystal, in cubic angstroms, if one unit on the scientist's scale is one nanometre. Give your answer to two significant figures. **(6 marks)**

(E/P) **13** The diagram shows a parallelepiped $ABCEFDHG$.

M is the midpoint of EG. The point N lies on AB such that $AN:NB = 2:1$.

a Find the ratio of the volume of the parallelepiped to the volume of the tetrahedron $NCME$. **(6 marks)**

b State, with justification, how this ratio varies as N moves along the line segment AB. **(2 marks)**

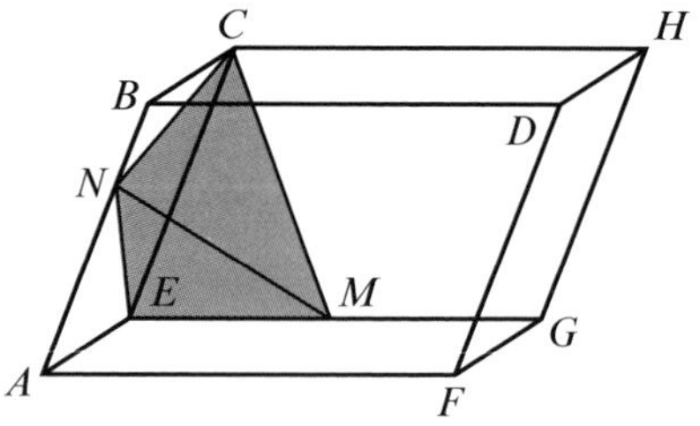

(E/P) **14** The diagram shows a pyramid with base vertices $A(-1, 0, 0)$, $B(0, 2, 1)$, $C(1, 2, 3)$ and $D(0, 0, 2)$. The vertex of the pyramid is at $E(3, 0, 1)$.

Find the exact volume of the pyramid. **(8 marks)**

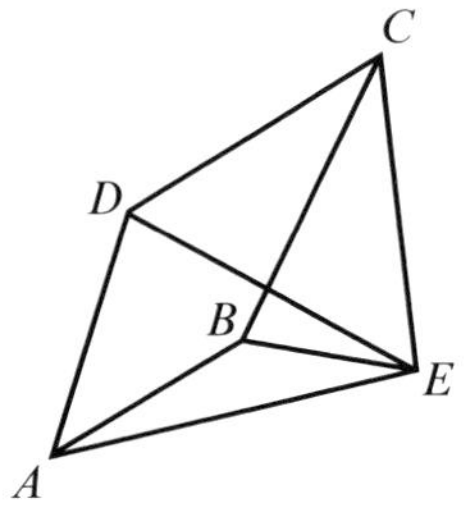

Problem-solving

Split the pyramid into two tetrahedrons.

Challenge

a Explain why $\mathbf{a}.(\mathbf{b} \times \mathbf{c}) = (\mathbf{a} \times \mathbf{b}).\mathbf{c}$.

b Use the result from part **a** to show that $\mathbf{d}.(\mathbf{a} \times \mathbf{b} + \mathbf{a} \times \mathbf{c}) = \mathbf{d}.(\mathbf{a} \times (\mathbf{b} + \mathbf{c}))$.

c Hence deduce that $\mathbf{a} \times \mathbf{b} + \mathbf{a} \times \mathbf{c} = \mathbf{a} \times (\mathbf{b} + \mathbf{c})$.

1.4 Straight lines

A

You can use the vector product to write a vector equation of a line in a form that doesn't require a parameter. Suppose that **a** is the position vector of a point on a line, and that the line is parallel to the vector **b**.

Let **r** be the position vector of a general point on the line.

$\overrightarrow{AR} = \overrightarrow{OR} - \overrightarrow{OA}$

$= \mathbf{r} - \mathbf{a}$

Since $\overrightarrow{AR}$ is parallel to **b**, $\overrightarrow{AR} \times \mathbf{b} = \mathbf{0}$.

So $(\mathbf{r} - \mathbf{a}) \times \mathbf{b} = \mathbf{0}$

- **$(\mathbf{r} - \mathbf{a}) \times \mathbf{b} = \mathbf{0}$ is an alternative form of the vector equation of a line passing through the point A with position vector a, and parallel to the vector b.**

  This may also be written as $\mathbf{r} \times \mathbf{b} = \mathbf{a} \times \mathbf{b}$.

Links A vector equation of a straight line passing through a point A with position vector **a**, and parallel to the vector **b**, is $\mathbf{r} = \mathbf{a} + \lambda\mathbf{b}$, where λ is a scalar parameter.
← **Core Pure Book 1, Chapter 9**

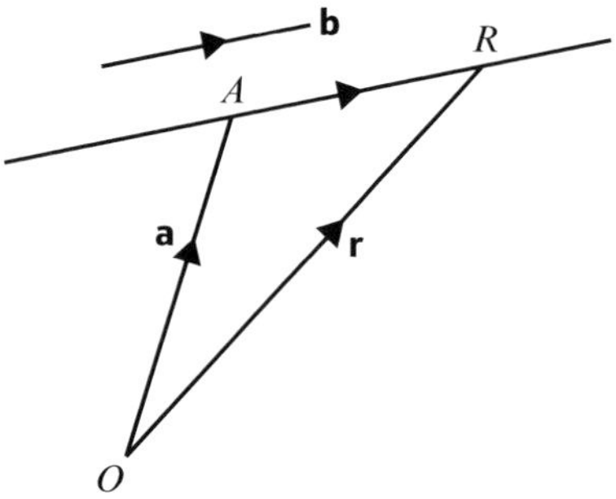

Online Explore the vector equation of a line, written using a cross product, with GeoGebra.

Example 16

Find the vector equation of the line through the points (1, 2, −1) and (3, −2, 2) in the form $(\mathbf{r} - \mathbf{a}) \times \mathbf{b} = \mathbf{0}$.

The line is in the direction $\begin{pmatrix} 3 \\ -2 \\ 2 \end{pmatrix} - \begin{pmatrix} 1 \\ 2 \\ -1 \end{pmatrix} = \begin{pmatrix} 2 \\ -4 \\ 3 \end{pmatrix}$

Any multiple of this vector is also parallel to the direction of the line.

So the equation is $\left(\mathbf{r} - \begin{pmatrix} 3 \\ -2 \\ 2 \end{pmatrix}\right) \times \begin{pmatrix} 2 \\ -4 \\ 3 \end{pmatrix} = \mathbf{0}$

You could use the position vector $\begin{pmatrix} 1 \\ 2 \\ -1 \end{pmatrix}$ instead of $\begin{pmatrix} 3 \\ -2 \\ 2 \end{pmatrix}$ in this equation.

You can use the direction vector of a straight line to find the angles α, β and γ that the line makes with the positive x-, y- and z-axes respectively. The angles α, β and γ lie in the range $0 \leqslant \alpha, \beta, \gamma \leqslant 180°$.

- **If a line is parallel to the vector $\mathbf{a} = x\mathbf{i} + y\mathbf{j} + z\mathbf{k}$, the direction ratios of the line are $x : y : z$, and the direction cosines of the line are**

$$\cos\alpha = \frac{x}{|\mathbf{a}|}, \cos\beta = \frac{y}{|\mathbf{a}|} \text{ and } \cos\gamma = \frac{z}{|\mathbf{a}|},$$

and are written as l, m and n respectively.

Links For the vector $\mathbf{a} = x\mathbf{i} + y\mathbf{j} + z\mathbf{k}$, the angle made with the positive x-axis is given by $\cos\alpha = \frac{x}{|\mathbf{a}|}$

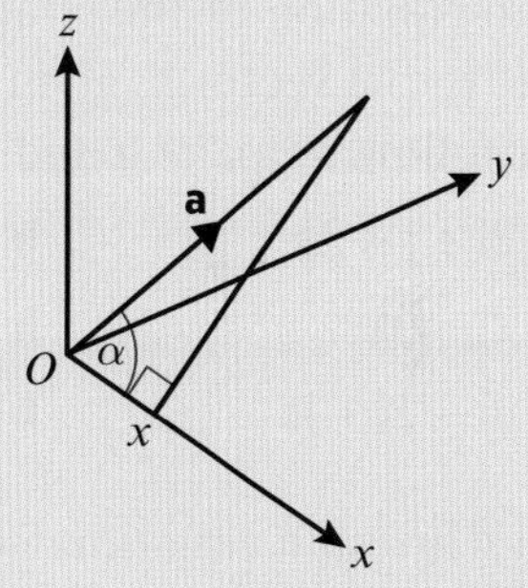

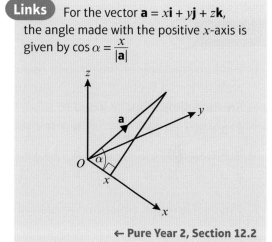

← Pure Year 2, Section 12.2

The sum of the squares of the direction cosines is always 1:

$$l^2 + m^2 + n^2 = \frac{x^2 + y^2 + z^2}{|\mathbf{a}|^2} = \frac{|\mathbf{a}|^2}{|\mathbf{a}|^2} = 1$$

- **A line with direction ratios $x : y : z$ has direction cosines l, m and n such that $l^2 + m^2 + n^2 = 1$.**

Example 17

A line has vector equation $\left(\mathbf{r} - \begin{pmatrix} 1 \\ 2 \\ -1 \end{pmatrix}\right) \times \begin{pmatrix} 4 \\ -3 \\ 2 \end{pmatrix} = \mathbf{0}$.

a Find the direction cosines of the line, l, m and n.

b Show that the Cartesian equation of the line can be written as $\frac{x-1}{l} = \frac{y-2}{m} = \frac{z+1}{n}$

A

a $l = \frac{4}{\sqrt{4^2 + (-3)^2 + 2^2}} = \frac{4}{\sqrt{29}}$

$m = \frac{-3}{\sqrt{4^2 + (-3)^2 + 2^2}} = -\frac{3}{\sqrt{29}}$

$n = \frac{2}{\sqrt{4^2 + (-3)^2 + 2^2}} = \frac{2}{\sqrt{29}}$

Use the **direction vector** of the line in the formulae for l, m and n.

b $\frac{x-1}{4} = \frac{y-2}{-3} = \frac{z+1}{2}$

Write the Cartesian equation of the line using the standard formula $\frac{x-x_1}{a} = \frac{y-y_1}{b} = \frac{z-z_1}{c}$

← Core Pure Book 1, Section 9.1

Multiplying each expression by $\sqrt{29}$,

$$\frac{x-1}{\frac{4}{\sqrt{29}}} = \frac{y-2}{\frac{-3}{\sqrt{29}}} = \frac{z+1}{\frac{2}{\sqrt{29}}}$$

Which is $\frac{x-1}{l} = \frac{y-2}{m} = \frac{z+1}{n}$

Problem-solving

The direction cosines are in the same ratio as the direction ratios.
$l:m:n = x:y:z$

Exercise 1D

1 Find an equation of the straight line passing through the point with position vector **a** which is parallel to the vector **b**, giving your answer in the form **r** × **b** = **c**, where **c** is a vector to be found for the following pairs **a** and **b**:

a **a** = 2**i** + **j** + 2**k**   **b** = 3**i** + **j** − 2**k**
b **a** = 2**i** − 3**k**   **b** = **i** + **j** + 5**k**
c **a** = 4**i** − 2**j** + **k**   **b** = −**i** − 2**j** + 3**k**

2 Find a Cartesian equation for each of the lines given in question **1**.

3 Find, in the form (**r** − **a**) × **b** = **0**, an equation of the straight line passing through the points with coordinates:

a (1, 3, 5), (6, 4, 2)   **b** (3, 4, 12), (4, 3, 5)
c (−2, 2, 6), (3, 7, 11)   **d** (4, 2, −4), (1, 1, 1)

4 Find a Cartesian equation for each of the lines given in question **3**.

5 Find, in the form (**r** − **a**) × **b** = **0**, an equation of the straight line given by the following equations, where λ is a scalar parameter.

a **r** = **i** + **j** − 2**k** + λ(2**i** − **k**)   **b** **r** = **i** + 4**j** + λ(3**i** + **j** − 5**k**)   **c** **r** = 3**i** + 4**j** − 4**k** + λ(2**i** − 2**j** − 3**k**)

6 Find the equation of the straight line with Cartesian equation

$$\frac{x-3}{2} = \frac{y+1}{5} = \frac{2z-3}{3}$$

in the form:

a **r** × **b** = **c**   **b** **r** = **a** + t**b**, where t is a scalar parameter.

A/P 7 Given that the point with coordinates $(p, q, 1)$ lies on the line with equation

$$\mathbf{r} \times \begin{pmatrix} 2 \\ 1 \\ 3 \end{pmatrix} = \begin{pmatrix} 8 \\ -7 \\ -3 \end{pmatrix}$$

find the values of p and q. **(4 marks)**

P 8 Given that the equation of a straight line is

$$\mathbf{r} \times \begin{pmatrix} 1 \\ 1 \\ -1 \end{pmatrix} = \begin{pmatrix} -1 \\ 2 \\ 1 \end{pmatrix}$$

Hint Let $\mathbf{a} = a_1\mathbf{i} + a_2\mathbf{j} + a_3\mathbf{k}$ and set up simultaneous equations.

find an equation for the line in the form $\mathbf{r} = \mathbf{a} + t\mathbf{b}$, where t is a scalar parameter. **(4 marks)**

E 9 A line L passes through the points A and B with position vectors $-3\mathbf{i} + 2\mathbf{j} + 7\mathbf{k}$ and $3\mathbf{i} + 4\mathbf{j} - 5\mathbf{k}$ respectively.

a Find the direction cosines of L. **(3 marks)**

b Hence or otherwise write a Cartesian equation of the line. **(2 marks)**

10 Write down the direction cosines of:

a the x-axis **b** the y-axis **c** the z-axis **d** the line $x = y = z$

E/P 11 Lines L_1 and L_2 intersect and have direction vectors $\mathbf{i} + 2\mathbf{j} + 3\mathbf{k}$ and $3\mathbf{i} + 2\mathbf{j} + \mathbf{k}$ respectively.

a Find the direction cosines l_1, m_1 and n_1 of line L_1. **(3 marks)**

b Find the direction cosines l_2, m_2 and n_2 of line L_2. **(3 marks)**

c Verify that $l_1l_2 + m_1m_2 + n_1n_2 = \cos\theta$ where θ is the angle between the two lines. **(4 marks)**

d Prove that the above result is true for any two intersecting lines. **(6 marks)**

P 12 The direction cosines of two lines L_1 and L_2 are $l_1 = -\frac{1}{\sqrt{11}}, m_1 = \frac{1}{\sqrt{11}}, n_1 = -\frac{3}{\sqrt{11}}$ and $l_2 = \frac{3}{\sqrt{14}}$, $m_2 = -\frac{2}{\sqrt{14}}, n_2 = -\frac{1}{\sqrt{14}}$ respectively.

Find, in radians correct to three significant figures, the acute angle between the two lines.

P 13 A line L makes angles of α, β and γ with the x, y and z-axes respectively.

Show that $\cos 2\alpha + \cos 2\beta + \cos 2\gamma = -1$.

P 14 Find, in degrees correct to one decimal place, the angles that the line segment $\overrightarrow{OP}$ makes with each of the axes given that P has coordinates $(2, 3, 4)$.

P 15 A straight line passes through the origin and makes angles of 45° to the x-axis and 60° with the z-axis. Find two possible equations of the line.

E/P 16 A line L passes through the point $(1, 2, -1)$ and makes equal angles with the axes.

a Find the direction cosines of L. **(3 marks)**

b Hence find the equation of the line in the form $\frac{x-a}{l} = \frac{y-b}{m} = \frac{z-c}{n}$ **(2 marks)**

A E/P **17** A telephone wire is modelled as a straight line in 3D space. **i** and **j** are the horizontal vectors due east and north respectively, and **k** is the vertical unit vector. The units are metres.

An engineer inspects the wire at the point with position vector 6**k**, and finds that it is horizontal, and directed on a bearing of 015°.

a Find a vector equation of the wire, giving your answer in the form $(\mathbf{r} - \mathbf{a}) \times \mathbf{b} = \mathbf{0}$. **(4 marks)**

b Hence show that the wire will intersect with a second wire with vector equation

$$\left(\mathbf{r} - \begin{pmatrix} 5 \\ 2 \\ 1 \end{pmatrix}\right) \times \begin{pmatrix} 5 - 2(\sqrt{6} - \sqrt{2}) \\ 2 - 2(\sqrt{6} + \sqrt{2}) \\ -5 \end{pmatrix} = \mathbf{0}$$ **(3 marks)**

c Give a possible criticism of this model. **(1 mark)**

Challenge

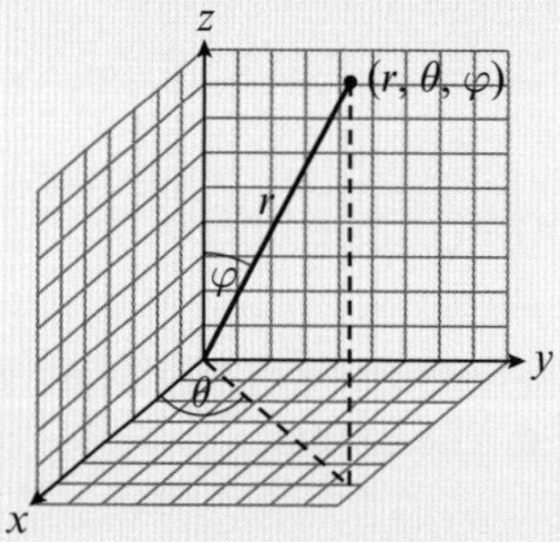

Spherical polar coordinates are defined by the distance from the origin, r, the 'azimuthal angle' (measured anti-clockwise from the x-axis in the xy-plane), θ, and the 'polar angle' (measured from the positive z-axis), φ.

A line L passes through the origin and the point with spherical polar coordinates $\left(3, \frac{\pi}{4}, \frac{\pi}{3}\right)$.

a Find, in their simplest form, the direction cosines of L.

b Find, in terms of θ and φ, expressions for the direction cosines of the line which passes through the origin and the point with spherical coordinates (r, θ, φ).

1.5 Solving geometrical problems

You can use the fact that the vector product **a** × **b** is perpendicular to both **a** and **b** to solve problems involving planes and lines in three dimensions.

Example 18

a Find, in the form **r.n** $= p$, an equation of the plane which contains the line l and the point with position vector **a** where l has equation $\mathbf{r} = 3\mathbf{i} + 5\mathbf{j} - 2\mathbf{k} + \lambda(-\mathbf{i} + 2\mathbf{j} - \mathbf{k})$ and $\mathbf{a} = 4\mathbf{i} + 3\mathbf{j} + \mathbf{k}$.

b Give the equation of the plane in Cartesian form.

A

a The vector $-\mathbf{i} + 2\mathbf{j} - \mathbf{k}$ is perpendicular to $\mathbf{n}$.

The vector $4\mathbf{i} + 3\mathbf{j} + \mathbf{k} - (3\mathbf{i} + 5\mathbf{j} - 2\mathbf{k})$ also lies in the plane and is also perpendicular to $\mathbf{n}$, i.e. $\mathbf{i} - 2\mathbf{j} + 3\mathbf{k}$ is perpendicular to $\mathbf{n}$.

So $\mathbf{n} = \begin{vmatrix} \mathbf{i} & \mathbf{j} & \mathbf{k} \\ -1 & 2 & -1 \\ 1 & -2 & 3 \end{vmatrix}$

$= 4\mathbf{i} + 2\mathbf{j}$

So the equation of the required plane is

$\mathbf{r}.(4\mathbf{i} + 2\mathbf{j}) = (4\mathbf{i} + 3\mathbf{j} + \mathbf{k}).(4\mathbf{i} + 2\mathbf{j})$

$\Rightarrow \mathbf{r}.(4\mathbf{i} + 2\mathbf{j}) = 16 + 6$

An equation of the plane is $\mathbf{r}.(4\mathbf{i} + 2\mathbf{j}) = 22$

b In Cartesian form this may be written as

$4x + 2y = 22$

$\Rightarrow 2x + y = 11$

Line l lies in the plane. The direction of l is $-\mathbf{i} + 2\mathbf{j} - \mathbf{k}$, and so this vector is perpendicular to $\mathbf{n}$.

The point (4, 3, 1) lies in the plane, and the point (3, 5, −2) lies on the line and so also in the plane, so the vector joining these two points also lies in the plane.

This vector $\mathbf{i} - 2\mathbf{j} + 3\mathbf{k}$ is also perpendicular to $\mathbf{n}$.

$\mathbf{n}$ is in the direction of the vector product of $-\mathbf{i} + 2\mathbf{j} - \mathbf{k}$ and $\mathbf{i} - 2\mathbf{j} + 3\mathbf{k}$.

Replace $\mathbf{r}$ with $x\mathbf{i} + y\mathbf{j} + z\mathbf{k}$ and perform the scalar product.

Example 19

Find a Cartesian equation of the plane that passes through the points $A(1, 0, -1)$, $B(2, 1, 0)$ and $C(2, 16, 6)$.

$\overrightarrow{AB} = \overrightarrow{OB} - \overrightarrow{OA} = \mathbf{i} + \mathbf{j} + \mathbf{k}$

$\overrightarrow{AC} = \overrightarrow{OC} - \overrightarrow{OA} = \mathbf{i} + 16\mathbf{j} + 7\mathbf{k}$

$\overrightarrow{AB} \times \overrightarrow{AC} = \begin{vmatrix} \mathbf{i} & \mathbf{j} & \mathbf{k} \\ 1 & 1 & 1 \\ 1 & 16 & 7 \end{vmatrix}$

$= -9\mathbf{i} - 6\mathbf{j} + 15\mathbf{k}$

So $\mathbf{r}.(-9\mathbf{i} - 6\mathbf{j} + 15\mathbf{k}) = (\mathbf{i} - \mathbf{k}).(-9\mathbf{i} - 6\mathbf{j} + 15\mathbf{k})$

$\Rightarrow \mathbf{r}.(-9\mathbf{i} - 6\mathbf{j} + 15\mathbf{k}) = -9 - 15 = -24$

So the equation of the plane may be written as

$\mathbf{r}.(3\mathbf{i} + 2\mathbf{j} - 5\mathbf{k}) = 8$

$\Rightarrow (x\mathbf{i} + y\mathbf{j} + z\mathbf{k}).(3\mathbf{i} + 2\mathbf{j} - 5\mathbf{k}) = 8$

$\Rightarrow 3x + 2y - 5z = 8$, which is a Cartesian equation of the plane.

This is the direction of the normal to the plane.

Use $\mathbf{r.n} = \mathbf{a.n}$, where $\mathbf{a} = \mathbf{i} - \mathbf{k}$

Replace $\mathbf{r}$ by $x\mathbf{i} + y\mathbf{j} + z\mathbf{k}$ to obtain the Cartesian equation.

You may wish to check that each point lies on this plane.

Example 20

A

Find the equation of the line of intersection of the planes Π_1 and Π_2 where Π_1 has equation $\mathbf{r}.(2\mathbf{i} - 2\mathbf{j} - \mathbf{k}) = 2$ and Π_2 has equation $\mathbf{r}.(\mathbf{i} - 3\mathbf{j} + \mathbf{k}) = 5$.

Direction vector of line is given by

$$\begin{vmatrix} \mathbf{i} & \mathbf{j} & \mathbf{k} \\ 2 & -2 & -1 \\ 1 & -3 & 1 \end{vmatrix} = \begin{pmatrix} -5 \\ -3 \\ -4 \end{pmatrix}$$

$\begin{pmatrix} 2 \\ -2 \\ -1 \end{pmatrix}$ is normal to Π_1 and $\begin{pmatrix} 1 \\ -3 \\ 1 \end{pmatrix}$ is normal to Π_2. The line must be perpendicular to both normal vectors, so you can use the vector product to find its direction vector.

Π_1: $2x - 2y - z = 2$

Π_2: $x - 3y + z = 5$

Write Cartesian equations of both planes. Fix the value of one variable and solve simultaneously to find a point on the line. Setting $z = 0$ simplifies the calculation.

Set $z = 0$ and solve simultaneously:

$$\left.\begin{aligned} 2x - 2y &= 2 \\ x - 3y &= 5 \end{aligned}\right\} \Rightarrow x = -1,\ y = -2$$

So $(-1, -2, 0)$ lies on the line, and the equation for the line is

$$\mathbf{r} = \begin{pmatrix} -1 \\ -2 \\ 0 \end{pmatrix} + \lambda\begin{pmatrix} 5 \\ 3 \\ 4 \end{pmatrix}$$

Problem-solving

You could also find two points on the line by setting $z = 0$, and also setting $x = 0$ (for example), then use these to find an equation for the line.

Example 21

Show that the shortest distance between the two skew lines with equations $\mathbf{r} = \mathbf{a} + \lambda\mathbf{b}$ and $\mathbf{r} = \mathbf{c} + \mu\mathbf{d}$, where λ and μ are scalars, is given by the formula $\left|\dfrac{(\mathbf{a} - \mathbf{c}).(\mathbf{b} \times \mathbf{d})}{|\mathbf{b} \times \mathbf{d}|}\right|$.

The shortest distance between the lines is XY where XY is perpendicular to both lines.

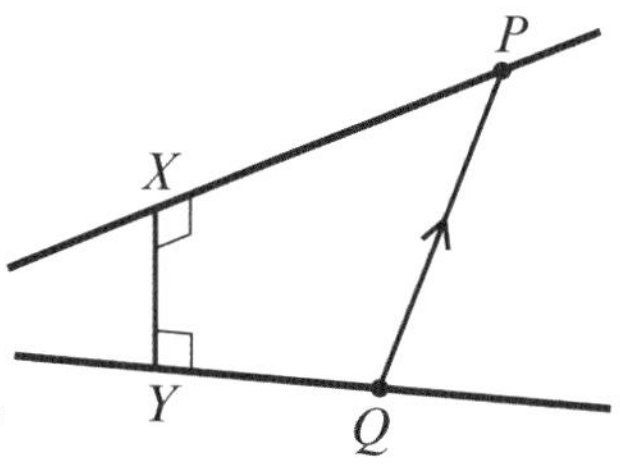

The common perpendicular to the two skew lines is in the direction $\mathbf{b} \times \mathbf{d}$ and a unit vector in that direction is $\dfrac{\mathbf{b} \times \mathbf{d}}{|\mathbf{b} \times \mathbf{d}|}$

If P is a point on the line with equation $\mathbf{r} = \mathbf{a} + \lambda\mathbf{b}$ and Q is a point on the line with equation $\mathbf{r} = \mathbf{c} + \mu\mathbf{d}$ then

$$\overrightarrow{QP} = \mathbf{a} - \mathbf{c} + \lambda\mathbf{b} - \mu\mathbf{d}$$

The projection of PQ in the direction of the common perpendicular is

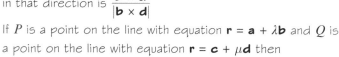

$$(\mathbf{a} - \mathbf{c} + \lambda\mathbf{b} - \mu\mathbf{d}).\frac{\mathbf{b} \times \mathbf{d}}{|\mathbf{b} \times \mathbf{d}|}$$

This gives $PQ\cos\theta$, where θ is the angle between PQ and the common perpendicular.

$= \quad (\mathbf{a} - \mathbf{c}).\dfrac{\mathbf{b} \times \mathbf{d}}{|\mathbf{b} \times \mathbf{d}|} + \lambda\mathbf{b}.\dfrac{\mathbf{b} \times \mathbf{d}}{|\mathbf{b} \times \mathbf{d}|} - \mu\mathbf{d}.\dfrac{\mathbf{b} \times \mathbf{d}}{|\mathbf{b} \times \mathbf{d}|}$ — Using the distributive property.

But $\mathbf{b}.(\mathbf{b} \times \mathbf{d}) = \mathbf{d}.(\mathbf{b} \times \mathbf{d}) = 0$ and the shortest distance must be a positive quantity, so the shortest distance is — $\mathbf{b} \times \mathbf{d}$ is perpendicular to both **b** and **d**.

given by $\left|\dfrac{(\mathbf{a} - \mathbf{c}).(\mathbf{b} \times \mathbf{d})}{|\mathbf{b} \times \mathbf{d}|}\right|$. — Use the modulus to ensure that the result is positive.

- **The shortest distance between the two skew lines with equations $\mathbf{r} = \mathbf{a} + \lambda\mathbf{b}$ and $\mathbf{r} = \mathbf{c} + \mu\mathbf{d}$, where λ and μ are scalars, is given by the formula**

$$\left|\frac{(\mathbf{a} - \mathbf{c}).(\mathbf{b} \times \mathbf{d})}{|\mathbf{b} \times \mathbf{d}|}\right|$$

Example 22

Find the shortest distance between the two skew lines with equations $\mathbf{r} = \mathbf{i} + \lambda(\mathbf{j} + \mathbf{k})$ and $\mathbf{r} = -\mathbf{i} + 3\mathbf{j} - \mathbf{k} + \mu(2\mathbf{i} - \mathbf{j} - \mathbf{k})$, where λ and μ are scalars.

$\mathbf{a} - \mathbf{c} = \mathbf{i} - (-\mathbf{i} + 3\mathbf{j} - \mathbf{k}) = 2\mathbf{i} - 3\mathbf{j} + \mathbf{k}$ — Use $\mathbf{a} = \mathbf{i}$ and $\mathbf{c} = -\mathbf{i} + 3\mathbf{j} - \mathbf{k}$.

$\mathbf{b} \times \mathbf{d} = \begin{vmatrix} \mathbf{i} & \mathbf{j} & \mathbf{k} \\ 0 & 1 & 1 \\ 2 & -1 & -1 \end{vmatrix} = 2\mathbf{j} - 2\mathbf{k}$ — Take the vector product of the two direction vectors.

So the shortest distance is $\left|\dfrac{(2\mathbf{i} - 3\mathbf{j} + \mathbf{k}).(2\mathbf{j} - 2\mathbf{k})}{\sqrt{2^2 + (-2)^2}}\right| = \left|\dfrac{-8}{\sqrt{8}}\right| = \sqrt{8}$ — Use the formula for shortest distance.

$= 2\sqrt{2}$

Exercise 1E

1 Find a Cartesian equation of the plane that passes through the points:

a (0, 4, 2), (1, 1, 2) and (−1, 5, 0)
b (1, 1, 0), (2, 3, −3) and (3, 7, −2)
c (3, 0, 0), (2, 0, −1) and (4, 1, 3)
d (1, −1, 6), (3, 1, −2) and (4, 1, 0)

2 Find, in the form $\mathbf{r}.\mathbf{n} = p$, an equation of the plane which contains the line l and the point with position vector **a** where:

a l has equation $\mathbf{r} = \mathbf{i} + \mathbf{j} - 2\mathbf{k} + \lambda(2\mathbf{i} - \mathbf{k})$ and $\mathbf{a} = 4\mathbf{i} + 3\mathbf{j} + \mathbf{k}$

b l has equation $\mathbf{r} = \mathbf{i} + 2\mathbf{j} + 2\mathbf{k} + \lambda(2\mathbf{i} + \mathbf{j} - 3\mathbf{k})$ and $\mathbf{a} = 3\mathbf{i} + 5\mathbf{j} + \mathbf{k}$

c l has equation $\mathbf{r} = 2\mathbf{i} - \mathbf{j} + \mathbf{k} + \lambda(\mathbf{i} + 2\mathbf{j} + 2\mathbf{k})$ and $\mathbf{a} = 7\mathbf{i} + 8\mathbf{j} + 6\mathbf{k}$

3 Find the equation of the line of intersection of the planes Π_1 and Π_2 where:

a Π_1 has equation $\mathbf{r}.(3\mathbf{i} - 2\mathbf{j} - \mathbf{k}) = 5$ and Π_2 has equation $\mathbf{r}.(4\mathbf{i} - \mathbf{j} - 2\mathbf{k}) = 5$

b Π_1 has equation $\mathbf{r}.(5\mathbf{i} - \mathbf{j} - 2\mathbf{k}) = 16$ and Π_2 has equation $\mathbf{r}.(16\mathbf{i} - 5\mathbf{j} - 4\mathbf{k}) = 53$

c Π_1 has equation $\mathbf{r}.(\mathbf{i} - 3\mathbf{j} + \mathbf{k}) = 10$ and Π_2 has equation $\mathbf{r}.(4\mathbf{i} - 3\mathbf{j} - 2\mathbf{k}) = 1$

A

4 Find the acute angle between the line with equation $(\mathbf{r} - 3\mathbf{j}) \times (-4\mathbf{i} - 7\mathbf{j} + 4\mathbf{k}) = \mathbf{0}$ and the plane with equation $\mathbf{r} = \lambda(4\mathbf{i} - \mathbf{j} - \mathbf{k}) + \mu(4\mathbf{i} - 5\mathbf{j} + 3\mathbf{k})$.

5 Find the shortest distance between the two skew lines with equations
$\mathbf{r} = \mathbf{i} + \lambda(-3\mathbf{i} - 12\mathbf{j} + 11\mathbf{k})$ and $\mathbf{r} = 3\mathbf{i} - \mathbf{j} + \mathbf{k} + \mu(2\mathbf{i} + 6\mathbf{j} - 5\mathbf{k})$, where λ and μ are scalars.

6 The plane Π has equation $\mathbf{r}.(\mathbf{i} + \mathbf{j} - \mathbf{k}) = 4$.

a Show that the line with equation $\mathbf{r} = 2\mathbf{i} + 3\mathbf{j} + \mathbf{k} + \lambda(-\mathbf{i} + 2\mathbf{j} + \mathbf{k})$ lies in the plane Π.

b Show that the line with equation $\mathbf{r} = -\mathbf{i} + 2\mathbf{j} + 4\mathbf{k} + \lambda(-\mathbf{i} + 2\mathbf{j} + \mathbf{k})$ is parallel to the plane Π and find the shortest distance from the line to the plane.

E **7** A tetrahedron has vertices at $A(1, 2, 3)$, $B(0, 1, -2)$, $C(3, 6, 1)$ and $D(5, -2, 4)$. Find:

a the Cartesian equation of the plane ABC **(3 marks)**

b the volume of the tetrahedron $ABCD$. **(3 marks)**

The normal to the plane ABC through point D intersects the plane at point E.

c Find the angle CDE, giving your answer in radians correct to three decimal places. **(5 marks)**

E **8** The lines L_1 and L_2 have equations

$$L_1: \mathbf{r} = \begin{pmatrix} -1 \\ 0 \\ 1 \end{pmatrix} + \lambda\begin{pmatrix} 3 \\ 3 \\ -2 \end{pmatrix}$$

$$L_2: \mathbf{r} = \begin{pmatrix} a \\ 4 \\ -4 \end{pmatrix} + \mu\begin{pmatrix} 2 \\ 1 \\ -3 \end{pmatrix}$$

If the lines L_1 and L_2 intersect, find:

a the value of a **(4 marks)**

b an equation for the plane containing the lines L_1 and L_2, giving your answer in the form $ax + by + cz + d = 0$, where a, b, c and d are integer constants. **(4 marks)**

For other values of a, the lines L_1 and L_2 do not intersect and are skew lines.

c Given that $a = 1$, find the shortest distance between the lines L_1 and L_2. **(3 marks)**

E **9** The plane Π has equation

$$\mathbf{r} = \begin{pmatrix} 1 \\ -2 \\ 1 \end{pmatrix} + \lambda\begin{pmatrix} -1 \\ 2 \\ -2 \end{pmatrix} + \mu\begin{pmatrix} 0 \\ 2 \\ -1 \end{pmatrix}$$

a Find a unit vector perpendicular to the plane Π. **(3 marks)**

The line l passes through the point $A(2, 3, 2)$ and meets Π at $(1, -2, 1)$.

The acute angle between the plane Π and the line l is α.

b Find α to the nearest degree. **(4 marks)**

c Find the perpendicular distance from A to the plane Π. **(4 marks)**

A E **10** The plane Π_1 has Cartesian equation $2x - y + 3z - 1 = 0$.

a Find the perpendicular distance from the point $(3, -3, 2)$ to the plane Π_1. **(3 marks)**

The plane Π_2 has vector equation

$$\mathbf{r} = \lambda\begin{pmatrix}-2\\-4\\0\end{pmatrix} + \mu\begin{pmatrix}-1\\4\\3\end{pmatrix},$$

where λ and μ are scalar parameters.

b Find the acute angle between Π_1 and Π_2 giving your answer in radians to three significant figures. **(5 marks)**

c Find a vector equation of the line of intersection of the two planes. **(6 marks)**

P **11** The plane Π_1 has vector equation

$$\mathbf{r} = \begin{pmatrix}2\\1\\3\end{pmatrix} + \lambda\begin{pmatrix}-1\\-2\\0\end{pmatrix} + \mu\begin{pmatrix}4\\-2\\3\end{pmatrix}$$

where λ and μ are real parameters.

Π_1 is transformed to the plane Π_2 by the transformation represented by the matrix $\mathbf{T}$, where

$$\mathbf{T} = \begin{pmatrix}1 & 0 & 2\\0 & 1 & -3\\0 & 2 & 1\end{pmatrix}$$

Find an equation of Π_2 in the form $\mathbf{r.n} = p$. **(9 marks)**

P **12** Four planes have Cartesian equations

Π_1: $2x - y + 3z = 1$ $\quad$ Π_2: $x + y - 3z = 2$ $\quad$ Π_3: $3x - 2y - z = 4$ $\quad$ Π_4: $x + y = 0$

Find the volume of the finite space enclosed by all four planes.

Challenge

a Show that the plane $x + y + z = 0$ is invariant under the linear transformation represented by the matrix $\begin{pmatrix}2 & -1 & 2\\2 & 2 & -1\\-1 & 2 & 2\end{pmatrix}$.

b Show that the only invariant point in this plane is the origin.

Mixed exercise 1

E **1** The points A, B and C have position vectors **a**, **b** and **c** respectively, relative to a fixed origin O, as shown in the diagram.

$\mathbf{a} = 2\mathbf{i} + 3\mathbf{j}$ $\qquad$ $\mathbf{b} = \mathbf{i} - 2\mathbf{j} + 2\mathbf{k}$ $\qquad$ $\mathbf{c} = 3\mathbf{i} + 2\mathbf{j} - 4\mathbf{k}$

Calculate:

a $\mathbf{b} \times \mathbf{c}$ **(3 marks)**

b $\mathbf{a}.(\mathbf{b} \times \mathbf{c})$ **(2 marks)**

c the area of triangle OBC **(2 marks)**

d the volume of tetrahedron $OABC$. **(1 mark)**

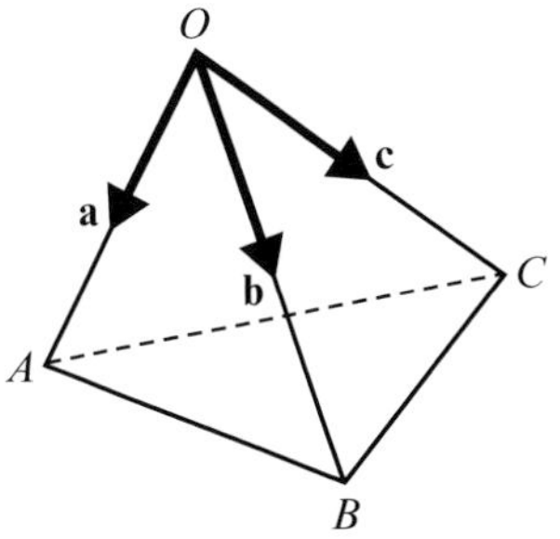

(E/P) **2** A soft drinks manufacturer is designing a package in the shape of a tetrahedron. He designs it in 3D software with the origin as his starting point. The position vectors of vertices A, B and C from the origin are $2\mathbf{i} + \mathbf{j} + 3\mathbf{k}$, $\mathbf{i} - 4\mathbf{j} - 3\mathbf{k}$ and $-\mathbf{i} + 3\mathbf{j} - \mathbf{k}$ respectively.

a Find $\overrightarrow{OB} \times \overrightarrow{OC}$. **(3 marks)**

He prints prototype packages using a 3D printer and a scale of 1 unit in the design representing 4 cm on the model.

b Given that the thickness of the plastic can be considered negligible, find, in cm^3, the volume of one prototype package. **(4 marks)**

(E/P) **3** The diagram shows a parallelepiped $ABCEFDHG$ with vertices $A(0, 0, 0)$, $E(3, -1, 2)$, $C(4, 1, -2)$, and $F(2, -5, 1)$.

A tetrahedron is formed by joining vertices A, C and E to the point M on side EF such that the ratio $EM : MF$ is $2 : 1$.

Show that the volume of the tetrahedron is $\frac{1}{9}$ of the volume of the parallelepiped. **(8 marks)**

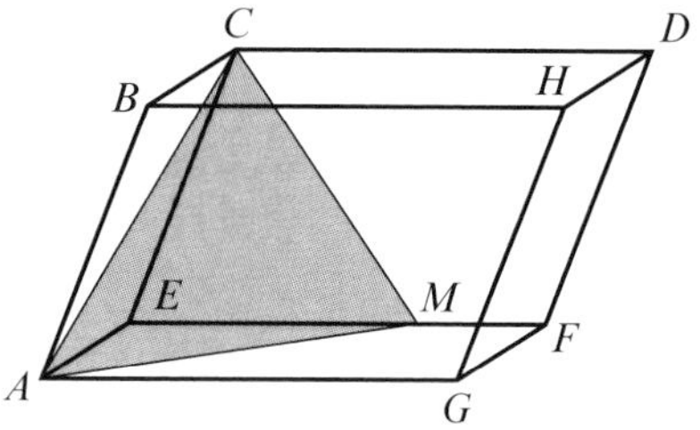

(E/P) **4** Relative to an origin O, the points A and B have position vectors $\mathbf{a}$ metres and $\mathbf{b}$ metres respectively, where

$$\mathbf{a} = 5\mathbf{i} + 2\mathbf{j} \qquad \mathbf{b} = 2\mathbf{i} - \mathbf{j} - 3\mathbf{k}$$

The point C moves such that the volume of the tetrahedron $OABC$ is always $5\,\text{m}^3$.

Determine Cartesian equations of the locus of possible positions of point C. **(6 marks)**

(E/P) **5** The lines L_1 and L_2 have equations $\mathbf{r} = \mathbf{a}_1 + s\mathbf{b}_1$ and $\mathbf{r} = \mathbf{a}_2 + t\mathbf{b}_2$ respectively, where

$$\mathbf{a}_1 = 3\mathbf{i} - 3\mathbf{j} - 2\mathbf{k} \qquad \mathbf{b}_1 = \mathbf{j} + 2\mathbf{k}$$
$$\mathbf{a}_2 = 8\mathbf{i} + 3\mathbf{j} \qquad \mathbf{b}_2 = 5\mathbf{i} + 4\mathbf{j} - 2\mathbf{k}.$$

a Verify that the point P with position vector $3\mathbf{i} - \mathbf{j} + 2\mathbf{k}$ lies on both L_1 and L_2. **(2 marks)**

b Find $\mathbf{b}_1 \times \mathbf{b}_2$. **(3 marks)**

c Find a Cartesian equation of the plane containing L_1 and L_2. **(4 marks)**

The points with position vectors $\mathbf{a}_1$ and $\mathbf{a}_2$ are A_1 and A_2 respectively.

d By expressing $\overrightarrow{A_1P}$ and $\overrightarrow{A_2P}$ as multiples of $\mathbf{b}_1$ and $\mathbf{b}_2$ respectively, or otherwise, find the area of the triangle PA_1A_2. **(3 marks)**

(A) (E) **6** The position vectors of the points A, B, C and D relative to a fixed origin O, are $-\mathbf{j} + 2\mathbf{k}$, $\mathbf{i} - 3\mathbf{j} + 5\mathbf{k}$, $2\mathbf{i} - 2\mathbf{j} + 7\mathbf{k}$ and $\mathbf{j} + 2\mathbf{k}$ respectively.

a Find $\mathbf{p} = \overrightarrow{AB} \times \overrightarrow{CD}$. **(3 marks)**

b Calculate $\overrightarrow{AC}.\mathbf{p}$. **(2 marks)**

c Hence determine the shortest distance between the line containing AB and the line containing CD. **(3 marks)**

(E) **7** Relative to a fixed origin O, the point M has position vector $-4\mathbf{i} + \mathbf{j} - 2\mathbf{k}$.

The straight line l has equation $\mathbf{r} \times \overrightarrow{OM} = 5\mathbf{i} - 10\mathbf{k}$.

a Express the equation of the line l in the form $\mathbf{r} = \mathbf{a} + t\mathbf{b}$, where $\mathbf{a}$ and $\mathbf{b}$ are constant vectors and t is a parameter. **(3 marks)**

A **b** Verify that the point N with coordinates (2, −3, 1) lies on l and find the area of triangle OMN. **(4 marks)**

E **8** A plane passes through the three points A, B, C, whose position vectors, referred to an origin O, are $(\mathbf{i} + 3\mathbf{j} + 3\mathbf{k})$, $(3\mathbf{i} + \mathbf{j} + 4\mathbf{k})$, $(2\mathbf{i} + 4\mathbf{j} + \mathbf{k})$ respectively.

a Find, in the form $l\mathbf{i} + m\mathbf{j} + n\mathbf{k}$, a unit normal vector to this plane. **(4 marks)**

b Find also a Cartesian equation of the plane. **(3 marks)**

c Find the perpendicular distance from the origin to this plane. **(3 marks)**

E **9** **a** Show that the vector $\mathbf{i} + \mathbf{k}$ is perpendicular to the plane with vector equation $\mathbf{r} = \mathbf{i} + s\mathbf{j} + t(\mathbf{i} - \mathbf{k})$. **(2 marks)**

b Find the perpendicular distance from the origin to this plane. **(3 marks)**

c Hence or otherwise obtain a Cartesian equation of the plane. **(3 marks)**

E **10** The points A, B and C have position vectors $\mathbf{i} + \mathbf{j} + \mathbf{k}$, $5\mathbf{i} - 2\mathbf{j} + \mathbf{k}$ and $3\mathbf{i} + 2\mathbf{j} + 6\mathbf{k}$ respectively, referred to an origin O.

a Find a vector perpendicular to the plane containing the points A, B and C. **(3 marks)**

b Hence, or otherwise, find an equation for the plane which contains the points A, B and C, in the form $ax + by + cz + d = 0$. **(3 marks)**

The point D has coordinates (1, 5, 6).

c Find the volume of the tetrahedron $ABCD$. **(4 marks)**

E **11** The plane Π passes through $A(3, -5, -1)$, $B(-1, 5, 7)$ and $C(2, -3, 0)$.

a Find $\overrightarrow{AC} \times \overrightarrow{BC}$. **(3 marks)**

b Hence, or otherwise, find the equation, in the form $\mathbf{r}.\mathbf{n} = p$, of the plane Π. **(3 marks)**

c The perpendicular from the point (2, 3, −2) to Π meets the plane at P. Find the coordinates of P. **(4 marks)**

E/P **12** Given that P and Q are the points with position vectors $\mathbf{p}$ and $\mathbf{q}$ respectively, relative to an origin O, and that $\mathbf{p} = 3\mathbf{i} - \mathbf{j} + 2\mathbf{k}$ and $\mathbf{q} = 2\mathbf{i} + \mathbf{j} - \mathbf{k}$,

a find $\mathbf{p} \times \mathbf{q}$. **(3 marks)**

b Hence, or otherwise, find an equation of the plane containing O, P and Q in the form $ax + by + cz = d$. **(3 marks)**

The line with equation $(\mathbf{r} - \mathbf{p}) \times \mathbf{q} = \mathbf{0}$ meets the plane with equation $\mathbf{r}.(\mathbf{i} + \mathbf{j} + \mathbf{k}) = 2$ at the point T.

c Find the coordinates of the point T. **(4 marks)**

E **13** The planes Π_1 and Π_2 are defined by the equations $2x + 2y - z = 9$ and $x - 2y = 7$ respectively.

a Find the acute angle between Π_1 and Π_2, giving your answer to the nearest degree. **(3 marks)**

b Find in the form $\mathbf{r} \times \mathbf{u} = \mathbf{v}$ an equation of the line of intersection of Π_1 and Π_2. **(4 marks)**

A E/P **14** The plane Π has vector equation

$$\mathbf{r} = \begin{pmatrix} 1 \\ 3 \\ 4 \end{pmatrix} + u\begin{pmatrix} 4 \\ 1 \\ 2 \end{pmatrix} + v\begin{pmatrix} 3 \\ 2 \\ -1 \end{pmatrix}$$

where u and v are parameters.

The line L has vector equation

$$\mathbf{r} = \begin{pmatrix} 2 \\ 1 \\ -3 \end{pmatrix} + t\begin{pmatrix} 2 \\ 3 \\ -4 \end{pmatrix}$$

where t is a parameter.

a Show that L is parallel to Π. **(4 marks)**

b Find the shortest distance between L and Π. **(3 marks)**

E **15** The plane Π has equation $2x + y + 3z = 21$ and the origin is O. The line l passes through the point $P(1, 2, 1)$ and is perpendicular to Π.

a Find a vector equation of l. **(3 marks)**

The line l meets the plane Π at the point M.

b Find the coordinates of M. **(3 marks)**

c Find $\overrightarrow{OP} \times \overrightarrow{OM}$. **(3 marks)**

d Hence, or otherwise, find the distance from P to the line OM, giving your answer in surd form.

The point Q is the reflection of P in Π. **(3 marks)**

e Find the coordinates of Q. **(3 marks)**

E/P **16** In a tetrahedron $ABCD$ the coordinates of the vertices B, C, D are $(1, 2, 3)$, $(2, 3, 3)$ and $(3, 2, 4)$ respectively. Find:

a the Cartesian equation of the plane BCD **(4 marks)**

b the sine of the angle between BC and the plane $x + 2y + 3z = 4$. **(3 marks)**

c If AC and AD are perpendicular to BD and BC respectively and if $AB = \sqrt{26}$, find the coordinates of the two possible positions of A. **(4 marks)**

E **17** Points A and B have position vectors $-2\mathbf{i} + \mathbf{j} + 5\mathbf{k}$ and $4\mathbf{i} + 2\mathbf{j} - 3\mathbf{k}$ respectively.

a Find the direction ratios of $\overrightarrow{AB}$. **(3 marks)**

b Find the direction cosines l, m and n of $\overrightarrow{AB}$. **(3 marks)**

c Write down the Cartesian equation of the line through A and B in the form

$$\frac{x - x_1}{l} = \frac{y - y_1}{m} = \frac{z - z_1}{n}$$ **(2 marks)**

P **18** A line L makes angles α, β and γ with the x-, y- and z-axes respectively.

Prove that $\sin^2\alpha + \sin^2\beta + \sin^2\gamma = 2$.

P **19** Two lines L_1 and L_2 have direction cosines equal to l_1, m_1, n_1 and l_2, m_2, n_2 respectively.

Show that if the two lines are parallel, then $\frac{l_1}{l_2} = \frac{m_1}{m_2} = \frac{n_1}{n_2}$

A P **20** A radio mast is modelled as a straight rod in 3D space. It is supported by guide wires W_1 and W_2 which are modelled as straight lines. W_1 passes through the origin and makes angles of 45°, 60° and 60° with the x-, y- and z-axes respectively.

The wire attaches to the pylon at point A.

a W_2 has vector equation $\mathbf{r} = \begin{pmatrix} \frac{8+3\sqrt{2}}{4} \\ 0 \\ \frac{5\sqrt{2}}{4} \end{pmatrix} + \lambda\begin{pmatrix} 3 \\ -4 \\ 1 \end{pmatrix}$.

Show that W_2 also passes through A and find the coordinates of A. **(7 marks)**

b The base of the pylon, B, lies in the xy-plane and the pylon is perpendicular to the xy-plane. Given that each unit in the model represents 10 m, find the distance that B is from the origin. **(4 marks)**

c Give one criticism of the model. **(1 mark)**

P **21** The plane Π_1 has vector equation

$$\mathbf{r} = \begin{pmatrix} 1 \\ 1 \\ -1 \end{pmatrix} + \lambda\begin{pmatrix} 0 \\ 3 \\ 2 \end{pmatrix} + \mu\begin{pmatrix} -4 \\ 1 \\ 2 \end{pmatrix}$$

where λ and μ are real parameters.

The plane Π_1 is transformed to the plane Π_2 by the transformation represented by the matrix $\mathbf{T}$, where

$$\mathbf{T} = \begin{pmatrix} 1 & 0 & 3 \\ 1 & -2 & -1 \\ -1 & 0 & 2 \end{pmatrix}$$

Show that the equation of the plane Π_2 can be written as $\mathbf{r}.\begin{pmatrix} 4 \\ 5 \\ 4 \end{pmatrix} = d$ where d is a constant to be found. **(9 marks)**

Challenge

The plane Π cuts the x-, y- and z-axes at the points $(p, 0, 0)$, $(0, q, 0)$ and $(0, 0, r)$ respectively. Given that the shortest distance between the plane and the origin is d, prove that

$$\frac{1}{p^2} + \frac{1}{q^2} + \frac{1}{r^2} = \frac{1}{d^2}$$

Summary of key points

1 The **scalar** (or **dot**) **product** of two vectors **a** and **b** is written as **a.b**, and defined as

$$\mathbf{a}.\mathbf{b} = |\mathbf{a}||\mathbf{b}|\cos\theta$$

where θ is the angle between **a** and **b**.

2 The **vector** (or **cross**) **product** of the vectors **a** and **b** is defined as

$$\mathbf{a} \times \mathbf{b} = |\mathbf{a}||\mathbf{b}|\sin\theta\,\hat{\mathbf{n}}$$

where θ is the angle between **a** and **b**.

3 $\mathbf{b} \times \mathbf{a} = -\mathbf{a} \times \mathbf{b}$

4 If **i**, **j** and **k** are unit vectors along the x-, y- and z-axes respectively, then:

- $\mathbf{i} \times \mathbf{i} = \mathbf{0}$
- $\mathbf{j} \times \mathbf{j} = \mathbf{0}$
- $\mathbf{k} \times \mathbf{k} = \mathbf{0}$
- $\mathbf{i} \times \mathbf{j} = \mathbf{k}$ and $\mathbf{j} \times \mathbf{i} = -\mathbf{k}$
- $\mathbf{j} \times \mathbf{k} = \mathbf{i}$ and $\mathbf{k} \times \mathbf{j} = -\mathbf{i}$
- $\mathbf{k} \times \mathbf{i} = \mathbf{j}$ and $\mathbf{i} \times \mathbf{k} = -\mathbf{j}$

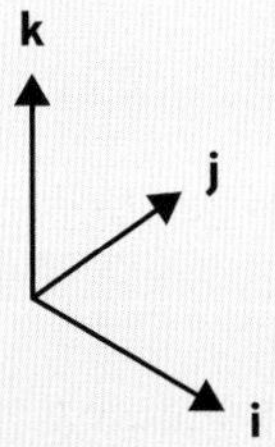

5 If $\mathbf{a} \times \mathbf{b} = \mathbf{0}$ then either $\mathbf{a} = \mathbf{0}$, $\mathbf{b} = \mathbf{0}$ or **a** and **b** are parallel.

6 $\mathbf{a} \times \mathbf{b} = (a_2b_3 - a_3b_2)\mathbf{i} + (a_3b_1 - a_1b_3)\mathbf{j} + (a_1b_2 - a_2b_1)\mathbf{k}$

$$= \begin{vmatrix} \mathbf{i} & \mathbf{j} & \mathbf{k} \\ a_1 & a_2 & a_3 \\ b_1 & b_2 & b_3 \end{vmatrix} = \mathbf{i}\begin{vmatrix} a_2 & a_3 \\ b_2 & b_3 \end{vmatrix} - \mathbf{j}\begin{vmatrix} a_1 & a_3 \\ b_1 & b_3 \end{vmatrix} + \mathbf{k}\begin{vmatrix} a_1 & a_2 \\ b_1 & b_2 \end{vmatrix}$$

7 If A and B have position vectors **a** and **b** respectively, then
Area of triangle $OAB = \frac{1}{2}|\mathbf{a} \times \mathbf{b}|$

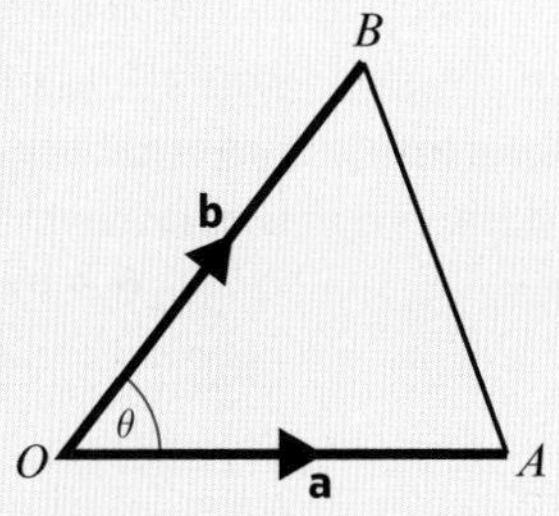

8 If A, B and C have position vectors **a**, **b** and **c** respectively, then

$$\text{Area of triangle } ABC = \tfrac{1}{2}\left|\overrightarrow{AB} \times \overrightarrow{AC}\right|$$
$$= \tfrac{1}{2}|(\mathbf{b} - \mathbf{a}) \times (\mathbf{c} - \mathbf{a})|$$
$$= \tfrac{1}{2}|(\mathbf{a} \times \mathbf{b}) + (\mathbf{b} \times \mathbf{c}) + (\mathbf{c} \times \mathbf{a})|$$

9 If A and B have position vectors **a** and **b** respectively, then

Area of parallelogram $OABC = |\mathbf{a} \times \mathbf{b}|$

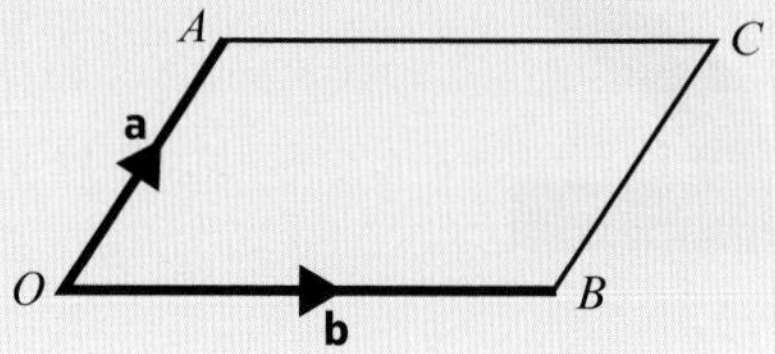

10 If A, B, C and D have position vectors **a**, **b**, **c** and **d** respectively, then

$$\text{Area of parallelogram } ABCD = \left|\overrightarrow{AB} \times \overrightarrow{AD}\right|$$
$$= |(\mathbf{b} - \mathbf{a}) \times (\mathbf{d} - \mathbf{a})|$$
$$= |(\mathbf{a} \times \mathbf{b}) + (\mathbf{b} \times \mathbf{d}) + (\mathbf{d} \times \mathbf{a})|$$

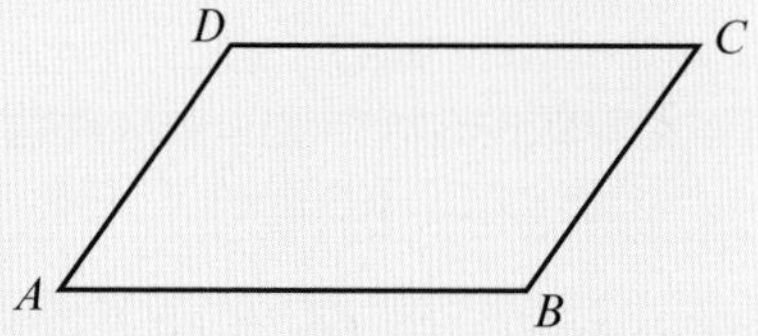

11 When $\mathbf{a} = (a_1\mathbf{i} + a_2\mathbf{j} + a_3\mathbf{k})$, $\mathbf{b} = (b_1\mathbf{i} + b_2\mathbf{j} + b_3\mathbf{k})$ and $\mathbf{c} = (c_1\mathbf{i} + c_2\mathbf{j} + c_3\mathbf{k})$,

$$\mathbf{a}.(\mathbf{b} \times \mathbf{c}) = a_1(b_2c_3 - b_3c_2) + a_2(b_3c_1 - b_1c_3) + a_3(b_1c_2 - b_2c_1)$$

This can also be written as

$$\mathbf{a}.(\mathbf{b} \times \mathbf{c}) = \begin{vmatrix} a_1 & a_2 & a_3 \\ b_1 & b_2 & b_3 \\ c_1 & c_2 & c_3 \end{vmatrix}$$

$\mathbf{a}.(\mathbf{b} \times \mathbf{c})$ is known as the **scalar triple product**.

12 $\mathbf{a}.(\mathbf{b} \times \mathbf{c}) = \mathbf{b}.(\mathbf{c} \times \mathbf{a}) = \mathbf{c}.(\mathbf{a} \times \mathbf{b})$
$\mathbf{a}.(\mathbf{a} \times \mathbf{p}) = \mathbf{a}.(\mathbf{p} \times \mathbf{a}) = 0$ for any vector **p**.

13 If three sides of a parallelepiped are given by vectors **a**, **b** and **c** as shown in the diagram, then the volume of the parallelepiped is given by $|\mathbf{a}.(\mathbf{b} \times \mathbf{c})|$.

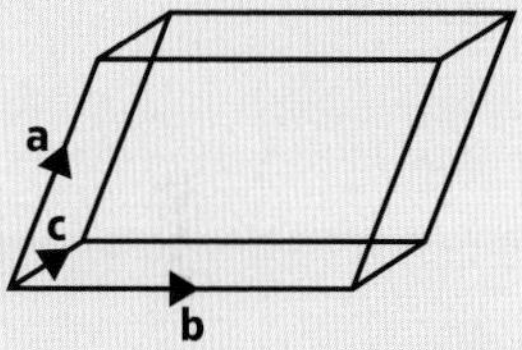

14 If three sides of a tetrahedron are given by vectors **a**, **b** and **c** as shown in the diagram, then the volume of the tetrahedron is given by $\frac{1}{6}|\mathbf{a}.(\mathbf{b} \times \mathbf{c})|$.

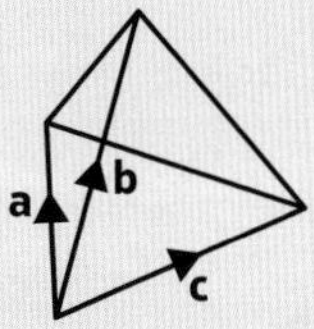

A

15 $(\mathbf{r} - \mathbf{a}) \times \mathbf{b} = \mathbf{0}$ is an alternative form of the vector equation of a line passing through the point A with position vector **a**, and parallel to the vector **b**.

This may also be written as $\mathbf{r} \times \mathbf{b} = \mathbf{a} \times \mathbf{b}$.

16 If a line is parallel to the vector $\mathbf{a} = x\mathbf{i} + y\mathbf{j} + z\mathbf{k}$, the direction ratios of the line are $x:y:z$, and the direction cosines of the line are

$$\cos\alpha = \frac{x}{|\mathbf{a}|}, \cos\beta = \frac{y}{|\mathbf{a}|}, \cos\gamma = \frac{z}{|\mathbf{a}|}$$

and are written as l, m and n respectively.

17 A line with direction ratios $x:y:z$ has direction cosines l, m and n such that $l^2 + m^2 + n^2 = 1$.

18 The shortest distance between the two skew lines with equations $\mathbf{r} = \mathbf{a} + \lambda\mathbf{b}$ and $\mathbf{r} = \mathbf{c} + \mu\mathbf{d}$, where λ and μ are scalars, is given by the formula $\left|\dfrac{(\mathbf{a} - \mathbf{c}).(\mathbf{b} \times \mathbf{d})}{|(\mathbf{b} \times \mathbf{d})|}\right|$

2 Conic sections 1

Objectives

After completing this chapter you should be able to:

- Plot and sketch a curve expressed parametrically → pages 33–35
- Work with the Cartesian equation and parametric equations of a parabola and a rectangular hyperbola → pages 35–45
- Find the equation of tangents and normals to parabolas and rectangular hyperbolas → pages 45–54
- Understand the focus–directrix property of a parabola → pages 54–56
- Solve locus problems involving the parabola and rectangular hyperbola → pages 55–56

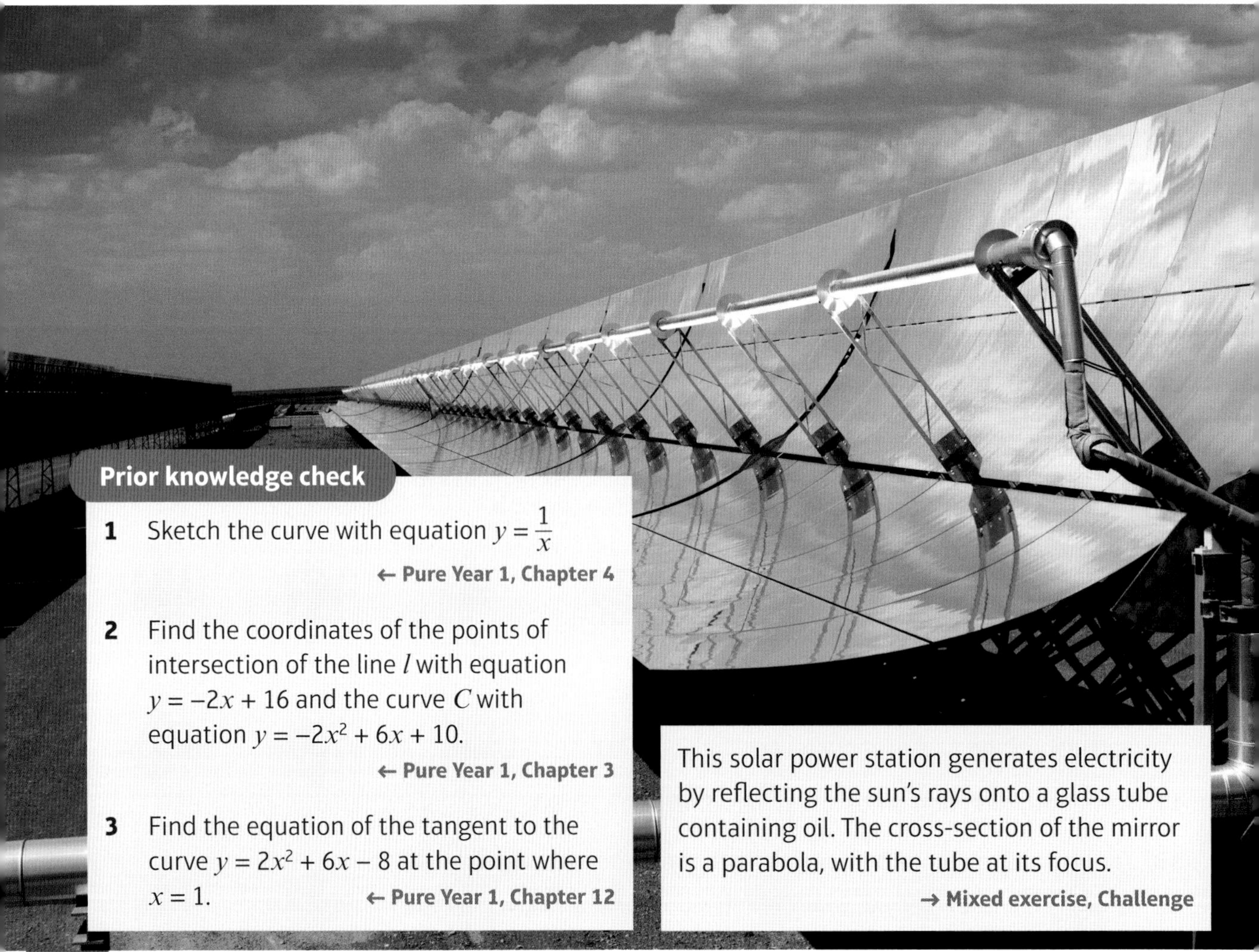

Prior knowledge check

1 Sketch the curve with equation $y = \frac{1}{x}$

← Pure Year 1, Chapter 4

2 Find the coordinates of the points of intersection of the line l with equation $y = -2x + 16$ and the curve C with equation $y = -2x^2 + 6x + 10$.

← Pure Year 1, Chapter 3

3 Find the equation of the tangent to the curve $y = 2x^2 + 6x - 8$ at the point where $x = 1$.

← Pure Year 1, Chapter 12

This solar power station generates electricity by reflecting the sun's rays onto a glass tube containing oil. The cross-section of the mirror is a parabola, with the tube at its focus.

→ Mixed exercise, Challenge

2.1 Parametric equations

You can define a curve using **parametric equations**, where the x- and y- coordinates of each point on the curve are given in terms of an independent variable (such as t) which is called a **parameter**. The parametric equations of a curve are written in the form

$$x = \text{p}(t),\ y = \text{q}(t)$$

Each value of t within the domain of the functions p and q generates a unique point on the curve.

- **To find the Cartesian equation of a curve given parametrically you eliminate the parameter t between the equations.**

Links

A Cartesian equation is an equation in terms of x and y only. ← **Pure Year 2, Chapter 8**

Example 1

A curve has parametric equations $x = at^2$, $y = 2at$, $t \in \mathbb{R}$ where a is a positive constant. Find the Cartesian equation of the curve.

$y = 2at$

So $t = \dfrac{y}{2a}$ (1) — Rearrange one equation into the form $t = \ldots$

$x = at^2$ (2)

Substitute (1) into (2):

$x = a\left(\dfrac{y}{2a}\right)^2$ — Substitute $t = \dfrac{y}{2a}$ into $x = at^2$.

So $x = \dfrac{ay^2}{4a^2}$ which simplifies to

$x = \dfrac{y^2}{4a}$

Hence, the Cartesian equation is

$y^2 = 4ax$ — This equation now involves x and y and not t. Note that a is a constant.

Example 2

A curve has parametric equations $x = ct$, $y = \frac{c}{t}$, $t \in \mathbb{R}$, $t \neq 0$, where c is a positive constant.

a Find the Cartesian equation of the curve.

b Hence sketch this curve.

a Method 1 — To obtain the Cartesian equation, eliminate t from the given parametric equations.

$x = ct$

So $t = \dfrac{x}{c}$ (1) — Rearrange one equation into the form $t = \ldots$

$y = \dfrac{c}{t}$ (2)

Substitute (1) into (2):

$y = \frac{c}{\left(\frac{x}{c}\right)}$ — Substitute $t = \frac{x}{c}$ into $y = \frac{c}{t}$

So $y = c \times \frac{c}{x}$ — This simplifies to $y = \frac{c^2}{x}$

Hence, the Cartesian equation is

$y = \frac{c^2}{x}$ — This equation now involves x and y. Note that c is a constant.

Method 2

$xy = ct \times \left(\frac{c}{t}\right)$ — Alternatively, you can multiply x by y on this occasion to eliminate t.

$xy = \frac{c^2 t}{t}$

Hence, the Cartesian equation is

$xy = c^2$ — This equation now involves x and y. Note that c is a constant.

This also may be expressed as

$y = \frac{c^2}{x}$

b

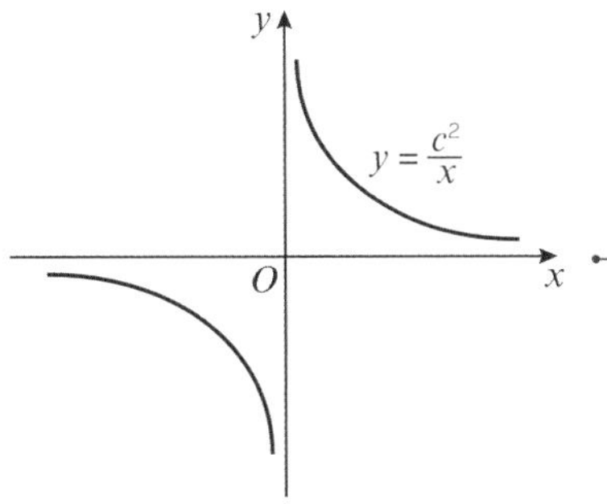

- As c is a positive constant, then c^2 is also a positive constant, which may be denoted by another constant, k.
- Hence the Cartesian equation represents a curve of the form $y = \frac{k}{x}$, $k > 0$.

← **Pure Year 1, Chapter 4**

Exercise 2A

1 Find the Cartesian equations of the curves given by these pairs of parametric equations.

a $x = 5t^2, y = 10t$ **b** $x = \frac{1}{2}t^2, y = t$ **c** $x = 50t^2, y = 100t$

d $x = \frac{1}{5}t^2, y = \frac{2}{5}t$ **e** $x = \frac{5}{2}t^2, y = 5t$ **f** $x = \sqrt{3}t^2, y = 2\sqrt{3}t$

g $x = 4t, y = 2t^2$ **h** $x = 6t, y = 3t^2$

2 Find the Cartesian equations of the curves given by these pairs of parametric equations.

a $x = t, y = \frac{1}{t}, t \neq 0$ **b** $x = 7t, y = \frac{7}{t}, t \neq 0$

c $x = 3\sqrt{5}t, y = \frac{3\sqrt{5}}{t}, t \neq 0$ **d** $x = \frac{t}{5}, y = \frac{1}{5t}, t \neq 0$

3 A curve has parametric equations $x = 3t, y = \frac{3}{t}, t \in \mathbb{R}, t \neq 0$.

a Find the Cartesian equation of the curve.

b Hence sketch this curve.

4 A curve has parametric equations $x = \sqrt{2}t$, $y = \frac{\sqrt{2}}{t}$, $t \in \mathbb{R}$, $t \neq 0$.

a Find the Cartesian equation of the curve. **b** Hence sketch this curve.

2.2 Parabolas

You have previously encountered parabolas in the form of quadratic curves, such as $y = x^2$. The parabola is one member of a family of curves known as the conic sections. These curves can be obtained by slicing a cone.
The parabola is obtained by slicing the cone parallel to its slope.

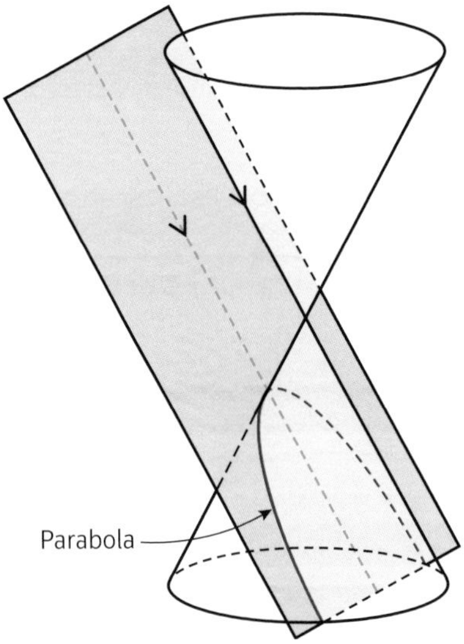

Links The circle is another example of a conic section, obtained by slicing a cone horizontally. You can learn about other conic sections later in this chapter and in the next chapter. → **Section 2.5, Chapter 3**

You need to be able to recognise and work with the parametric form of the equation for a parabola.

- **The curve opposite is a parabola with Cartesian equation $y^2 = 4ax$, where a is a positive constant.**
  - **The curve has parametric equations**

    $x = at^2$, $y = 2at$, $t \in \mathbb{R}$
  - **The curve is symmetrical about the x-axis.**
  - **A general point P on this curve has coordinates (x, y) or $(at^2, 2at)$.**

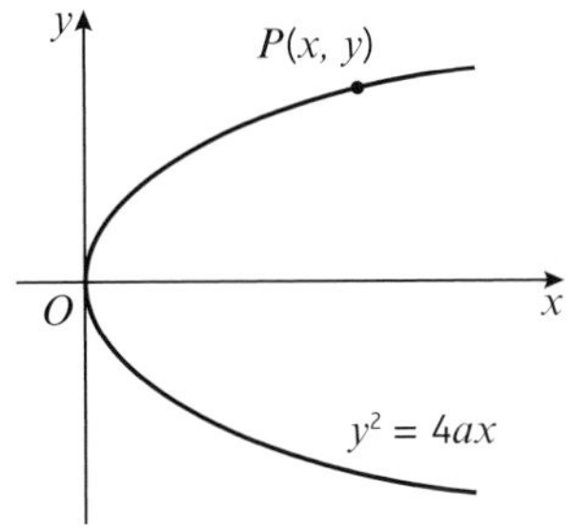

You also need to be able to define a parabola in terms of its **focus–directrix** properties.

- **A parabola is the locus of points that are the same distance from a fixed point S, called the focus, and a fixed straight line called the directrix. In the diagram on the right, $SP = PX$ for all points P on the parabola. For the parabola with Cartesian equation $y^2 = 4ax$,**
  - **the focus, S, has coordinates $(a, 0)$**
  - **the directrix has equation $x + a = 0$**
  - **the vertex is at the point (0, 0).**

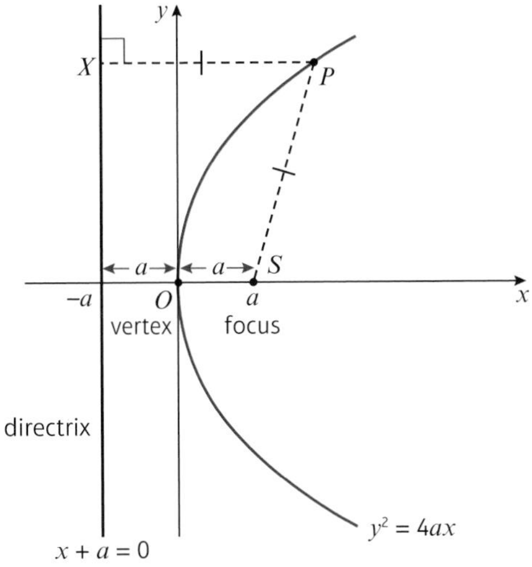

Online Explore the focus–directrix properties of a parabola using GeoGebra.

Example 3

Find an equation of the parabola with:

a focus (7, 0) and directrix $x + 7 = 0$

b focus $\left(\frac{\sqrt{3}}{4}, 0\right)$ and directrix $x = -\frac{\sqrt{3}}{4}$

a Focus (7, 0) and directrix $x + 7 = 0$

The focus and directrix are in the form $(a, 0)$ and $x + a = 0$.

So $a = 7$

So parabola has equation $y^2 = 28x$

Write equation in the form $y^2 = 4ax$ with $a = 7$.

b Focus $\left(\frac{\sqrt{3}}{4}, 0\right)$ and directrix $x = -\frac{\sqrt{3}}{4}$

$x + \frac{\sqrt{3}}{4} = 0$

Rearrange the directrix to the form $x + a = 0$.

So $a = \frac{\sqrt{3}}{4}$

So parabola has equation $y^2 = \sqrt{3}x$.

With $a = \frac{\sqrt{3}}{4}$, $y^2 = 4\left(\frac{\sqrt{3}}{4}\right)x$.

Example 4

Find the coordinates of the focus and an equation for the directrix of a parabola with equation:

a $y^2 = 24x$ **b** $y^2 = \sqrt{32}x$.

a $y^2 = 24x$

This is in the form $y^2 = 4ax$ with $a = 6$.

So the focus has coordinates (6, 0)

Focus has coordinates $(a, 0)$.

and the directrix has equation $x + 6 = 0$.

Directrix has equation $x + a = 0$.

b $y^2 = \sqrt{32}x$

$\sqrt{32} = 4\sqrt{2}$ so this is in the form $y^2 = 4ax$ with $a = \sqrt{2}$.

So the focus has coordinates $(\sqrt{2}, 0)$

and the directrix has equation $x + \sqrt{2} = 0$.

Exercise 2B

1 Find an equation of the parabola with:

a focus (5, 0) and directrix $x + 5 = 0$

b focus (8, 0) and directrix $x + 8 = 0$

c focus (1, 0) and directrix $x = -1$

d focus $\left(\frac{3}{2}, 0\right)$ and directrix $x = -\frac{3}{2}$

e focus $\left(\frac{\sqrt{3}}{2}, 0\right)$ and directrix $x + \frac{\sqrt{3}}{2} = 0$

2 Find the coordinates of the focus, and an equation for the directrix of each of the following parabolas.

a $y^2 = 12x$ **b** $y^2 = 20x$ **c** $y^2 = 10x$

d $y^2 = 4\sqrt{3}x$ **e** $y^2 = \sqrt{2}x$ **f** $y^2 = 5\sqrt{2}x$

3 Find the coordinates of the focus, and an equation of the parabola that passes through the general point:

a $(6t^2, 12t)$ **b** $(3\sqrt{2}\,t^2, 6\sqrt{2}\,t)$

Hint The parabola with general point $(6t^2, 12t)$ has parametric equations $x = 6t^2$, $y = 12t$.

Challenge

1 Find a Cartesian equation of the parabola with:

a focus (0, 4) and directrix $y = -4$

b focus (3, 3) and directrix $y = 0$

c focus (8, 0) and directrix $x = 2$

2 The parabola C has focus (2, 2) and directrix $x + y + 4 = 0$. Show that a Cartesian equation for C is $x + y = \frac{1}{16}(x - y)^2$.

Problem-solving Use a matrix transformation to rotate the general point $(at^2, 2at)$, for a suitable value of a.

Example 5

The point $P(8, -8)$ lies on the parabola C with equation $y^2 = 8x$. The point S is the focus of the parabola. The line l passes through S and P.

a Find the coordinates of S.

b Find an equation for l, giving your answer in the form $ax + by + c = 0$, where a, b and c are integers.

The line l meets the parabola C again at the point Q. The point M is the midpoint of PQ.

c Find the coordinates of Q.

d Find the coordinates of M.

e Draw a sketch showing the parabola C, the line l and the points P, Q, S and M.

a $y^2 = 8x$

This is in the form $y^2 = 4ax$ with $a = 2$.

The focus, S, has coordinates (2, 0)

The focus has coordinates $(a, 0)$.

b $m = \frac{-8 - 0}{8 - 2} = \frac{-8}{6}$

Use $m = \frac{y_2 - y_1}{x_2 - x_1}$, where $(x_1, y_1) = (2, 0)$ and $(x_2, y_2) = (8, -8)$.

So $m = -\frac{4}{3}$

$y - 0 = -\frac{4}{3}(x - 2)$

$3y = -4(x - 2)$

$3y = -4x + 8$

Use $y - y_1 = m(x - x_1)$. Here $m = -\frac{4}{3}$ and $(x_1, y_1) = (2, 0)$.

$4x + 3y - 8 = 0$

The line l has equation $4x + 3y - 8 = 0$.

Rearrange into the form $ax + by + c = 0$.

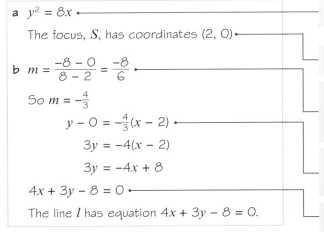

c l: $4x + 3y - 8 = 0$ (1)

C: $y^2 = 8x$

As the line l meets the curve C, solve these equations simultaneously.

$8x + 6y - 16 = 0$

Multiply (1) by 2.

$y^2 + 6y - 16 = 0$

Use $y^2 = 8x$.

$(y + 8)(y - 2) = 0$

So $y = -8$ or $y = 2$.

$y = -8$ corresponds to point P.

When $y = 2$, $x = \frac{1}{2}$ so Q has coordinates $\left(\frac{1}{2}, 2\right)$.

Use $y = 2$ to find the x-coordinate of Q.

d The midpoint is $\left(\frac{8 + \frac{1}{2}}{2}, \frac{-8 + 2}{2}\right)$

Use $\left(\frac{x_1 + x_2}{2}, \frac{y_1 + y_2}{2}\right)$, where $P = (x_1, y_1) = (8, -8)$ and $Q = (x_2, y_2) = \left(\frac{1}{2}, 2\right)$.

The point M has coordinates $\left(\frac{17}{4}, -3\right)$.

Simplify.

e The parabola C has equation

$y^2 = 8x$

The line l has equation

$4x + 3y - 8 = 0$

The line l cuts the parabola at the points $P(8, -8)$ and $Q\left(\frac{1}{2}, 2\right)$.

The points $S(2, 0)$ and $M\left(\frac{17}{4}, -3\right)$ also lie on the line l.

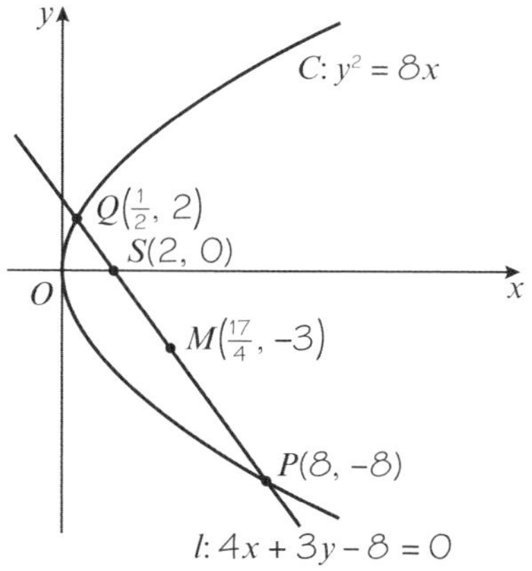

Notation

The line segment PQ is a **chord** of the parabola. A chord which passes through the focus is sometimes called a **focal chord**.

Example 6

The parabola C has general point $(at^2, 2at)$. The line $x = k$ intersects C at the points P and Q. Find, in terms of a and k, the length of the chord PQ.

If $x = k$, then $k = at^2$.

So $t^2 = \frac{k}{a}$

$t = \pm\sqrt{\frac{k}{a}}$

If $t = \sqrt{\frac{k}{a}}$, then $y = 2a\sqrt{\frac{k}{a}} = 2\sqrt{ak}$

Similarly, if $t = -\sqrt{\frac{k}{a}}$, then $y = -2\sqrt{ak}$

The coordinates of P and Q are $(k, 2\sqrt{ak})$ and $(k, -2\sqrt{ak})$.

The length of the chord PQ is $4\sqrt{ak}$.

The x-coordinate of a point on the curve is $x = at^2$.

Solve the equation for t.

Substitute each value for t separately to obtain the two y-values.

This is a vertical line segment, so the distance from P to Q can be found by subtracting the y-coordinates.

Problem-solving

You could also solve this problem by finding a Cartesian equation of C and substituting $x = k$ to find two corresponding values of y.

Exercise 2C

1 The line $y = 2x - 3$ meets the parabola $y^2 = 3x$ at the points P and Q. Find the coordinates of P and Q.

2 The line $y = x + 6$ meets the parabola $y^2 = 32x$ at the points A and B. Find the exact length of AB, giving your answer as a surd in its simplest form.

Hint Use the distance formula $d = \sqrt{(x_2 - x_1)^2 + (y_2 - y_1)^2}$ ← Pure Year 1, Chapter 5

3 The line $y = x - 20$ meets the parabola $y^2 = 10x$ at the points A and B. The midpoint of AB is the point M. Find the coordinates of M.

P 4 The parabola C has parametric equations $x = 6t^2$, $y = 12t$. The focus of C is at the point S.

a State the coordinates of S and the equation of the directrix of C.

b Sketch the graph of C.

The points P and Q on the parabola are both at a distance 9 units away from the directrix of the parabola.

c State the distance PS.

d Find the exact length PQ, giving your answer as a surd in its simplest form.

e Find the area of the triangle PQS, giving your answer in the form $k\sqrt{2}$, where k is an integer.

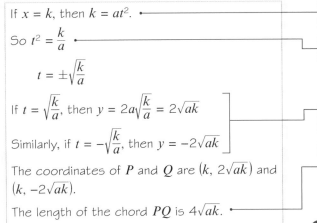

5 The parabola C has equation $y^2 = 4ax$, where a is a constant. The point $\left(\frac{5}{4}t^2, \frac{5}{2}t\right)$ is a general point on C.

a Find a Cartesian equation of C.

The point P lies on C and has y-coordinate 5.

b Find the x-coordinate of P.

The point Q lies on the directrix of C where $y = 3$. The line l passes through the points P and Q.

c Find the coordinates of Q.

d Find an equation for l, giving your answer in the form $ax + by + c = 0$, where a, b and c are integers.

(E) **6** A parabola C has equation $y^2 = 4x$. The point S is the focus of C.

a Find the coordinates of S. **(1 mark)**

The point P with y-coordinate 4 lies on C.

b Find the x-coordinate of P. **(1 mark)**

The line l passes through S and P.

c Find an equation for l, giving your answer in the form $ax + by + c = 0$, where a, b and c are integers. **(2 marks)**

The line l meets C again at the point Q.

d Find the coordinates of Q. **(3 marks)**

e Find the distance of the directrix of C to the point Q. **(2 marks)**

(E/P) **7** The diagram shows the point P which lies on the parabola C with equation $y^2 = 12x$.

The point S is the focus of C. The points Q and R lie on the directrix to C. The line segment PQ is parallel to the line segment RS as shown in the diagram. The length of PS is 12 units.

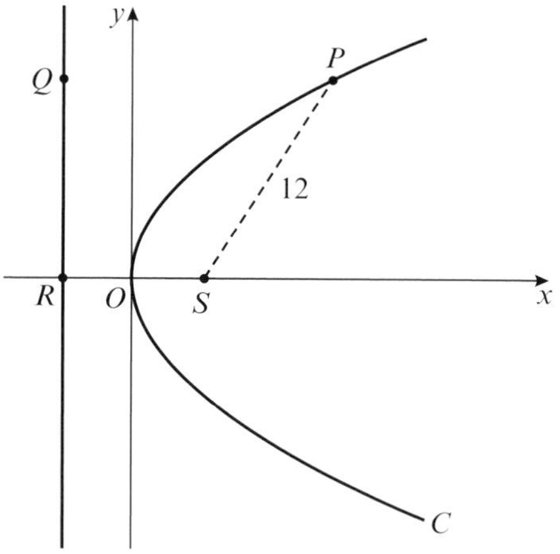

a Find the coordinates of R and S. **(2 marks)**

b Hence find the exact coordinates of P and Q. **(2 marks)**

c Find the area of the quadrilateral $PQRS$, giving your answer in the form $k\sqrt{3}$, where k is an integer. **(2 marks)**

(E/P) **8** The points $P(16, 8)$ and $Q(4, b)$, where $b < 0$ lie on the parabola C with equation $y^2 = 4ax$.

a Find the values of a and b. **(2 marks)**

P and Q also lie on the line l_1. The midpoint of PQ is the point R.

b Find an equation of l_1, giving your answer in the form $y = mx + c$, where m and c are constants to be determined. **(3 marks)**

c Find the coordinates of R. **(1 mark)**

The line l_2 is perpendicular to l, and passes through R.

d Find an equation of l_2, giving your answer in the form $y = mx + c$, where m and c are constants to be determined. **(3 marks)**

The line l_2 meets the parabola C at two points.

e Show that the x-coordinates of these two points can be written in the form $x = \lambda \pm \mu\sqrt{13}$, where λ and μ are integers to be determined. **(4 marks)**

P **9** The point $P(at^2, 2at)$ lies on the parabola C with equation $y^2 = 4ax$. The line l passes through P and the focus of the parabola, S.

a Find an expression for the gradient of l in terms of t. **(2 marks)**

The line intersects the parabola again at a point Q.

b Find the coordinates of Q, giving your answer in terms of a and t. **(4 marks)**

P **10** The diagram shows the parabola with equation $y^2 = 36x$. The region R is bounded by the parabola, the x-axis and the line $x = 10$. Find the exact area of R.

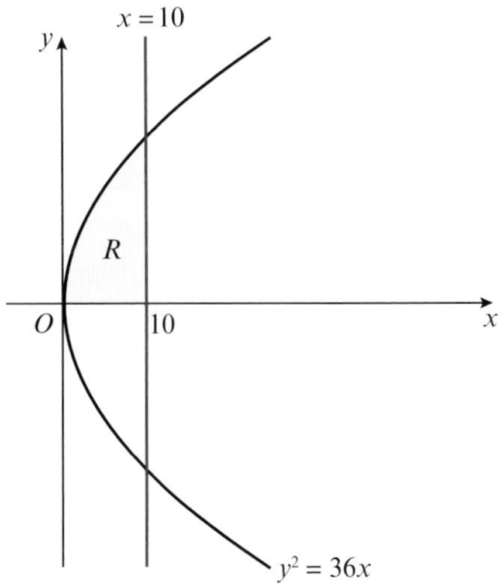

Problem-solving

The equation $y = \sqrt{4ax}$ represents the **top half** of the parabola $y^2 = 4ax$. Use integration to find the area under this curve between $x = 0$ and $x = 10$.

P **11** The diagram shows the parabola C with equation $y^2 = \frac{1}{2}x$. The straight line l with equation $y = \frac{1}{8}x$ cuts C at the points O and P. Find the area of the shaded region R. **(4 marks)**

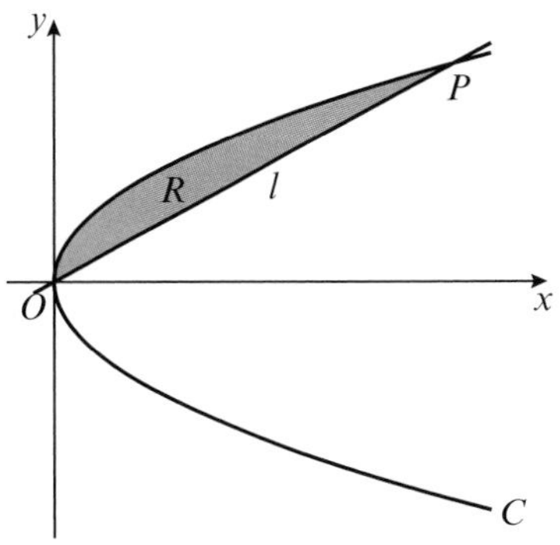

E/P **12** The diagram shows the points $P(2, a)$ and $Q(2, b)$ which lie on the parabola C with equation $y^2 = 8x$. The point T lies on the directrix to C.

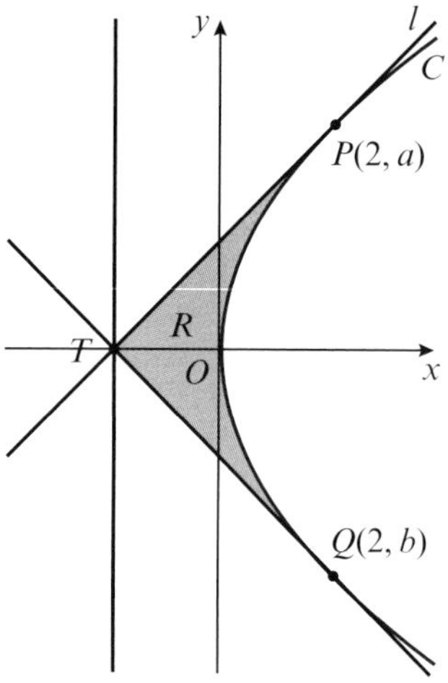

a Find the values of a and b. **(1 mark)**

T and P lie on the line l.

b Find an equation of l, giving your answer in the form $y = mx + c$, where m and c are constants to be determined. **(2 marks)**

c Find the area of the shaded region R. **(4 marks)**

E/P **13** A parabola C has equation $y^2 = 16x$.
The point S is the focus to C.

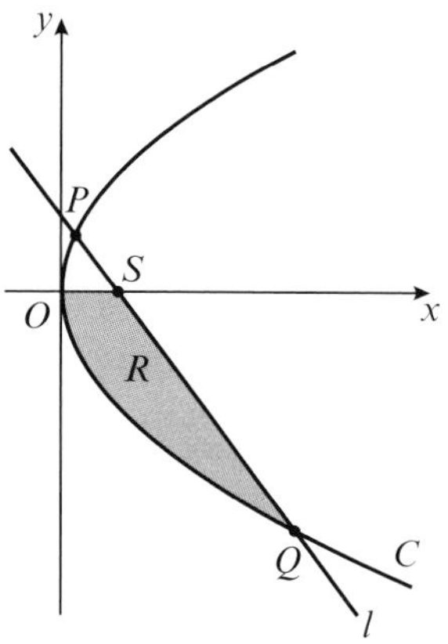

a Find the coordinates of S. **(1 mark)**

The point P with y-coordinate 4 lies on C.

b Find the x-coordinate of P. **(1 mark)**

The straight line l passes through S and P.

c Find an equation for l giving your answer in the form $y = mx + c$, where m and c are constants to be found. **(2 marks)**

The line l meets C again at Q. The shaded region R is bounded by the curve C, the line l and the x-axis.

d Find the area of the shaded region R. **(6 marks)**

2.3 Rectangular hyperbolas

If you slice through a cone in such a way that the slice intersects both halves, you obtain a curve called a hyperbola.

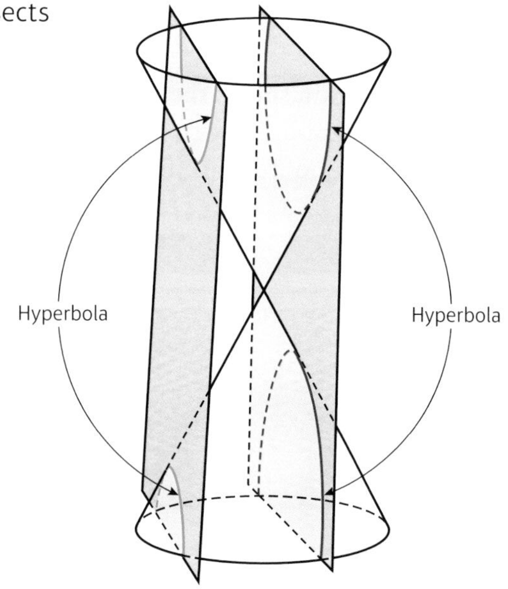

Notation
A hyperbola has two sections. These are sometimes called different **branches** of the hyperbola.

In this chapter you will consider one specific type of hyperbola called a **rectangular hyperbola**.
This curve has two asymptotes which meet at right angles.

- **The curve opposite is a rectangular hyperbola with Cartesian equation $xy = c^2$, where c is a positive constant.**
  - **The curve has parametric equations**

    $$x = ct,\ y = \frac{c}{t},\ t \in \mathbb{R},\ t \neq 0$$
  - **The curve has asymptotes with equations $x = 0$ (the y-axis) and $y = 0$ (the x-axis).**
  - **A general point P on this curve has coordinates (x, y) or $\left(ct, \frac{c}{t}\right)$.**

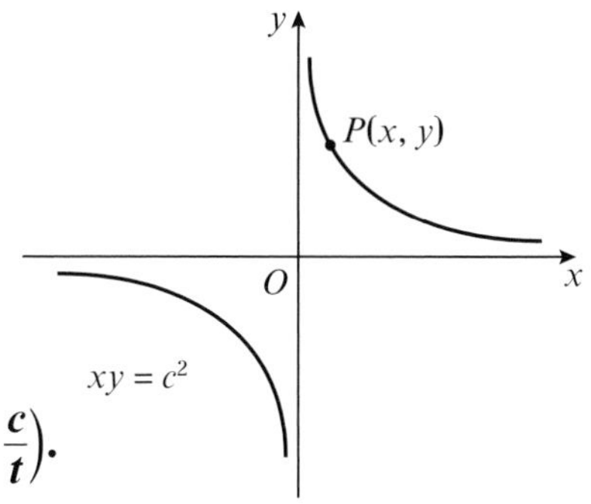

Example 7

The rectangular hyperbola H has Cartesian equation $xy = 64$. The line l with equation $x + 2y - 36 = 0$ intersects the curve at the points P and Q.

a Find the coordinates of P and Q.

b Find the equation of the perpendicular bisector of PQ in the form $y = mx + c$.

a $x + 2y - 36 = 0 \Rightarrow x = -2y + 36$ — Rearrange to obtain $x = \ldots$

$(-2y + 36)y = 64$ — Substitute into $xy = 64$.

$-2y^2 + 36y - 64 = 0$

$y^2 - 18y + 32 = 0$

$(y - 16)(y - 2) = 0$ — Expand and then factorise the quadratic.

$y = 2 \Rightarrow x = 32 \Rightarrow P(32, 2)$

$y = 16 \Rightarrow x = 4 \Rightarrow Q(4, 16)$ — Substitute the y-coordinates into either equation to calculate the x-coordinates.

b Midpoint of PQ is $(18, 9)$ — The midpoint of (x_1, y_1) and (x_2, y_2) is $\left(\frac{x_1 + x_2}{2}, \frac{y_1 + y_2}{2}\right)$

Gradient of PQ is $-\frac{1}{2}$ — Rearrange $x + 2y - 36 = 0$ to obtain $y = -\frac{1}{2}x + 18$.

Gradient of perpendicular bisector is 2. — The gradients of perpendicular lines multiply to equal -1.

$y - 9 = 2(x - 18)$ — Use $y - y_1 = m(x - x_1)$ with $m = 2$ and $(x_1, y_1) = (18, 9)$.

$\Rightarrow y = 2x - 27$

Exercise 2D

1 A rectangular hyperbola has equation $xy = 12$.

a Sketch the curve.

The line l with equation $y = -3x + 15$ intersects the curve at the points P and Q.

b Find the coordinates of P and Q.

c Find the equation of the perpendicular bisector of PQ.

d Find the x-coordinates of the points where the perpendicular bisector intersects the rectangular hyperbola.

2 The rectangular hyperbola with equation $xy = 9$ and the straight line with equation $y = x$ intersect at the points P and Q.

a Find the coordinates of the points P and Q.

The lines $3x - y + 6 = 0$ and $x - 3y - 6 = 0$ intersect the rectangular hyperbola at P and also at the points S and T respectively.

b Find the length of ST.

c Show that the midpoint of ST lies on the straight line $y = x$.

(P) **3** The straight line $3x + 4y + 48 = 0$ intersects the rectangular hyperbola with parametric equations $x = 6t$, $y = \frac{6}{t}$, $t \neq 0$, at the points P and Q. The straight line $4x - 3y - 11 = 0$ intersects the rectangular hyperbola with equation $xy = 36$ at the points Q and R. Find the area of the triangle PQR.

(P) **4** The points $P\left(cp, \frac{c}{p}\right)$ and $Q\left(cq, \frac{c}{q}\right)$ both lie on the hyperbola with equation $xy = c^2$.

Show that the chord PQ has equation $x + pqy = c(p + q)$.

(P) **5** The parabola C has equation $y^2 = 4ax$ and the rectangular hyperbola H has equation $xy = c^2$, where $a > 0$ and $c > 0$. Show that C and H intersect exactly once, and find the coordinates of the point of intersection, giving your answer in terms of a and c.

(E/P) **6** The rectangular hyperbola with equation $xy = c^2$ contains point P with x-coordinate $\frac{c}{2}$ and point Q with x-coordinate $-4c$. Find, in terms of c, the exact length of the chord PQ. **(5 marks)**

(E) **7** A rectangular hyperbola H has parametric equations $x = 9t$, $y = \frac{9}{t}$, $t \neq 0$. The straight line l with equation $4x - 3y + 69 = 0$ intersects H at the points P and Q.

a Show that l intersects H where $12t^2 + 23t - 9 = 0$. **(3 marks)**

b Hence, or otherwise, find the coordinates of P and Q. **(4 marks)**

8 The rectangular hyperbola H has parametric equations $x = 12t$, $y = \frac{12}{t}$, $t \neq 0$.

a Write the Cartesian equation of H in the form $xy = c^2$. **(1 mark)**

P and Q are points on the hyperbola such that $t = \frac{1}{2}$ and $t = 6$ respectively.

b Find the length of the line segment PQ, giving your answer in the form $a\sqrt{10}$. **(3 marks)**

c Find the equation of the perpendicular bisector of PQ. **(3 marks)**

9 The diagram shows the straight line with equation $x + 2y - 10 = 0$ that intersects the rectangular hyperbola with equation $xy = 8$ at the points P and Q.

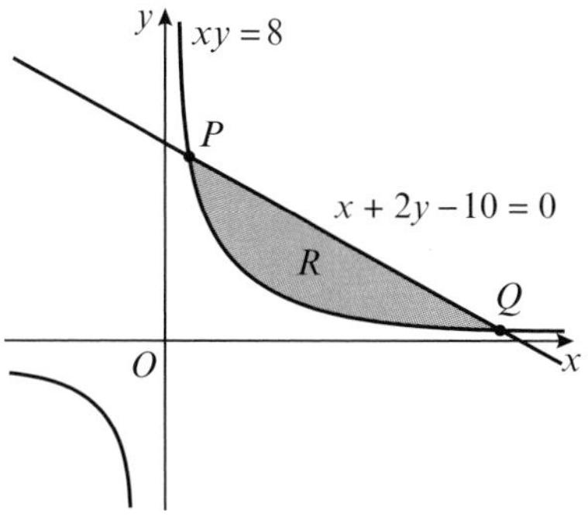

a Find the coordinates of P and Q. **(2 marks)**

b Find the exact area of the shaded region, R, bounded by the hyperbola and the line. Give your answer in the form $a + b \ln c$, where a, b and c are constants to be found. **(5 marks)**

Challenge

The rectangular hyperbola with equation $xy = c^2$ is rotated through 45° anticlockwise about the origin. Show that the resulting curve can be written in the form $y^2 - x^2 = k^2$, where $k > 0$, giving k in terms of c.

Problem-solving

The resulting curve is a rectangular hyperbola with asymptotes $y = x$ and $y = -x$.

2.4 Tangents and normals

You can use parametric differentiation or implicit differentiation to find the gradient of any point on a parabola. You do not need to be able to use either of these techniques if you are studying for AS level Further Maths only.

Links

Parametric and implicit differentiation are covered in Pure Year 2.

← Pure Year 2, Sections 9.7, 9.8

Parametric differentiation	**Implicit differentiation**
$x = at^2 \Rightarrow \frac{dx}{dt} = 2at$	$y^2 = 4ax$
$y = 2at \Rightarrow \frac{dy}{dt} = 2a$	$2y\frac{dy}{dx} = 4a$
$\frac{dy}{dx} = \frac{dy}{dt} \div \frac{dx}{dt} = \frac{1}{t}$	$\frac{dy}{dx} = \frac{2a}{y}$

These two expressions are equivalent, since $t = \frac{y}{2a}$. However, it is sometimes useful to find the gradient in terms of the parameter.

- **For the general parabola $y^2 = 4ax$, the gradient is given by $\frac{dy}{dx} = \frac{2a}{y}$**

Watch out If you need to use this result in an AS exam, it will be given with the question. In an A level exam you would be expected to derive this result if the question says 'prove' or 'use calculus'.

You can find the gradient at any point on a rectangular hyperbola by rearranging the equation into the form $y = \frac{c^2}{x}$ and differentiating.

Example 8

The point P, with x-coordinate 2, lies on the rectangular hyperbola H with equation $xy = 8$.
Find:

a the equation of the tangent, T, to H at point P

b the equation of the normal, N to H at the point P
giving your answers in the form $ax + by + c = 0$, where a, b and c are integers.

a $H: xy = 8$

$y = \frac{8}{x} \Rightarrow y = 8x^{-1}$

$\frac{dy}{dx} = -8x^{-2} = -\frac{8}{x^2}$

When $x = 2$, $m_T = \frac{dy}{dx} = -\frac{8}{2^2} = -2$ — Substitute $x = 2$ to calculate the gradient of the tangent to H at P.

When $x = 2$, $y = \frac{8}{2} = 4$ — Find the y-coordinate when $x = 2$. Hence P has coordinates (2, 4).

T: $y - 4 = -2(x - 2)$ — Use $y - y_1 = m_T(x - x_1)$ to find the equation of the tangent, T. Here $m_T = -2$ and $(x_1, y_1) = (2, 4)$.

T: $2x + y - 8 = 0$ — Then rearrange into the required form.

Therefore, the equation of the tangent to H at P is $2x + y - 8 = 0$.

b Gradient of tangent at P is $m_T = -2$.

So gradient of normal is $m_N = \frac{1}{2}$ — Use $m_N = \frac{-1}{m_T} = \frac{-1}{-2} = \frac{1}{2}$

N: $y - 4 = \frac{1}{2}(x - 2)$ — Use $y - y_1 = m_N(x - x_1)$ to find the equation of the normal, N. Here $m_N = \frac{1}{2}$ and $(x_1, y_1) = (2, 4)$.

N: $2y - 8 = x - 2$

N: $x - 2y + 6 = 0$ — Rearrange into the required form.

Therefore, the equation of the normal to H at P is $x - 2y + 6 = 0$.

Example 9

The point P with coordinates (75, 30) lies on the parabola C with equation $y^2 = 12x$.
Find the equation of the tangent to C at P, giving your answer in the form $y = mx + c$, where m and c are constants.

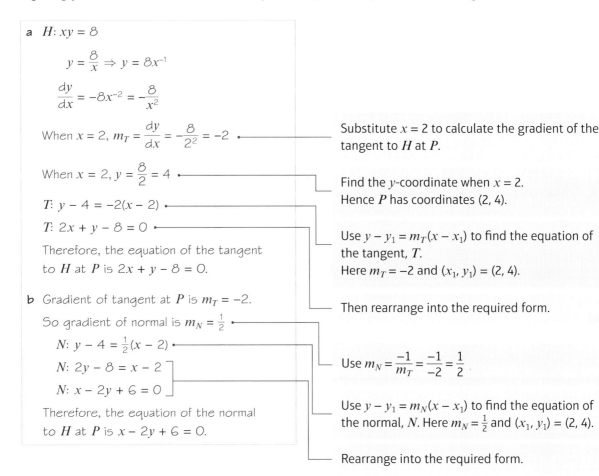

$y^2 = 12x$

$2y\frac{dy}{dx} = 12$

$\frac{dy}{dx} = \frac{6}{y}$

When $y = 30$, $\frac{dy}{dx} = \frac{6}{30} = \frac{1}{5}$ so $m = \frac{1}{5}$

$y - 30 = \frac{1}{5}(x - 75)$

$\Rightarrow y = \frac{1}{5}x + 15$

Therefore, the equation of the tangent to C at P is $y = \frac{1}{5}x + 15$.

Use implicit differentiation to find $\frac{dy}{dx}$.
Students who are only studying for AS Further Maths could use the result $\frac{dy}{dx} = \frac{2a}{y}$ with $a = 3$.

Use $\frac{dy}{dx} = \frac{6}{y}$ to find the gradient of the tangent.

Use $y - y_1 = m(x - x_1)$ to find the equation of the tangent. Here $m = \frac{1}{5}$ and $(x_1, y_1) = (75, 30)$.

Example 10

The point $P(4, 8)$ lies on the parabola C with equation $y^2 = 4ax$. Find:

a the value of a

b an equation of the normal to C at P.

The normal to C at P cuts the parabola again at the point Q. Find:

c the coordinates of Q

d the length PQ, giving your answer as a simplified surd.

a $8^2 = 4a \times 4 \Rightarrow a = 4$

b $y^2 = 16x$

$2y\frac{dy}{dx} = 16$ so $\frac{dy}{dx} = \frac{8}{y}$

When $y = 8$, $m_T = \frac{8}{8} = 1$

So gradient of normal is $m_N = -1$.

$y - 8 = -1(x - 4)$

$\Rightarrow y = -x + 12$

Therefore, the equation of the normal to C at P is $y = -x + 12$.

c When the normal cuts the curve,

$(-x + 12)^2 = 16x$

$x^2 - 24x + 144 = 16x$

$x^2 - 40x + 144 = 0$

$(x - 4)(x - 36) = 0$

So $x = 4$ or $x = 36$

When $x = 36$, $y = -36 + 12 = -24$.

So Q has coordinates $(36, -24)$.

d $PQ = \sqrt{(36 - 4)^2 + (-24 - 8)^2}$

$PQ = \sqrt{32^2 + (-32)^2} = \sqrt{2048} = 32\sqrt{2}$

Substitute $(x, y) = (4, 8)$ into $y^2 = 4ax$ and simplify to find a.

Use $\frac{dy}{dx} = \frac{2a}{y}$ or implicit differentiation.

Use m_T for the gradient of the tangent and m_N for the gradient of the normal.
$m_N = \frac{-1}{m_T}$

Use $y - y_1 = m_N(x - x_1)$ to find the equation of the tangent. Here $m_N = -1$ and $(x_1, y_1) = (4, 8)$.

Substitute $y = -x + 12$ into $y^2 = 16x$.

Multiply out and solve the quadratic.

$x = 4$ corresponds to point P.

Use the distance formula to find the length of PQ, and give your answer as a simplified surd.

Exercise 2E

In this exercise, AS students may use, without proof, the result that, for the general parabola $y^2 = 4ax$, $\frac{dy}{dx} = \frac{2a}{y}$

1 Find the equation of the tangent to the curve:

a $y^2 = 4x$ at the point $(16, 8)$

b $y^2 = 8x$ at the point $(4, 4\sqrt{2})$

c $xy = 25$ at the point $(5, 5)$

d $xy = 4$ at the point where $x = \frac{1}{2}$

e $y^2 = 7x$ at the point $(7, -7)$

f $xy = 16$ at the point where $x = 2\sqrt{2}$.

Give your answers in the form $ax + by + c = 0$.

2 Find the equation of the normal to the curve:

a $y^2 = 20x$ at the point where $y = 10$

b $xy = 9$ at the point $\left(-\frac{3}{2}, -6\right)$.

Give your answers in the form $ax + by + c = 0$, where a, b and c are integers.

3 The point $A(-2, -16)$ lies on the rectangular hyperbola H with equation $xy = 32$.

a Find an equation of the normal to H at A.

The normal to H at A meets H again at the point B.

b Find the coordinates of B.

(P) 4 The points $P(4, 12)$ and $Q(-8, -6)$ lie on the rectangular hyperbola H with equation $xy = 48$.

a Show that an equation of the line PQ is $3x - 2y + 12 = 0$.

The point A lies on H. The normal to H at A is parallel to the chord PQ.

b Find the exact coordinates of the two possible positions of A.

5 The distinct points A and B, where $x = 3$, lie on the parabola C with equation $y^2 = 27x$.

a Find the coordinates of A and B.

Line l_1 is the tangent to C at A and line l_2 is the tangent to C at B. Given that at A, $y > 0$,

b draw a sketch showing the parabola C. Indicate on your sketch the points A and B and the lines l_1 and l_2.

c Find:

i an equation for l_1

ii an equation for l_2

giving your answers in the form $ax + by + c = 0$, where a, b and c are integers.

(E) 6 The rectangular hyperbola H is defined by the equations $x = \sqrt{3}t$, $y = \frac{\sqrt{3}}{t}$, $t \in \mathbb{R}$, $t \neq 0$.

The point P lies on H with x-coordinate $2\sqrt{3}$. Find:

a a Cartesian equation for the curve H **(2 marks)**

b an equation of the normal to H at P. **(4 marks)**

The normal to H at P meets H again at the point Q.

c Find the exact coordinates of Q. **(3 marks)**

7 The point $P(4t^2, 8t)$ lies on the parabola C with equation $y^2 = 16x$. The point P also lies on the rectangular hyperbola H with equation $xy = 4$.

a Find the value of t, and hence find the coordinates of P. **(3 marks)**

The normal to H at P meets the x-axis at the point N.

b Find the coordinates of N. **(4 marks)**

The tangent to C at P meets the x-axis at the point T.

c Find the coordinates of T. **(3 marks)**

d Hence, find the area of the triangle NPT. **(2 marks)**

Example 11

The point $P(at^2, 2at)$, lies on the parabola C with equation $y^2 = 4ax$ where a is a positive constant. Show that an equation of the normal to C at P is $y + tx = 2at + at^3$.

$2y\frac{dy}{dx} = 4a$ so $\frac{dy}{dx} = \frac{2a}{y}$

If $y = 2at$, then $\frac{dy}{dx} = \frac{2a}{2at} = \frac{1}{t}$

Substitute $y = 2at$ into $\frac{dy}{dx} = \frac{2a}{y}$

Gradient of tangent at P is $m_T = \frac{1}{t}$

So gradient of normal is $m_N = -t$.

P has coordinates $(at^2, 2at)$.

N: $y - 2at = -t(x - at^2)$

Use $y - y_1 = m_N(x - x_1)$ to find the equation of the normal, N. Here $m_N = -t$ and $(x_1, y_1) = (at^2, 2at)$.

N: $y - 2at = -tx + at^3$

N: $y + tx = 2at + at^3$

Rearrange into the required form.

Therefore, the equation of the normal to C at P is $y + tx = 2at + at^3$

- **An equation of the normal to the parabola with equation $y^2 = 4ax$ at the point $P(at^2, 2at)$ is $y + tx = 2at + at^3$**

You can use a similar method to find an equation for a tangent to a parabola.

- **An equation of the tangent to the parabola with equation $y^2 = 4ax$ at the point $P(at^2, 2at)$ is $ty = x + at^2$**

Links
The derivation of this result is left as an exercise. → **Exercise 2F Q4**

Example 12

The point $P\left(ct, \frac{c}{t}\right)$, $t \neq 0$, lies on the rectangular hyperbola H with equation $xy = c^2$ where c is a positive constant.

a Show that an equation of the tangent to H at P is $x + t^2y = 2ct$.

A rectangular hyperbola G has equation $xy = 9$. The tangent to G at the point A and the tangent to G at the point B meet at the point $(-1, 7)$.

b Find the coordinates of A and B.

a $H: \quad xy = c^2$

$y = \frac{c^2}{x} \Rightarrow y = c^2x^{-1}$ — Rearrange the equation for H in the form $y = x^n$.

$\frac{dy}{dx} = -c^2x^{-2} = -\frac{c^2}{x^2}$ — Differentiate to determine the gradient of H.

At P, $x = ct$ and

$m_T = \frac{dy}{dx} = -\frac{c^2}{(ct)^2} = -\frac{c^2}{c^2t^2} = -\frac{1}{t^2}$ — Substitute $x = ct$, to calculate the gradient of the tangent to H.

Gradient of tangent at P is $m_T = -\frac{1}{t^2}$.

P has coordinates $\left(ct, \frac{c}{t}\right)$.

$T: \quad y - \frac{c}{t} = -\frac{1}{t^2}(x - ct)$ — Use $y - y_1 = m_T(x - x_1)$ to find the equation of the tangent, T. Here $m_T = -\frac{1}{t^2}$ and $(x_1, y_1) = \left(ct, \frac{c}{t}\right)$.

$T: \; t^2y - ct = -(x - ct)$

$T: \; t^2y - ct = -x + ct$

$T: \; x + t^2y = 2ct$ — Rearrange into the required form.

Therefore, the equation of the tangent to H at P is $x + t^2y = 2ct$.

b Compare $xy = 9$ with $xy = c^2$.

As c is positive, $c = 3$. — $c^2 = 9 \Rightarrow c = \sqrt{9} \Rightarrow c = 3$.

Tangent to G is $x + t^2y = 6t \quad (1)$ — Substitute $c = 3$ into the equation of the tangent derived in part **a**.

$-1 + t^2(7) = 6t$ — Substitute $x = -1$ and $y = 7$ in (1) as the tangent goes through point $(-1, 7)$.

$7t^2 - 6t - 1 = 0$

$(7t + 1)(t - 1) = 0$

$\Rightarrow t = -\frac{1}{7}$ or $t = 1$

P has coordinates $\left(ct, \frac{c}{t}\right) = \left(3t, \frac{3}{t}\right)$. — Substitute $c = 3$ into the general coordinates of P.

When $t = -\frac{1}{7}$, the coordinates are

$\left(3\left(-\frac{1}{7}\right), \frac{3}{-\frac{1}{7}}\right) = \left(-\frac{3}{7}, -21\right)$ — Substitute $t = -\frac{1}{7}$ into $P\left(3t, \frac{3}{t}\right)$.

When $t = 1$, the coordinates are

$\left(3 \times 1, \frac{3}{1}\right) = (3, 3)$. — Substitute $t = 1$ into $P\left(3t, \frac{3}{t}\right)$.

Therefore, the coordinates of A and B are $\left(-\frac{3}{7}, -21\right)$ and $(3, 3)$.

- **An equation of the tangent to the rectangular hyperbola with equation $xy = c^2$ at the point $P\left(ct, \frac{c}{t}\right)$ is $x + t^2y = 2ct$**

You can use a similar method to find an equation for a normal to a rectangular hyperbola.

- **An equation of the normal to the rectangular hyperbola with equation $xy = c^2$ at the point $P\left(ct, \frac{c}{t}\right)$ is $t^3x - ty = c(t^4 - 1)$**

Links

The derivation of this result is left as an exercise. → **Mixed exercise Q6**

Example 13

The parabola C has equation $y^2 = 20x$. The point $P(5p^2,10p)$ is a general point on C. The line l is normal to C at the point P.

a Show that an equation for l is $px + y = 10p + 5p^3$.

The point P lies on C. The normal to C at P passes through the point (30, 0) as shown on the diagram. The region R is bounded by this line, the curve C and the x-axis.

b Given that P lies in the first quadrant, show that the area of the shaded region R is $\frac{1100}{3}$

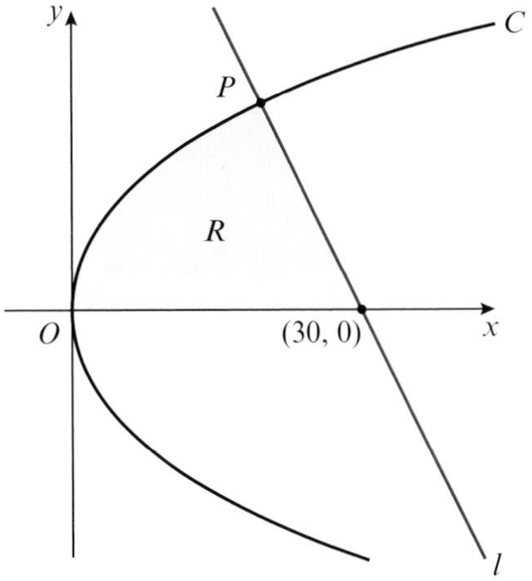

a $y^2 = 20x$

$2y\frac{dy}{dx} = 20$ so $\frac{dy}{dx} = \frac{10}{y}$

At $P(5p^2, 10p)$, $\frac{dy}{dx} = \frac{10}{10p} = \frac{1}{p}$

So, the gradient of the tangent at P is

$m_T = \frac{1}{p}$

Therefore, the gradient of the normal is

$m_N = -p$.

$y - 10p = -p(x - 5p^2)$

$y - 10p = -px + 5p^3$

$px + y = 10p + 5p^3$

Use the fact that $m_T \times m_N = -1$ to find the gradient of the normal.

Problem-solving

Since you know the gradient in terms of the parameter p, you can find an equation for the normal at P in terms of p.

Use $y - y_1 = m_N(x - x_1)$ with $m_N = -p$ and $(x_1, y_1) = (5p^2, 10p)$.

b At (30, 0), $30p = 10p + 5p^3$

$5p^3 - 20p = 0$

$p(p^2 - 4) = 0$

$p = 0$, $p = -2$ or $p = 2$

Discard $p = 0$ and $p = -2$, so $p = 2$.

Use the fact that the line passes through (30, 0) to find the value of p.

Problem-solving

The three solutions correspond to the three different normals to the curve that pass through the point (30, 0). You are interested in the one that lies in the first quadrant, so choose $p = 2$.

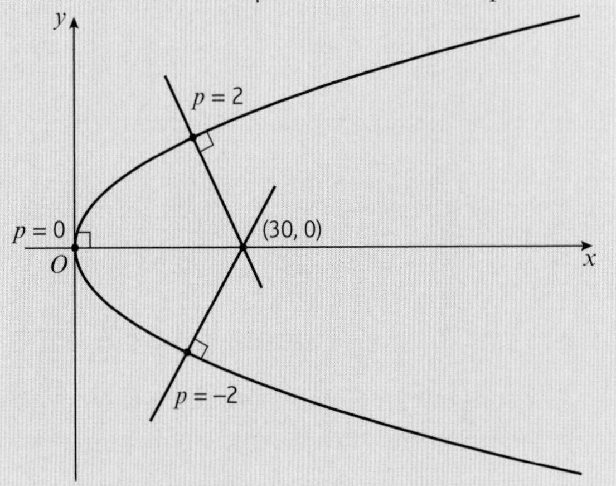

Coordinates of P are (20, 20). — Substitute $p = 2$ into $P(5p^2, 10p)$.

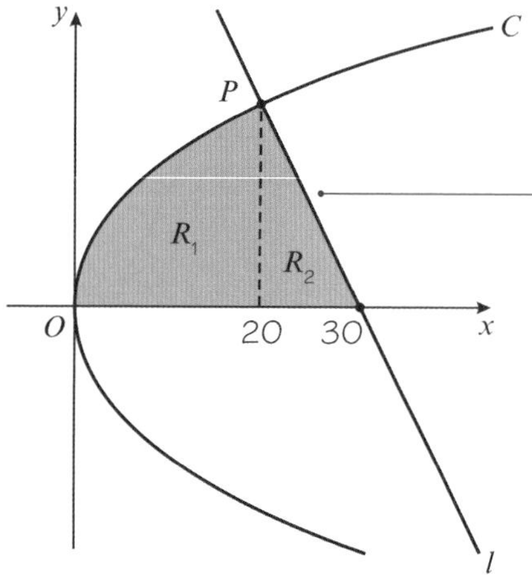

Split the area into two parts as shown. R_1 can be found using integration. R_2 is the area of a triangle.

$R_1 = 20^{\frac{1}{2}}\int_0^{20} x^{\frac{1}{2}}\,dx$

$= 20^{\frac{1}{2}}\left[\frac{2}{3}x^{\frac{3}{2}}\right]_0^{20}$

$= \frac{2}{3} \times 20^2 = \frac{800}{3}$

If $y^2 = 20x$, then $y = (20x)^{\frac{1}{2}} = 20^{\frac{1}{2}}x^{\frac{1}{2}}$.
$R_1 = \int_a^b y\,dx$ with $a = 0$, $b = 20$ and $y = 20^{\frac{1}{2}}x^{\frac{1}{2}}$.

$R_2 = \frac{1}{2}bh = \frac{1}{2} \times 10 \times 20 = 100$ — R_2 is a triangle with base 10 and height 20.

$R = R_1 + R_2 = 100 + \frac{800}{3} = \frac{1100}{3}$

Exercise 2F

In this exercise, AS students may use, without proof, the result that, for the general parabola $y^2 = 4ax$, $\frac{dy}{dx} = \frac{2a}{y}$

1 The point $P(3t^2, 6t)$ lies on the parabola C with equation $y^2 = 12x$.

a Show that an equation of the tangent to C at P is $yt = x + 3t^2$.

b Show that an equation of the normal to C at P is $xt + y = 3t^3 + 6t$.

2 The point $P\left(6t, \frac{6}{t}\right)$, $t \neq 0$, lies on the rectangular hyperbola H with equation $xy = 36$.

a Show that an equation of the tangent to H at P is $x + t^2y = 12t$.

b Show that an equation of the normal to H at P is $t^3x - ty = 6(t^4 - 1)$.

3 The point $P(5t^2, 10t)$ lies on the parabola C with equation $y^2 = 4ax$, where a is a constant and $t \neq 0$.

a Find the value of a.

b Show that an equation of the tangent to C at P is $yt = x + 5t^2$.

The tangent to C at P cuts the x-axis at the point X and the y-axis at the point Y. The point O is the origin of the coordinate system.

c Find, in terms of t, the area of the triangle OXY.

4 The point $P(at^2, 2at)$, lies on the parabola C with equation $y^2 = 4ax$, where a is a positive constant.

a Show that an equation of the tangent to C at P is $ty = x + at^2$.

The tangent to C at the point A and the tangent to C at the point B meet at the point with coordinates $(-4a, 3a)$.

b Find, in terms of a, the coordinates of A and B.

5 The point $P\left(4t, \frac{4}{t}\right)$, $t \neq 0$, lies on the rectangular hyperbola H with equation $xy = 16$.

a Show that an equation of the tangent to H at P is $x + t^2y = 8t$. **(4 marks)**

The tangent to H at the point A and the tangent to H at the point B meet at the point X with y-coordinate 5. X lies on the directrix of the parabola C with equation $y^2 = 16x$.

b Write down the coordinates of X. **(1 mark)**

c Find the coordinates of A and B. **(3 marks)**

d Deduce the equations of the tangents to H which pass through X. Give your answers in the form $ax + by + c = 0$, where a, b and c are integers. **(4 marks)**

6 The point $P(at^2, 2at)$ lies on the parabola C with equation $y^2 = 4ax$, where a is a constant and $t \neq 0$. The tangent to C at P cuts the x-axis at the point A.

a Find, in terms of a and t, the coordinates of A. **(4 marks)**

The normal to C at P cuts the x-axis at the point B.

b Find, in terms of a and t, the coordinates of B. **(4 marks)**

c Hence find, in terms of a and t, the area of the triangle APB. **(4 marks)**

7 The point $P(2t^2, 4t)$ lies on the parabola C with equation $y^2 = 8x$.

a Show that an equation of the normal to C at P is $xt + y = 2t^3 + 4t$. **(4 marks)**

The normals to C at the points R, S and T meet at the point $(12, 0)$.

b Find the coordinates of R, S and T. **(4 marks)**

c Deduce the equations of the normals to C which all pass through the point $(12, 0)$. **(4 marks)**

8 The point $P(at^2, 2at)$ lies on the parabola C with equation $y^2 = 4ax$, where a is a positive constant and $t \neq 0$. The tangent to C at P meets the y-axis at Q.

a Find in terms of a and t, the coordinates of Q. **(5 marks)**

The point S is the focus of the parabola.

b State the coordinates of S. **(1 mark)**

c Show that PQ is perpendicular to SQ. **(4 marks)**

9 The point $P(6t^2, 12t)$ lies on the parabola C with equation $y^2 = 24x$.

a Show that an equation of the tangent to the parabola at P is $ty = x + 6t^2$. **(4 marks)**

The point X has y-coordinate 9 and lies on the directrix of C.

b State the x-coordinate of X. **(1 marks)**

The tangent at the point B on C goes through point X.

c Find the possible coordinates of B. **(4 marks)**

(E/P) **10** The points $P(4p^2, 8p)$ and $Q(4q^2, 8q)$ lie on the parabola with equation $y^2 = 16x$. Prove that the normals to the parabola at points P and Q meet at $(8 + 4(p^2 + pq + q^2), -4pq(p + q))$. **(8 marks)**

(E/P) **11** The rectangular hyperbola, H, has Cartesian equation $xy = 64$. The points $P\left(8p, \frac{8}{p}\right)$ and $Q\left(8q, \frac{8}{q}\right)$ lie on H.

a Show that the equation of the tangent at point P is $p^2y + x = 16p$. **(4 marks)**

The tangents at P and Q meet at the point R.

b Given that the line OR is perpendicular to the line PQ, prove that $p^2q^2 = 1$. **(9 marks)**

(E/P) **12** A parabola is defined by the parametric equations $x = at^2$ and $y = 2at$.

a Show that the equation of the tangent to the parabola at the point $P(at^2, 2at)$ is $ty = x + at^2$. **(4 marks)**

b Show that the tangent intersects the x-axis at $T(-at^2, 0)$. **(4 marks)**

P is the point $(at^2, 2at)$ and S is the focus of the parabola.

c By considering gradients, or otherwise, show that PT can never be perpendicular to PS. **(4 marks)**

(E/P) **13** The point $P(p^2, 2p)$ lies on the parabola C with equation $y^2 = 4x$. The line l is tangent to C at the point P.

a Show that an equation for l is $py = x + p^2$. **(4 marks)**

b Find the area of the shaded region R. **(4 marks)**

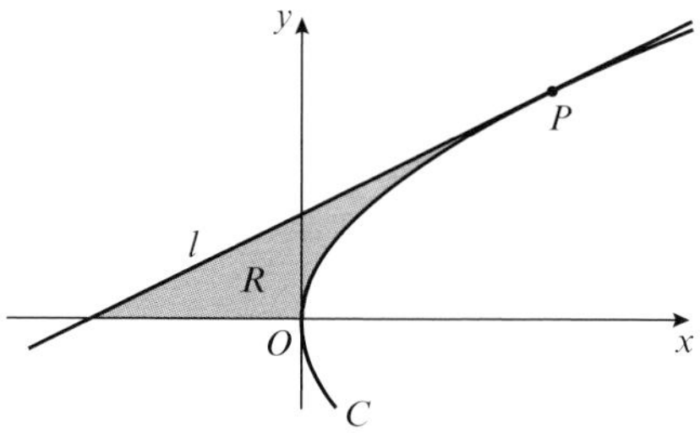

2.5 Loci

You can use the focus–directrix property of a parabola to derive its general equation.

Example 14

The curve C is the locus of points that are equidistant from the line with equation $x + 6 = 0$ and the point $(6, 0)$. Prove that C has Cartesian equation $y^2 = 4ax$, stating the value of a.

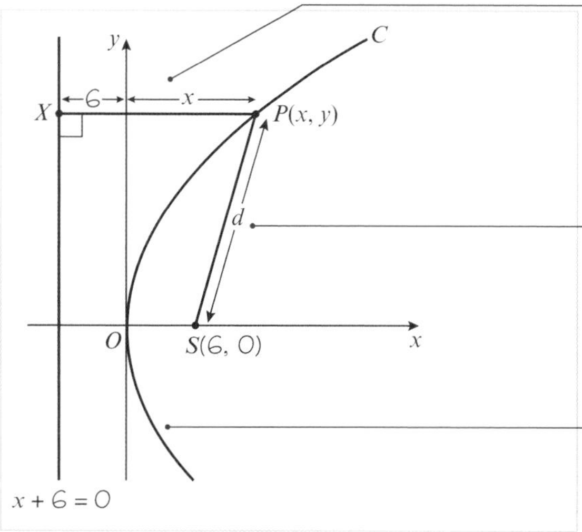

The (shortest) distance of P to the line $x + 6 = 0$ is the distance XP.

The line XP is horizontal and has distance $XP = x + 6$.

The distance SP is the same as the distance XP.

The locus of P is the curve shown.

From sketch, the locus satisfies

$SP = XP$.

Therefore, $SP^2 = XP^2$:

$(x - 6)^2 + (y - 0)^2 = (x + 6)^2$

$x^2 - 12x + 36 + y^2 = x^2 + 12x + 36$

$-12x + y^2 = 12x$

This simplifies to $y^2 = 24x$.

So the locus of P has an equation of the form $y^2 = 4ax$, where $a = 6$.

This means the distance SP is the same as the distance XP.

Use $d^2 = (x_2 - x_1)^2 + (y_2 - y_1)^2$ to find SP^2, where $S(6, 0)$ and $P(x, y)$.

This is in the form $y^2 = 4ax$.

So $4a = 24$, gives $a = 6$.

You can solve other locus problems involving the parabola and the rectangular hyperbola by considering general points on each curve.

Example 15

The point P lies on a parabola with equation $y^2 = 4ax$. Show that the locus of the midpoints of OP is a parabola.

The general point on the parabola $y^2 = 4ax$ has coordinates $(at^2, 2at)$.

Midpoint of $OP = \left(\frac{1}{2}at^2, at\right)$

$x = \frac{1}{2}at^2,\ y = at \Rightarrow y^2 = 2ax$

This is the equation of a parabola with focus $(\frac{1}{2}a, 0)$

Use the general point on the parabola to find the coordinates of the midpoint of OP in terms of the parameter. The locus can then be determined by considering the parametric equations for this general point.

Problem-solving

Any equation of the form $y^2 = kx$ is a parabola. You can find its focus by setting $k = 4a$.

Exercise 2G

P **1** A point P obeys a rule such that the distance of P to the point $(7, 0)$ is the same as the distance of P to the straight line $x + 7 = 0$. Prove that the locus of P has a Cartesian equation of the form $y^2 = 4ax$, stating the value of the constant a.

P **2** A point P obeys a rule such that the distance of P to the point $(2\sqrt{5}, 0)$ is the same as the distance of P to the straight line $x = -2\sqrt{5}$. Prove that the locus of P has an equation of the form $y^2 = 4ax$, stating the value of the constant a.

P **3** A point P obeys a rule such that the distance of P to the point $(0, 2)$ is the same as the distance of P to the straight line $y = -2$.

a Prove that the locus of P has an equation of the form $y = kx^2$, stating the value of the constant k.

Given that the locus of P is a parabola,

b state the coordinates of the focus of P, and an equation of the directrix to P

c sketch the locus of P with its focus and its directrix.

E/P **4** A point P is equidistant from the point $(a, 0)$ and the straight line $x + a = 0$. Prove that the locus of P is a parabola with equation $y^2 = 4ax$. **(4 marks)**

E/P **5** A point P is equidistant from the point $S(3, 0)$ and the line $x + 3 = 0$.

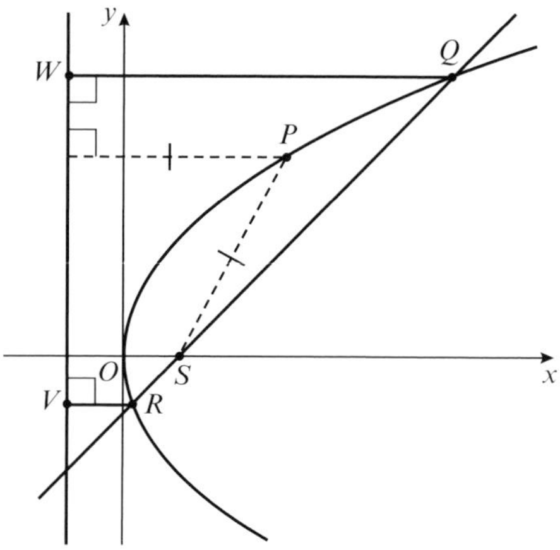

a Prove that the locus of P has an equation of the form $y^2 = kx$, where k is a constant to be found. **(4 marks)**

The point Q with y-coordinate $6\sqrt{6}$ lies on the locus.

b Show that the equation of the line through Q and S is $y = \frac{2\sqrt{6}}{5}x - \frac{6\sqrt{6}}{5}$ **(4 marks)**

The line also intersects the curve at the point R.

c Find the coordinates of the point R. **(3 marks)**

d Find the area of the trapezium $QRVW$. **(2 marks)**

E/P **6** Given that $P(x, y)$ is a general point on a rectangular hyperbola with equation $xy = c^2$, show that the locus of points $Q\left(x, \frac{1}{2}y\right)$ is also a rectangular hyperbola, stating its equation in the form $xy = k^2$, where k is given in terms of c. **(5 marks)**

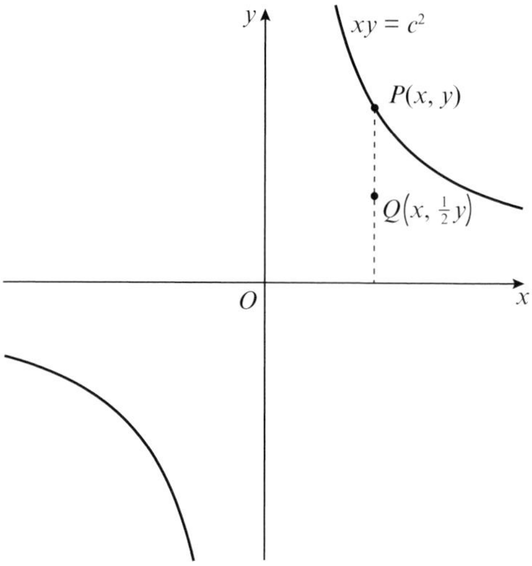

Hint Q is the midpoint of P and its 'foot' on the x-axis.

E/P **7** The points A and B lie on the x- and y-axes respectively. The point M is the midpoint of AB. A and B vary such that the area of triangle AOB is a constant value, q.

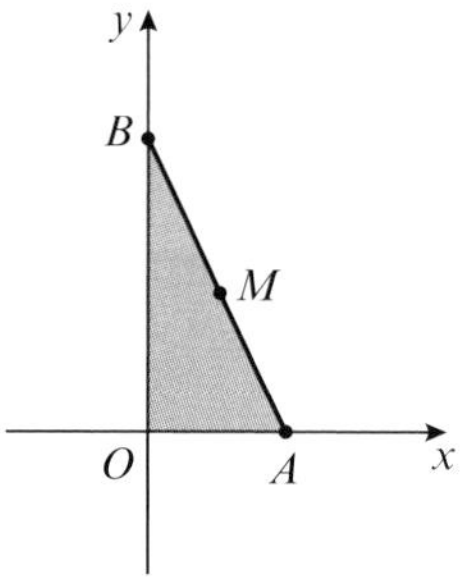

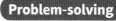

a Prove that the locus of M is a rectangular hyperbola. **(4 marks)**

b Give the equation of the locus from part **a** in the form $xy = c^2$, where c is given in terms of q. **(1 mark)**

Online Explore the locus of M using GeoGebra.

Challenge

A coordinate grid is drawn on a piece of paper. The point $(a, 0)$ and the line $x + a = 0$ are marked. The paper is then folded and creased in such a way that the point meets the line. Prove that the crease line is a tangent to the parabola with equation $y^2 = 4ax$.

Problem-solving

The parabola will form the **envelope** to the family of crease lines constructed in this way.

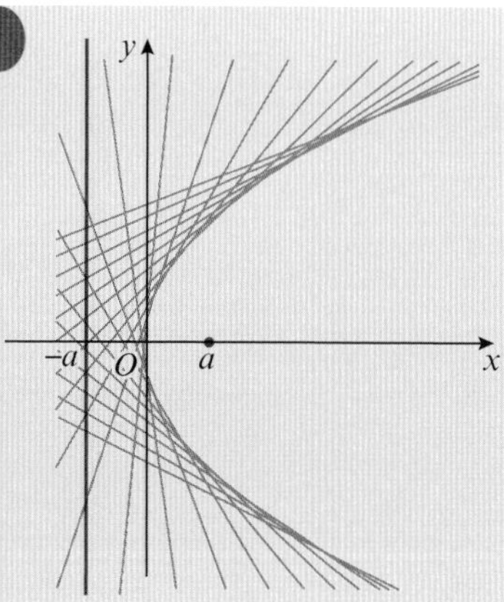

Mixed exercise 2

1 A parabola C has equation $y^2 = 12x$. The point S is the focus of C.

a Find the coordinates of S. **(1 mark)**

The line l with equation $y = 3x$ intersects C at the point P where $y > 0$.

b Find the coordinates of P. **(2 marks)**

c Find the area of the triangle OPS, where O is the origin. **(3 marks)**

2 A parabola C has equation $y^2 = 24x$. The point P with coordinates $(k, 6)$, where k is a constant, lies on C.

a Find the value of k. **(1 mark)**

The point S is the focus of C.

b Find the coordinates of S. **(1 mark)**

The line l passes through S and P and intersects the directrix of C at the point D.

c Show that an equation for l is $4x + 3y - 24 = 0$. **(2 marks)**

d Find the area of the triangle OPD, where O is the origin. **(3 marks)**

3 The parabola C has parametric equations $x = 12t^2$, $y = 24t$. The focus to C is at the point S.

a Find a Cartesian equation of C. **(2 marks)**

The point P lies on C where $y > 0$. P is 28 units from S.

b Find an equation of the directrix of C. **(1 mark)**

c Find the exact coordinates of the point P. **(3 marks)**

d Find the area of the triangle OSP, giving your answer in the form $k\sqrt{3}$, where k is an integer. **(3 marks)**

4 The point $(4t^2, 8t)$ lies on the parabola C with equation $y^2 = 16x$. The line l with equation $4x - 9y + 32 = 0$ intersects the curve at the points P and Q.

a Find the coordinates of P and Q. **(4 marks)**

b Show that an equation of the normal to C at $(4t^2, 8t)$ is $xt + y = 4t^3 + 8t$. **(4 marks)**

c Hence, find the equations of the normals to C at P and at Q. **(1 mark)**

The normal to C at P and the normal to C at Q meet at the point R.

d Find the coordinates of R and show that R lies on C. **(4 marks)**

e Find the distance OR, giving your answer in the form $k\sqrt{97}$, where k is an integer. **(2 marks)**

5 The point $P(at^2, 2at)$ lies on the parabola C with equation $y^2 = 4ax$, where a is a positive constant. The point Q lies on the directrix of C, and on the x-axis.

a State the coordinates of the focus of C and the coordinates of Q. **(2 marks)**

The tangent to C at P passes through the point Q.

b Find, in terms of a, the two sets of possible coordinates of P. **(5 marks)**

(E) **6** The point $P\left(ct, \frac{c}{t}\right)$, $c > 0$, $t \neq 0$, lies on the rectangular hyperbola H with equation $xy = c^2$.

a Show that the equation of the normal to H at P is $t^3x - ty = c(t^4 - 1)$. **(4 marks)**

b Hence, find the equation of the normal n to the curve J with the equation $xy = 36$ at the point (12, 3). Give your answer in the form $ax + by = d$, where a, b and d are integers. **(2 marks)**

The line n meets J again at the point Q.

c Find the coordinates of Q. **(4 marks)**

(E) **7** A rectangular hyperbola H has equation $xy = 9$. The lines l_1 and l_2 are distinct tangents to H. The gradients of l_1 and l_2 are both $-\frac{1}{4}$. Find the equations of l_1 and l_2. **(5 marks)**

(E) **8** The point P lies on the rectangular hyperbola $xy = c^2$, where $c > 0$. The tangent to the rectangular hyperbola at the point $P\left(ct, \frac{c}{t}\right)$, $t > 0$, cuts the x-axis at the point X and cuts the y-axis at the point Y.

a Find, in terms of c and t, the coordinates of X and Y. **(6 marks)**

b Given that the area of the triangle OXY is 144, find the exact value of c. **(3 marks)**

(E) **9** The points $P(4at^2, 4at)$ and $Q(16at^2, 8at)$ lie on the parabola C with equation $y^2 = 4ax$, where a is a positive constant.

a Show that an equation of the tangent to C at P is $2ty = x + 4at^2$. **(4 marks)**

b Hence, write down the equation of the tangent to C at Q. **(1 mark)**

The tangent to C at P meets the tangent to C at Q at the point R.

c Find, in terms of a and t, the coordinates of R. **(5 marks)**

(E/P) **10** A rectangular hyperbola H has Cartesian equation $xy = c^2$, $c > 0$. The point $\left(ct, \frac{c}{t}\right)$, where $t > 0$ is a general point on H.

a Show that an equation of the tangent to H at $\left(ct, \frac{c}{t}\right)$ is $x + t^2y = 2ct$. **(4 marks)**

The point P lies on H. The tangent to H at P cuts the x-axis at the point X with coordinates $(2a, 0)$, where a is a constant.

b Use the answer to part **a** to show that P has coordinates $\left(a, \frac{c^2}{a}\right)$. **(2 marks)**

The point Q, which lies on H, has x-coordinate $2a$.

c Find the y-coordinate of Q. **(2 marks)**

d Hence, find the equation of the line OQ, where O is the origin. **(2 marks)**

The lines OQ and XP meet at point R.

e Find, in terms of a, the x-coordinate of R. **(3 marks)**

Given that the line OQ is perpendicular to the line XP,

f show that $c^2 = 2a^2$ **(2 marks)**

g find, in terms of a, the y-coordinate of R. **(1 mark)**

11 The line with equation $2x - y - 12 = 0$ intersects the parabola C with equation $y^2 = 12x$ at the points P and Q.

a Find the coordinates of $P(a, b)$ and $Q(m, n)$. **(3 marks)**

b Find the area of the shaded region R bounded by the curve C and the lines $y = b$ and $x = m$. **(5 marks)**

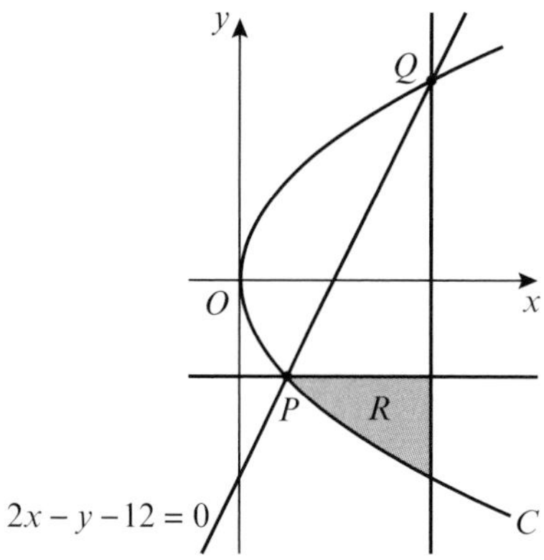

12 The point $P(9p^2, 18p)$ lies on the parabola with equation $y^2 = 36x$. The line l is normal to the parabola at P.

a Show that an equation for l is $y + px = 18p + 9p^3$. **(4 marks)**

Given that the line passes through the point $T(27, 0)$,

b find the coordinates of the three possible positions of P. **(3 marks)**

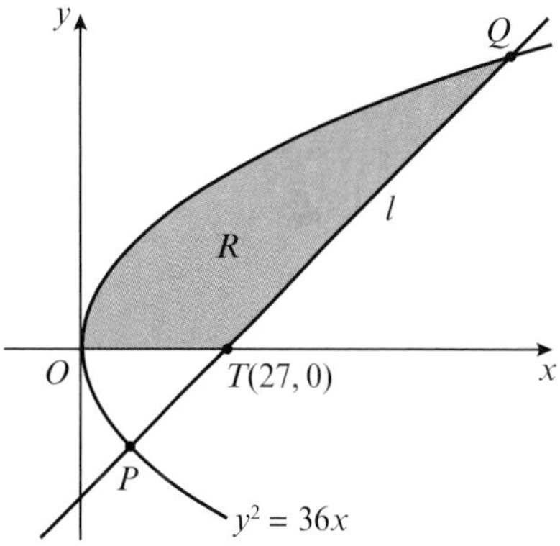

Given further that l has positive gradient, and that it intersects the parabola again at point Q, as shown in the diagram,

c find the coordinates of Q **(2 marks)**

d find the area of the shaded region R, bounded by l, the parabola and the x-axis. **(6 marks)**

13 Points $P(ap^2, 2ap)$ and $Q(aq^2, 2aq)$ lie on the parabola with equation $y^2 = 4ax$.

a Show that the equation of the line joining P and Q is $(p + q)y - 2x = 2apq$. **(4 marks)**

Given that the line PQ passes through the focus,

b show that $pq = -1$ **(2 marks)**

c find the coordinates of the point of intersection of the tangents to the parabola at the points P and Q **(3 marks)**

d show that this point of intersection lies on the directrix. **(2 marks)**

14 If P is a general point on a rectangular hyperbola, and the tangent at P cuts the x- and y-axes at A and B respectively, show that:

a $AP = PB$ **(3 marks)**

b the triangle AOB has constant area. **(3 marks)**

(E/P) **15** The chord PQ of a parabola with equation $y^2 = 4ax$ passes through the focus of the parabola as shown in the diagram. Show that:

a the tangents to the parabola at P and Q meet on the directrix **(7 marks)**

b the locus of the midpoint of PQ has equation $y^2 = 2a(x - a)$ **(8 marks)**

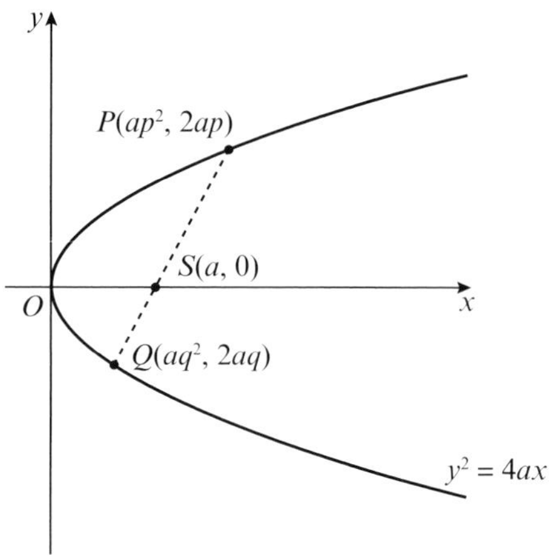

Challenge

When a ray of light is reflected, the angle between the incident ray and the normal at the point of contact with the surface is the same as the angle between the normal and the reflected ray.

The diagram below shows a parabolic mirror, with equation $y^2 = 4ax$. A ray of light parallel to the x-axis hits the mirror at the point $P(at^2, 2at)$. The line N is the normal to the mirror at the point P, and the angles of incidence and reflection, α, are shown on the diagram.

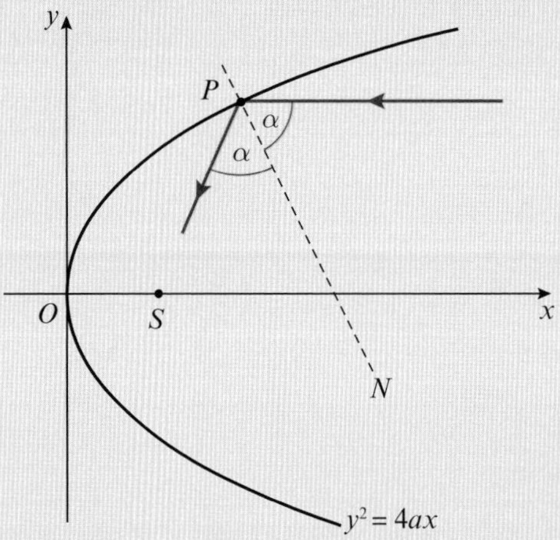

a Prove that $\tan \alpha = t$.

b Hence find an expression for $\tan 2\alpha$ in terms of t, and show that the gradient of the reflected ray is $\frac{2t}{t^2 - 1}$

c Hence show that the reflected ray passes through the focus of the parabola, S.

Summary of key points

1 To find the Cartesian equation of a curve given parametrically you eliminate the parameter t between the equations.

2 The curve opposite is a **parabola** with Cartesian equation $y^2 = 4ax$, where a is a positive constant.

- This curve has parametric equations $x = at^2$, $y = 2at$, $t \in \mathbb{R}$.
- The curve is symmetrical about the x-axis.
- A general point P on this curve has coordinates (x, y) or $(at^2, 2at)$.

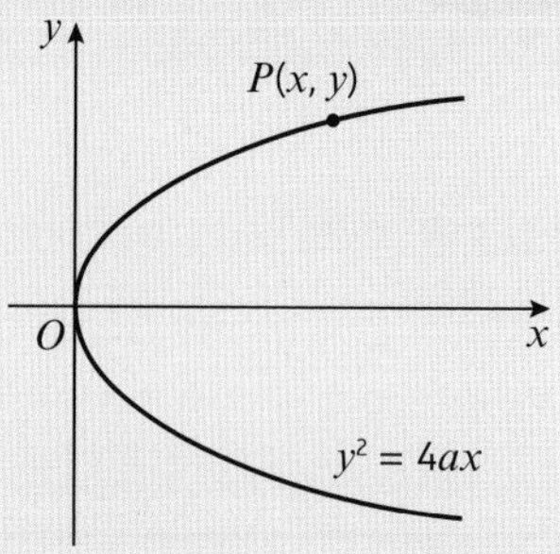

3 A parabola is the locus of points that are the same distance from a fixed point S, called the **focus**, and a fixed straight line called the **directrix**. In the diagram on the right, $SP = PX$ for all points P on the parabola.
For the parabola with Cartesian equation $y^2 = 4ax$,

- the **focus**, S, has coordinates $(a, 0)$
- the **directrix** has equation $x + a = 0$
- the **vertex** is at the point $(0, 0)$.

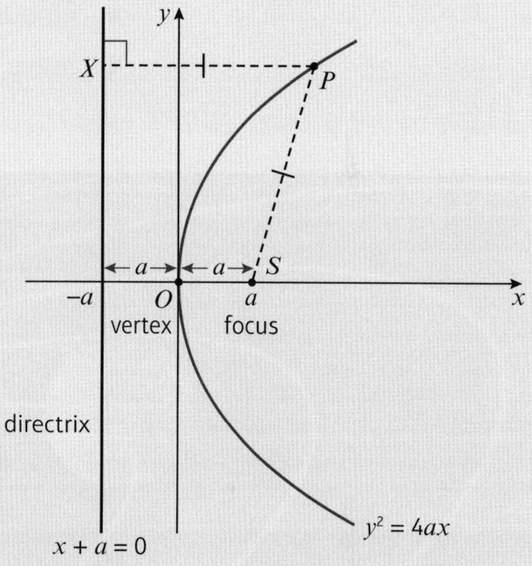

4 The curve opposite is a **rectangular hyperbola** with Cartesian equation $xy = c^2$, where c is a positive constant.

- This curve has parametric equations
  $$x = ct,\ y = \frac{c}{t},\ t \in \mathbb{R},\ t \neq 0$$
- The curve has asymptotes with equations $x = 0$ (the y-axis) and $y = 0$ (the x-axis).
- A general point P on this curve has coordinates (x, y) or $\left(ct, \frac{c}{t}\right)$.

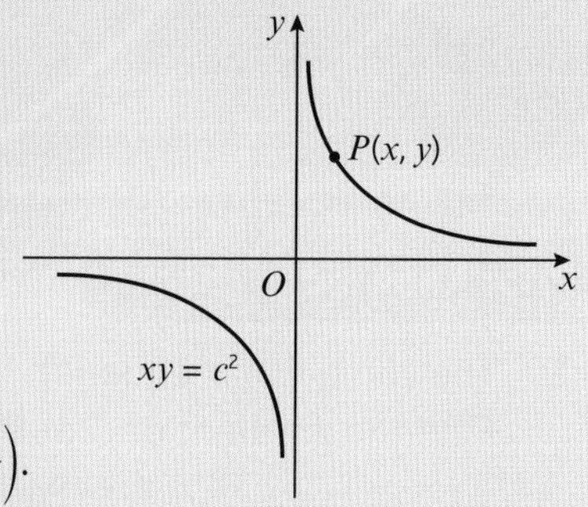

5 For the general parabola $y^2 = 4ax$, the gradient is given by $\frac{dy}{dx} = \frac{2a}{y}$.

6 An equation of the tangent to the parabola with equation $y^2 = 4ax$ at the point $P(at^2, 2at)$ is $ty = x + at^2$

An equation of the normal to the parabola with equation $y^2 = 4ax$ at the point $P(at^2, 2at)$ is $y + tx = 2at + at^3$

7 An equation of the tangent to the rectangular hyperbola with equation $xy = c^2$ at the point $P\left(ct, \frac{c}{t}\right)$ is $x + t^2y = 2ct$

An equation of the normal to the rectangular hyperbola with equation $xy = c^2$ at the point $P\left(ct, \frac{c}{t}\right)$ is $t^3x - ty = c(t^4 - 1)$

3 Conic sections 2

Objectives

After completing this chapter you should be able to:

- Identify an ellipse or a hyperbola from its Cartesian or parametric equations → pages 63–67
- Find the foci, directrices, and eccentricity for an ellipse or a hyperbola → pages 67–74
- Find tangents and normals to these curves → pages 74–83
- Solve simple loci questions → pages 83–87

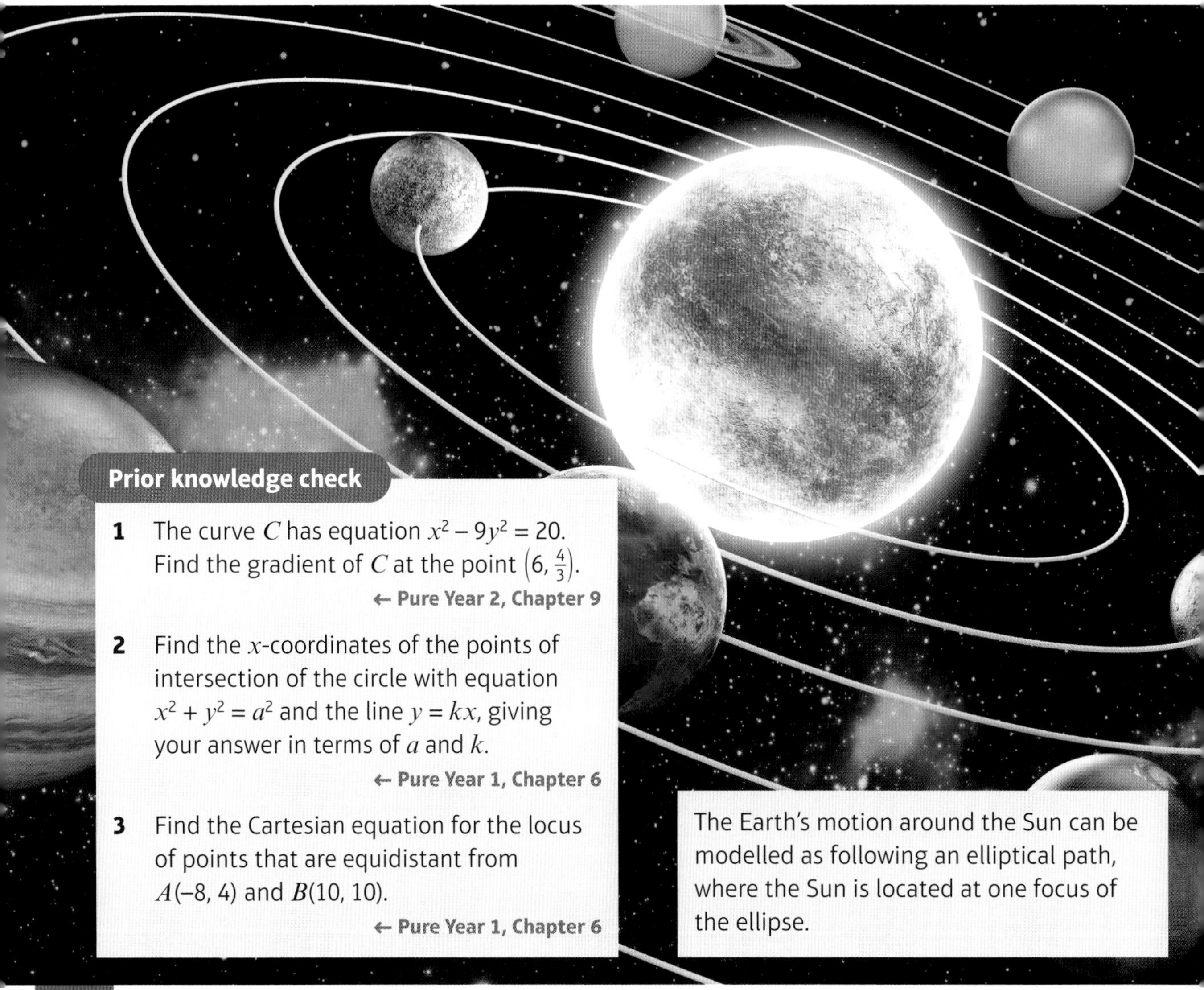

Prior knowledge check

1 The curve C has equation $x^2 - 9y^2 = 20$. Find the gradient of C at the point $\left(6, \frac{4}{3}\right)$.
← Pure Year 2, Chapter 9

2 Find the x-coordinates of the points of intersection of the circle with equation $x^2 + y^2 = a^2$ and the line $y = kx$, giving your answer in terms of a and k.
← Pure Year 1, Chapter 6

3 Find the Cartesian equation for the locus of points that are equidistant from $A(-8, 4)$ and $B(10, 10)$.
← Pure Year 1, Chapter 6

The Earth's motion around the Sun can be modelled as following an elliptical path, where the Sun is located at one focus of the ellipse.

3.1 Ellipses

A

In the previous chapter you encountered the **parabola** and the **rectangular hyperbola**, which are both examples of **conic sections**.

If you slice a cone in such a way as to produce a **closed** curve, the resulting curve is called an **ellipse**.

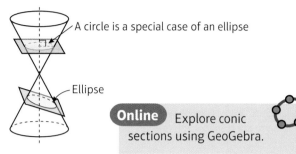

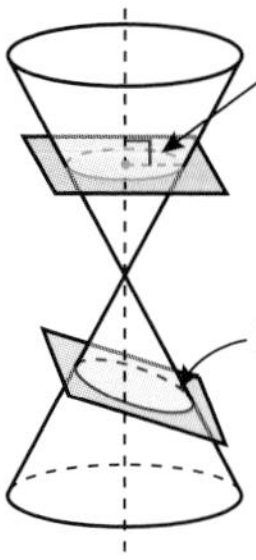

Online Explore conic sections using GeoGebra.

- **A standard ellipse has the Cartesian equation**

$$\frac{x^2}{a^2} + \frac{y^2}{b^2} = 1$$

When $x = 0$, $\frac{y^2}{b^2} = 1$ and so $y = \pm b$.

When $y = 0$, $\frac{x^2}{a^2} = 1$ and so $x = \pm a$.

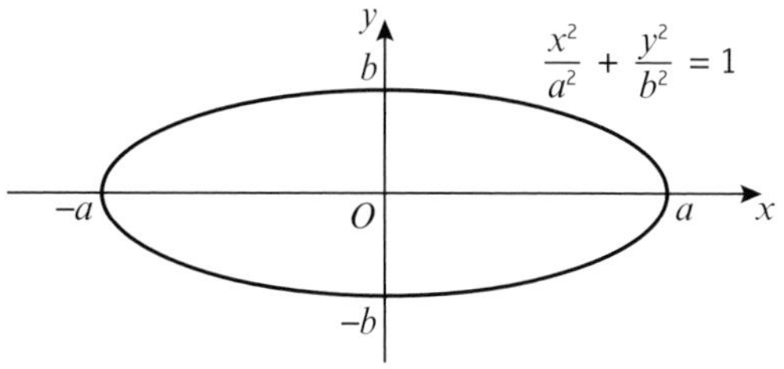

You can define a **general point** P on the ellipse in terms of a parameter, t.

- **The standard ellipse has parametric equations**

  $x = a\cos t,\ y = b\sin t,\ 0 \leqslant t < 2\pi$

- **A general point P on an ellipse has coordinates $(a\cos t, b\sin t)$.**

Note Substituting $x = a\cos t$ and $y = b\sin t$ into $\frac{x^2}{a^2} + \frac{y^2}{b^2}$ produces $\cos^2 t + \sin^2 t$ which is equal to 1. ← **Pure Year 1, Section 10.3**

Example 1

The ellipse E has equation $4x^2 + 9y^2 = 36$.

a Sketch E. **b** Write down parametric equations for E.

a $4x^2 + 9y^2 = 36$

$\frac{4x^2}{36} + \frac{9y^2}{36} = 1$ — First put the equation for E into standard form.

$\frac{x^2}{9} + \frac{y^2}{4} = 1$

So $a = 3$ and $b = 2$ — Identify the value of a and the value of b.

So sketch of E is

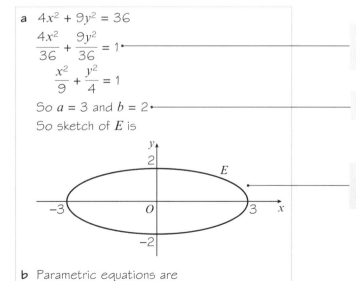

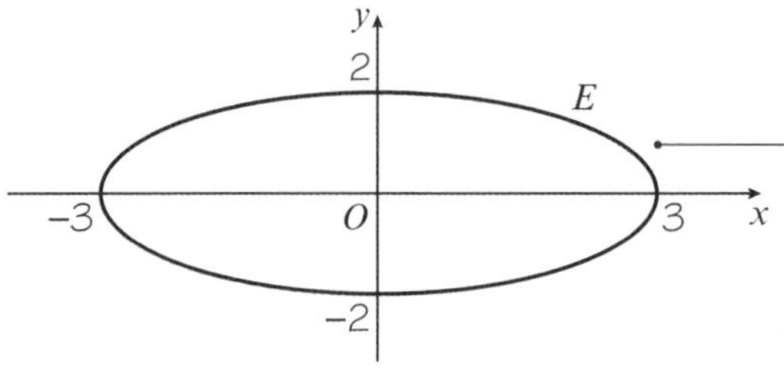

Draw the sketch: mark on intersections with the axes.

b Parametric equations are

$x = 3\cos t$, $y = 2\sin t$, $0 \leqslant t < 2\pi$

Example 2

The ellipse E has parametric equations

$x = 3\cos\theta,\ y = 5\sin\theta,\ 0 \leqslant \theta < 2\pi$

a Sketch E. **b** Find a Cartesian equation of E.

a $-5 \leqslant y \leqslant 5$

$-3 \leqslant x \leqslant 3$ — Since $\sin\theta$ and $\cos\theta$ are both between −1 and 1.

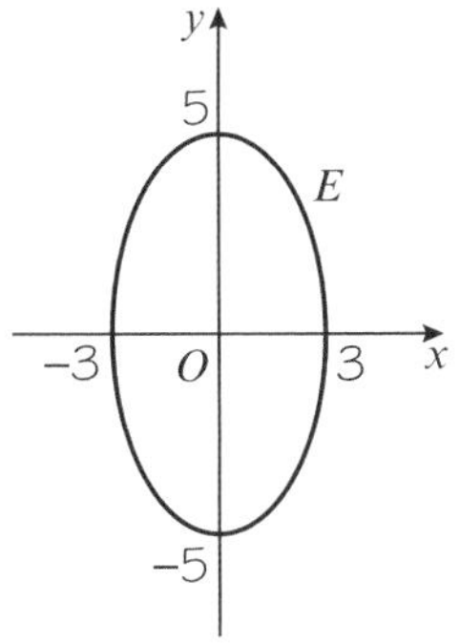

b $a = 3$ and $b = 5$, so the equation is $\frac{x^2}{9} + \frac{y^2}{25} = 1$ — Compare with the standard formula.

Exercise 3A

1 **a** Sketch the following ellipses showing clearly where the curves cross the coordinate axes.

i $x^2 + 4y^2 = 16$ **ii** $4x^2 + y^2 = 36$ **iii** $x^2 + 9y^2 = 25$

b Find parametric equations for these curves.

2 **a** Sketch ellipses with the following parametric equations.

i $x = 2\cos\theta,\ y = 3\sin\theta$ **ii** $x = 4\cos\theta,\ y = 5\sin\theta$

iii $x = \cos\theta,\ y = 5\sin\theta$ **iv** $x = 4\cos\theta,\ y = 3\sin\theta$

b Find a Cartesian equation for each ellipse.

(P) **3** The diagram shows the circles with equations $x^2 + y^2 = a^2$ and $x^2 + y^2 = b^2$. The line OS makes an angle θ with the positive x-axis and intersects the circles at points P and Q respectively. The point R has the same y-coordinate as P and the same x-coordinate as Q, as shown in the diagram.

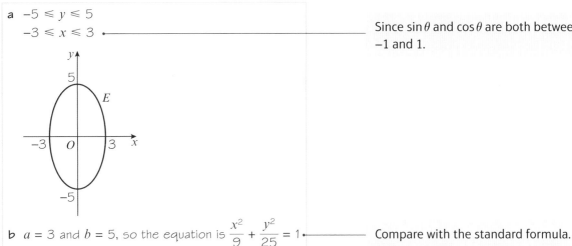

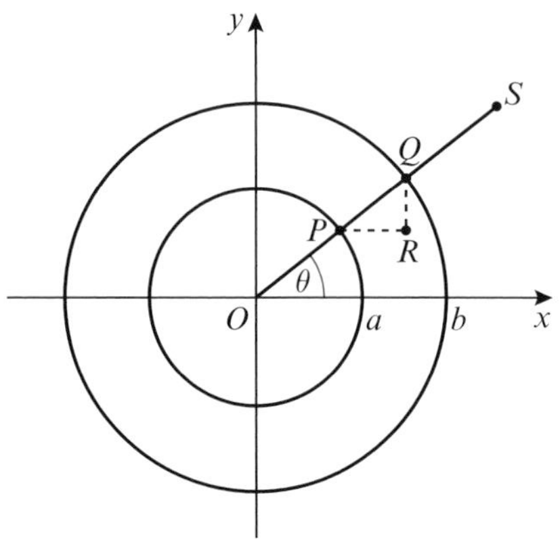

a Find the coordinates of R in terms of a, b and θ.

b Hence describe the locus of R as θ varies from 0 to 2π, and give its Cartesian equation.

c Sketch the curve with parametric equations

$x = 4\cos t,\ y = \sin t,\ \frac{\pi}{2} \leqslant t \leqslant \frac{3\pi}{2}$

showing clearly any points where the curve meets or intersects the coordinate axes.

Challenge

A The curve C is formed by rotating the ellipse with equation $\frac{x^2}{a^2} + \frac{y^2}{b^2} = 1$ through 45° anticlockwise about the origin.

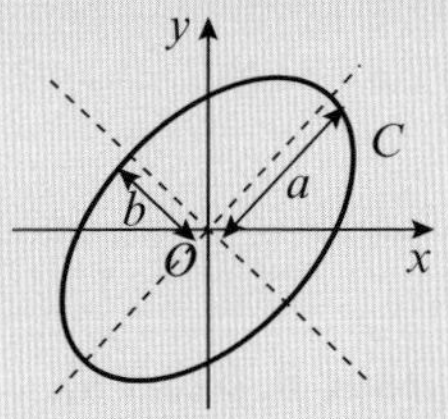

Show that C has equation $\frac{(x + y)^2}{2a^2} + \frac{(x - y)^2}{2b^2} = 1$

Problem-solving

Write the position vector of a general point on the original ellipse as $\begin{pmatrix} a\cos t \\ b\sin t \end{pmatrix}$ and then apply a suitable linear transformation.

3.2 Hyperbolas

In the previous chapter, you encountered rectangular hyperbolas with parametric equations $x = ct$, $y = \frac{c}{t}$, $t \in \mathbb{R}$, $t \neq 0$, where c is a positive constant. The Cartesian equation of this rectangular hyperbola is $xy = c^2$. This family of curves have perpendicular asymptotes with equations $x = 0$ (the y-axis) and $y = 0$ (the x-axis). A general point P on the curve has coordinates $P\left(ct, \frac{c}{t}\right)$.

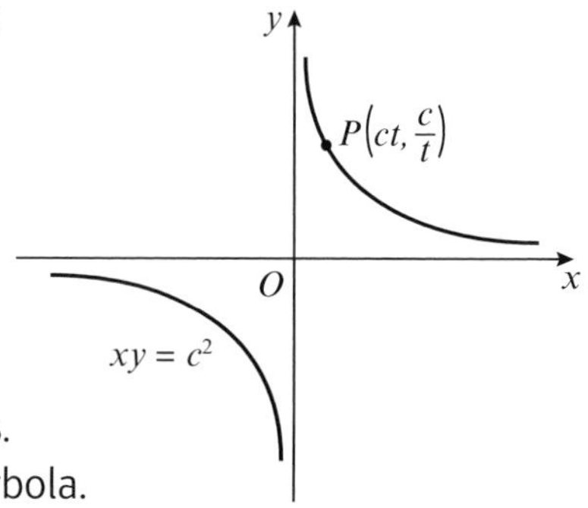

In general, hyperbolas do not need to have perpendicular asymptotes. You can find Cartesian and parametric equations for a standard hyperbola.

- **A standard hyperbola has Cartesian equation**

$$\frac{x^2}{a^2} - \frac{y^2}{b^2} = 1$$

When $y = 0$, $x^2 = a^2$ and so the curve crosses the x-axis at $(\pm a, 0)$. As x and y tend to infinity, $\frac{x^2}{a^2} \approx \frac{y^2}{b^2}$ and so the equations of the asymptotes are $y = \pm\frac{b}{a}x$.

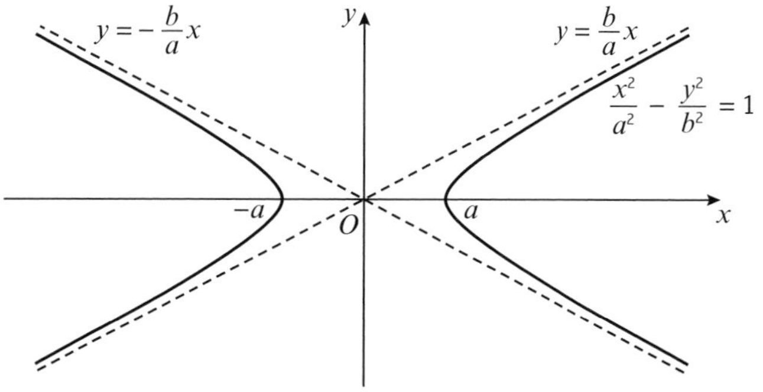

When $a = b$, this creates a rectangular hyperbola with equation $x^2 - y^2 = a^2$ with asymptotes at $y = x$ and $y = -x$. These asymptotes are perpendicular to one another.

Note The equations of the asymptotes are given in the formula booklet.

Watch out Although $x^2 - y^2 = a^2$ is an example of a rectangular hyperbola because its asymptotes are perpendicular, it is not part of the family of curves of the form $xy = c^2$ encountered in the previous chapter.

A In the previous section you saw that the parametric equations of the ellipse were connected to the trigonometric relationship $\cos^2\theta + \sin^2\theta \equiv 1$. You can use the corresponding relationship for the hyperbolic functions to find parametric equations for the hyperbola.

- **The standard hyperbola has parametric equations**

  $x = \pm a\cosh t,\ y = b\sinh t,\ t \in \mathbb{R}$

Links $\cosh^2 x - \sinh^2 x \equiv 1$ ← Core Pure Book 2, Chapter 6

- **The standard hyperbola has alternative parametric equations**

  $x = a\sec\theta,\ y = b\tan\theta,\ -\pi \leqslant \theta < \pi,\ \theta \neq \pm\frac{\pi}{2}$

- **A general point P on a hyperbola has coordinates $(\pm a\cosh t, b\sinh t)$ or $(a\sec\theta, b\tan\theta)$.**

Example 3

The hyperbola H has equation $9x^2 - 4y^2 = 36$.

a Sketch H.

b Write down the equations of the asymptotes of H.

c Find parametric equations for H.

a Rearrange the equation to get

$$\frac{x^2}{4} - \frac{y^2}{9} = 1$$

So $a = 2$ and $b = 3$

Write the equation in the form $\frac{x^2}{a^2} - \frac{y^2}{b^2} = 1$ and identify values for a and b.

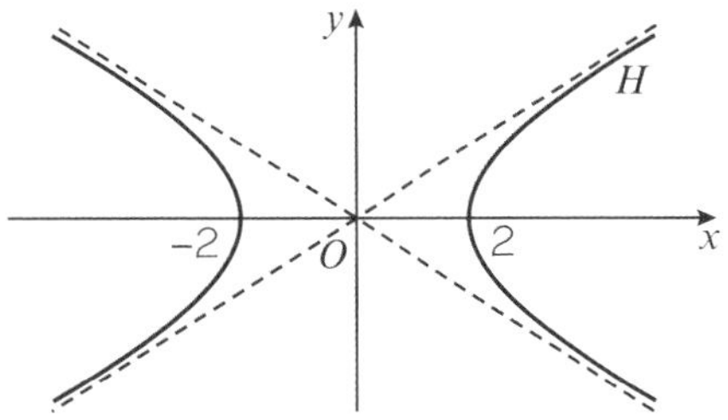

b Equations of the asymptotes are

$y = \frac{3}{2}x$ and $y = -\frac{3}{2}x$

The equations of the asymptotes are $y = \pm\frac{b}{a}x$.

c Parametric equations are

$x = \pm 2\cosh t,\ y = 3\sinh t,\ t \in \mathbb{R}$

Use $x = a\cosh t$ and $y = b\sinh t$.

Example 4

A hyperbola H has parametric equations

$$x = 4\sec t,\ y = \tan t,\ -\pi \leqslant t < \pi,\ t \neq \pm\frac{\pi}{2}$$

a Find a Cartesian equation for H.

b Sketch H.

c Write down the equations of the asymptotes of H.

a Using $\sec^2 t - \tan^2 t \equiv 1$,

$$\left(\frac{x}{4}\right)^2 - y^2 = 1$$

Cartesian equation is

$$\frac{x^2}{16} - y^2 = 1$$

Alternatively, compare with $x = a\sec\theta$ and $y = b\tan\theta$ and use the standard equation.

b $a = 4$ and $b = 1$

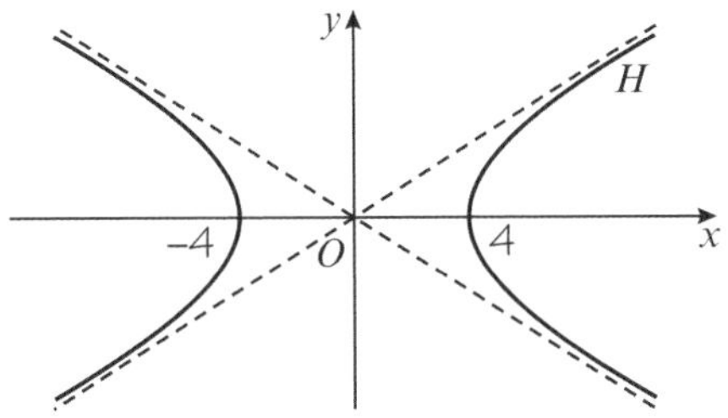

By comparing with $\frac{x^2}{a^2} - \frac{y^2}{b^2} = 1$ and using $a = 4$ and $b = 1$.

c Equations of asymptotes are

$$y = \pm\tfrac{1}{4}x$$

Use $y = \pm\frac{b}{a}x$.

Exercise 3B

1 Sketch the following hyperbolas showing clearly the intersections with the x-axis and the equations of the asymptotes.

a $x^2 - 4y^2 = 16$ **b** $4x^2 - 25y^2 = 100$ **c** $\frac{x^2}{8} - \frac{y^2}{2} = 1$

2 **a** Sketch the hyperbolas with the following parametric equations. Give the equations of the asymptotes and show points of intersection with the x-axis.

i $x = 2\sec\theta,\ y = 3\tan\theta,\ -\pi \leqslant \theta < \pi,\ \theta \neq \pm\frac{\pi}{2}$

ii $x = \pm 4\cosh t,\ y = 3\sinh t,\ t \in \mathbb{R}$

iii $x = \pm\cosh t,\ y = 2\sinh t,\ t \in \mathbb{R}$

iv $x = 5\sec\theta,\ y = 7\tan\theta,\ -\pi \leqslant \theta < \pi,\ \theta \neq \pm\frac{\pi}{2}$

b Find the Cartesian equation for each of the hyperbolas from part **a**.

Challenge

The rectangular hyperbola with equation $xy = c^2$ is rotated through 45° anticlockwise about the origin. Show that the resulting curve satisfies the equation $y^2 - x^2 = a^2$, and state the relationship between a and c in this case.

3.3 Eccentricity

You can define the ellipse and hyperbola in terms of their focus–directrix properties. In order to do this, you need to generalise the approach used for the parabola in the previous chapter. To do this you need to consider the **eccentricity** of a particular conic section.

Links The parabola with equation $y^2 = 4ax$ is the locus of all the points, P, that are equidistant from a fixed point, S, (the focus) and a fixed line (the directrix). ← **Section 2.2**

A

- **For all points, P, on a conic section, the ratio of the distance of P from a fixed point (called the focus) and a fixed straight line (called the directrix) is constant. This ratio, e, is known as the eccentricity of the curve.**

The diagram shows a fixed point, S, a fixed straight line, the directrix, and a point, P, on a conic section.

For all points, P, on the curve, the ratio $\frac{PS}{PM} = e$ is constant.

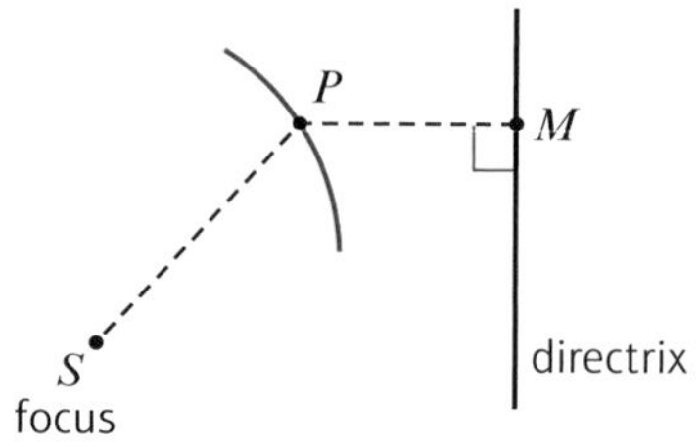

- **If $0 < e < 1$, the point P describes an ellipse.**
- **If $e = 1$, the point P describes a parabola.**
- **If $e > 1$, the point P describes a hyperbola.**

Watch out The special case where $e = 0$ represents a circle, and the special case where e is infinite represents a straight line. These are both examples of conic sections, but you will not need to consider them in this chapter.

Example 5

Show that, for $0 < e < 1$, the ellipse with focus $(ae, 0)$ and directrix $x = \frac{a}{e}$ has equation $\frac{x^2}{a^2} + \frac{y^2}{b^2} = 1$.

Let P be the point with coordinates (x, y).

$$\frac{PS}{PM} = e \Rightarrow PS^2 = e^2PM^2$$

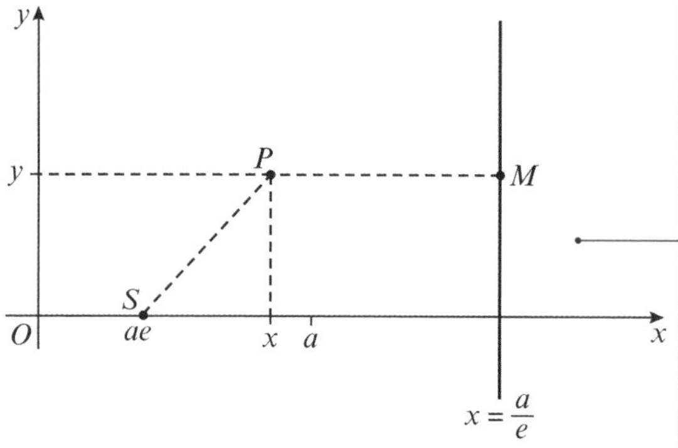

Draw a diagram.

$PS^2 = (x - ae)^2 + y^2$

$PM^2 = \left(\frac{a}{e} - x\right)^2 = \frac{(a - ex)^2}{e^2}$

Find expressions for PS^2 and PM^2 in terms of a, e and x, y.

So $PS^2 = e^2PM^2$ gives

$x^2 - 2aex + a^2e^2 + y^2 = a^2 - 2aex + e^2x^2$

$x^2(1 - e^2) + y^2 = a^2(1 - e^2)$

Simplify.

$$\frac{x^2}{a^2} + \frac{y^2}{a^2(1 - e^2)} = 1$$

So if $b^2 = a^2(1 - e^2)$ then you have the standard equation of the ellipse.

Problem-solving

This equation only produces an ellipse if $0 < e < 1$. If $e = 0$, then $1 - e^2 = 1$ and the equation reduces to the equation of a circle. If $e > 1$, then $1 - e^2$ is negative and the equation produces a hyperbola.

A Because the ellipse is symmetrical about the y-axis, the above derivation will also work for a focus $(-ae, 0)$ with a directrix $x = -\frac{a}{e}$

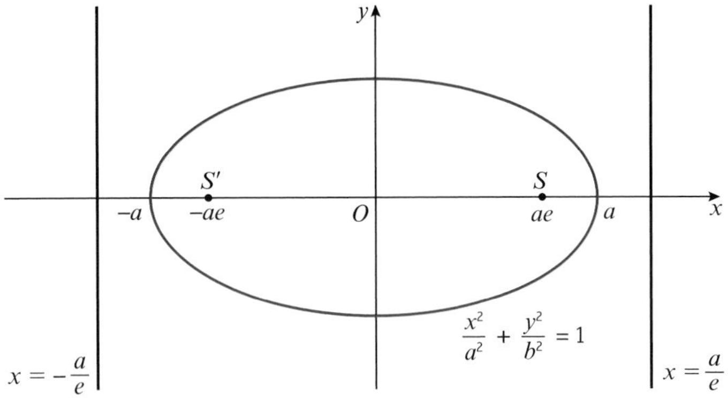

Online Explore the foci and directices of an ellipse using GeoGebra.

- **For an ellipse with equation $\frac{x^2}{a^2} + \frac{y^2}{b^2} = 1$, and $a > b$,**
  - **the eccentricity, $0 < e < 1$, is given by $b^2 = a^2(1 - e^2)$**
  - **the foci are at $(\pm ae, 0)$**
  - **the directrices are $x = \pm\frac{a}{e}$**

Notation **Foci** is the plural of focus and **directrices** is the plural of directrix.

Notice that the foci are on the **major axis** which in this case is the x-axis because $a > b$.

If the major axis is along the y-axis ($b > a$), then the foci will be on the y-axis at $(0, \pm be)$ and the directrices will have equations $y = \pm\frac{b}{e}$. The eccentricity will be given by $a^2 = b^2(1 - e^2)$.

Example 6

Find the foci of the ellipses with the following equations and give the equations of the directrices.

a $\frac{x^2}{9} + \frac{y^2}{4} = 1$ **b** $\frac{x^2}{16} + \frac{y^2}{25} = 1$

In each case sketch the ellipse, and show the directrices and foci.

a $\frac{x^2}{9} + \frac{y^2}{4} = 1$

$b^2 = a^2(1 - e^2)$ gives $4 = 9(1 - e^2)$ so $e^2 = \frac{5}{9}$

Note that $a = 3$ and $b = 2$.
Since $a > b$ use $b^2 = a^2(1 - e^2)$.

So $e = \frac{\sqrt{5}}{3}$

Use $(\pm ae, 0)$.

Foci are at $(\pm\sqrt{5}, 0)$.

Use $x = \pm\frac{a}{e}$

Directrices are $x = \pm\frac{9}{\sqrt{5}}$

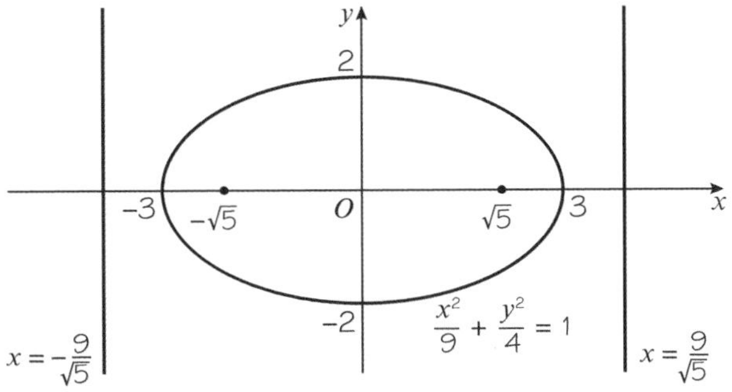

A

b $\frac{x^2}{16} + \frac{y^2}{25} = 1$

$a^2 = b^2(1 - e^2)$ gives $16 = 25(1 - e^2)$

So $e^2 = \frac{9}{25}$ and $e = \frac{3}{5}$

Foci are at $(0, \pm 3)$.

Directrices are $y = \pm\frac{25}{3}$

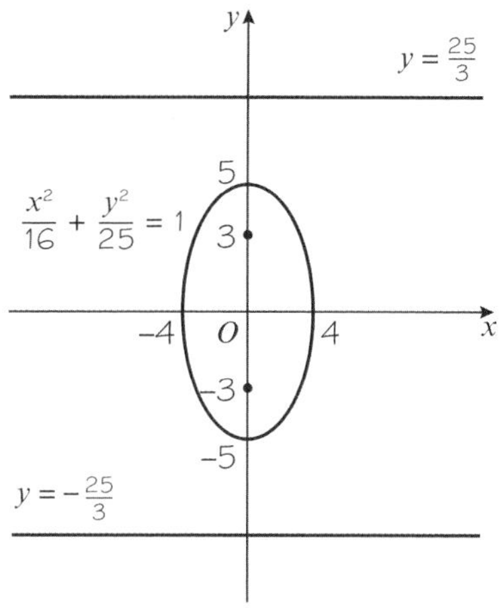

Note that $a = 4$ and $b = 5$. Since $b > a$ use $a^2 = b^2(1 - e^2)$.

Use $(0, \pm be)$.

Use $y = \pm\frac{b}{e}$

Example 7

The ellipse with equation $\frac{x^2}{a^2} + \frac{y^2}{b^2} = 1$ has foci at $S(ae, 0)$ and $S'(-ae, 0)$. Show that if P is any point on the ellipse then $PS + PS' = 2a$.

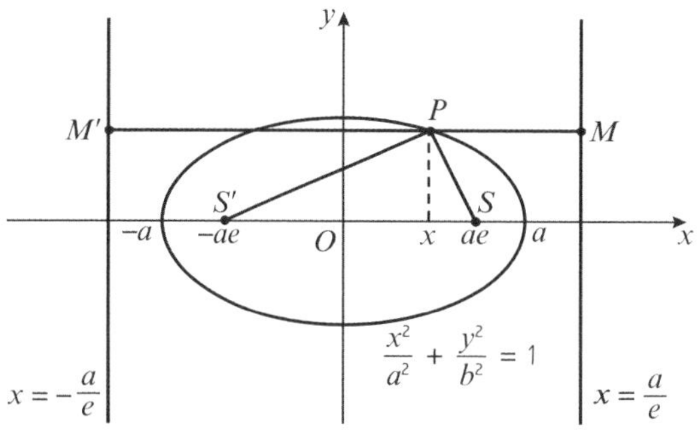

Let M be the point on the directrix $x = \frac{a}{e}$ where $PS = ePM$.

Let M' be the point on the directrix $x = -\frac{a}{e}$ where $PS' = ePM'$.

Let P be (x, y).

$PM = \frac{a}{e} - x$

$PM' = x + \frac{a}{e}$

So $PS + PS' = ePM + ePM'$

$= e\left(\frac{a}{e} - x\right) + e\left(\frac{a}{e} + x\right) = a - ex + a + ex$

$= 2a$

Use the focus and directrix definitions of an ellipse.

PM and PM' are parallel to the x-axis.

Note This is an important property of an ellipse.

Example 8

A Show that for $e > 1$ the hyperbola with foci at $(\pm ae, 0)$ and directrices at $x = \pm\frac{a}{e}$ has equation $\frac{x^2}{a^2} - \frac{y^2}{b^2} = 1$

Let $P(x, y)$ be a point on the hyperbola.

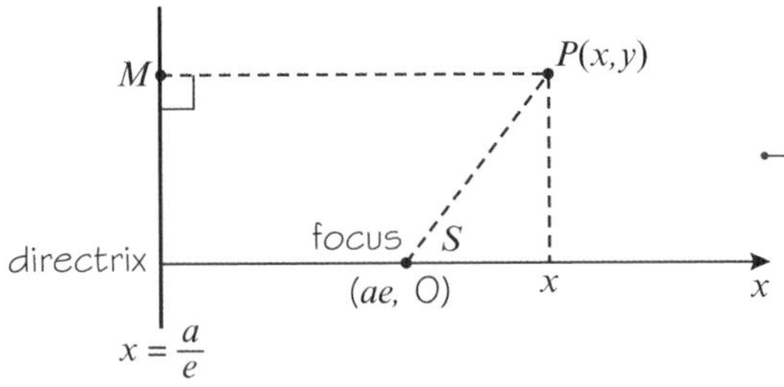

Draw a diagram.

$\frac{PS}{PM} = e \Rightarrow PS^2 = e^2PM^2$

$PS^2 = (x - ae)^2 + y^2$

$PM^2 = \left(x - \frac{a}{e}\right)^2 = \frac{(ex - a)^2}{e^2}$

Find expressions for PS^2 and PM^2 in terms of a, e and x, y.

So $PS^2 = e^2PM^2$ gives

$x^2 - \cancel{2aex} + a^2e^2 + y^2 = e^2x^2 - \cancel{2aex} + a^2$

Simplify

$a^2(e^2 - 1) = x^2(e^2 - 1) - y^2$

$1 = \frac{x^2}{a^2} - \frac{y^2}{a^2(e^2 - 1)}$

$e > 1$ so $a^2(e^2 - 1)$ will be positive.

So if $b^2 = a^2(e^2 - 1)$ you have the standard equation of a hyperbola.

- **For a hyperbola with equation $\frac{x^2}{a^2} - \frac{y^2}{b^2} = 1$,**
  - **the eccentricity, $e > 1$, is given by $b^2 = a^2(e^2 - 1)$**
  - **the foci are at $(\pm ae, 0)$**
  - **the directrices are $x = \pm\frac{a}{e}$**

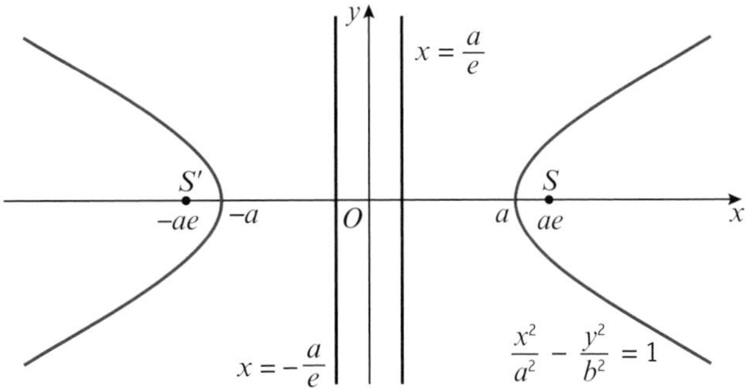

Example 9

Find foci of the following hyperbolas.

In each case, sketch the hyperbola and show the directrices.

a $\frac{x^2}{9} - \frac{y^2}{4} = 1$ **b** $\frac{x^2}{16} - \frac{y^2}{25} = 1$

A

a $\frac{x^2}{9} - \frac{y^2}{4} = 1$, so $a = 3$ and $b = 2$.

Compare the equation with $\frac{x^2}{a^2} - \frac{y^2}{b^2} = 1$ and identify a and b.

Eccentricity is given by $b^2 = a^2(e^2 - 1)$.

$4 = 9(e^2 - 1)$

Use $b^2 = a^2(e^2 - 1)$.

So $\frac{4}{9} + 1 = e^2$

$\Rightarrow e = \sqrt{\frac{13}{9}} = \frac{\sqrt{13}}{3}$

Foci are at $(\pm\sqrt{13}, 0)$.

Use $(\pm ae, 0)$.

Directrices are $x = \pm\frac{9}{\sqrt{13}}$

Use $x = \pm\frac{a}{e}$

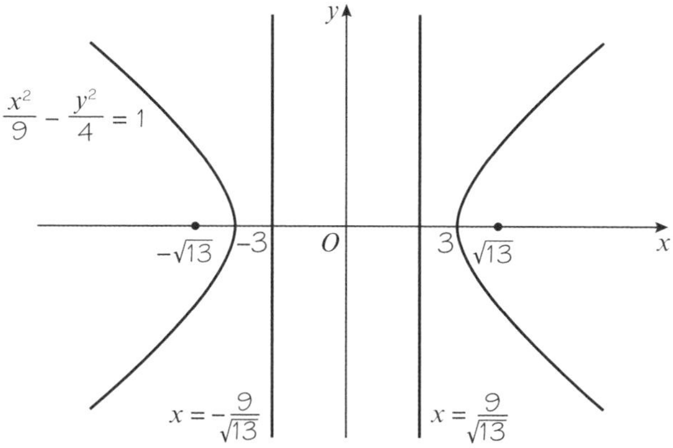

Online Explore the foci and directices of a hyperbola using GeoGebra.

b $\frac{x^2}{16} - \frac{y^2}{25} = 1$, so $a = 4$ and $b = 5$.

Compare the equation with $\frac{x^2}{a^2} - \frac{y^2}{b^2} = 1$ and identify a and b.

Eccentricity is given by $b^2 = a^2(e^2 - 1)$.

$25 = 16(e^2 - 1)$

Use $b^2 = a^2(e^2 - 1)$.

$\frac{25}{16} + 1 = e^2$ so $e = \sqrt{\frac{41}{16}} = \frac{\sqrt{41}}{4}$

Foci are at $(\pm\sqrt{41}, 0)$.

Use $(\pm ae, 0)$.

Directrices are $x = \pm\frac{16}{\sqrt{41}}$

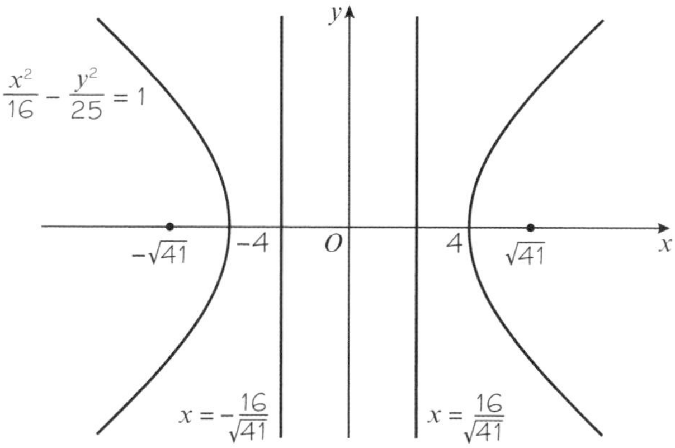

Problem-solving

In this example $b > a$. However, unlike with an ellipse, the foci do not move to the y-axis. Setting $x = 0$ in the general equation of a hyperbola would give $-\frac{y^2}{b^2} = 1$ which is never satisfied for real values of y.

Exercise 3C

A

1 Find the eccentricity of the following ellipses.

a $\frac{x^2}{9}+\frac{y^2}{5}=1$ **b** $\frac{x^2}{16}+\frac{y^2}{9}=1$ **c** $\frac{x^2}{4}+\frac{y^2}{8}=1$

2 Find the foci and directrices of the following ellipses.

a $\frac{x^2}{4}+\frac{y^2}{3}=1$ **b** $\frac{x^2}{16}+\frac{y^2}{7}=1$ **c** $\frac{x^2}{5}+\frac{y^2}{9}=1$

3 An ellipse with equation $\frac{x^2}{a^2}+\frac{y^2}{b^2}=1$ has focus (3, 0) and the equation of the directrix is $x = 12$.

a Explain why $a > b$.

b Find:

i the eccentricity of the ellipse **ii** the values of a and b.

c Sketch the ellipse, showing the directrices and any points of intersection with the coordinate axes.

4 An ellipse with equation $\frac{x^2}{a^2}+\frac{y^2}{b^2}=1$ has focus (0, 2) and the equation of the directrix is $y = 8$.

a Explain why $b > a$.

b Find:

i the eccentricity of the ellipse **ii** the values of a and b.

c Sketch the ellipse, showing the directrices and any points of intersection with the coordinate axes.

5 Find the eccentricities of the following hyperbolas.

a $\frac{x^2}{5}-\frac{y^2}{3}=1$ **b** $\frac{x^2}{9}-\frac{y^2}{7}=1$ **c** $\frac{x^2}{9}-\frac{y^2}{16}=1$

6 Sketch the following hyperbolas, showing clearly the positions of their foci and directrices.

a $\frac{x^2}{4}-\frac{y^2}{8}=1$ **b** $\frac{x^2}{16}-\frac{y^2}{9}=1$ **c** $\frac{x^2}{4}-\frac{y^2}{5}=1$

7 **a** For each of the following hyperbolas, find the eccentricity and show that the foci are at (±5, 0).

i $\frac{x^2}{24}-y^2=1$ **ii** $x^2-\frac{y^2}{24}=1$ **iii** $\frac{x^2}{16}-\frac{y^2}{9}=1$ **iv** $\frac{x^2}{9}-\frac{y^2}{16}=1$

b Hence sketch all four hyperbolas on the same graph, showing the foci and labelling each curve with its eccentricity.

E/P 8 The latus rectum of an ellipse is a chord perpendicular to the major axis that passes through a focus. Show that the length of the latus rectum of the ellipse with equation $\frac{x^2}{a^2}+\frac{y^2}{b^2}=1$, where $a > b$, is $\frac{2b^2}{a}$ **(5 marks)**

E/P 9 The distance between the foci of an ellipse is 16 and the distance between the directrices is 25.

a Find the eccentricity of the ellipse. **(3 marks)**

b Given that both the foci of the ellipse lie on the y-axis, find its equation in the form $\frac{x^2}{a^2}+\frac{y^2}{b^2}=1$. **(2 marks)**

E/P 10 The point P lies on the ellipse with equation $x^2 + 4y^2 = 36$, and A and B are the points $-3\sqrt{3},0$ and $3\sqrt{3},0$ respectively. Prove that $PA + PB = 12$. **(4 marks)**

A E/P **11** Ellipse E has equation $\frac{x^2}{a^2} + \frac{y^2}{b^2} = 1$, such that $a > b$. The foci of E are at S and S' and the point P is $(0, b)$.

Show that $\cos(PSS') = e$, the eccentricity of E. **(6 marks)**

E/P **12** The ellipse E has foci at S and S'. The point P on E is such that angle PSS' is a right angle and angle $PS'S = 30°$.

Show that the eccentricity of the ellipse, e, is $\frac{1}{\sqrt{3}}$ **(6 marks)**

3.4 Tangents and normals to an ellipse

You can use parametric differentiation or implicit differentiation to find the equations of the tangent and normal to an ellipse at a given point. It is often simpler to derive the equations rather than memorising formulae.

Watch out If you are asked to **prove** a result you will need to show enough working to demonstrate your process for finding the gradient.

Example 10

Find the equation of the tangent to the ellipse with equation $\frac{x^2}{9} + \frac{y^2}{4} = 1$ at the point $P(3\cos t, 2\sin t)$.

$y = 2\sin t,\ x = 3\cos t$

$$\frac{dy}{dx} = \frac{\frac{dy}{dt}}{\frac{dx}{dt}} = \frac{2\cos t}{-3\sin t}$$

Find the gradient.

Problem-solving

You could also differentiate the equation implicitly: $\frac{2}{9}x + \frac{1}{2}y\frac{dy}{dx} = 0$ and therefore $\frac{dy}{dx} = -\frac{4x}{9y}$

$$y - 2\sin t = \frac{2\cos t}{-3\sin t}(x - 3\cos t)$$

Write down the equation of the tangent using $y - y_1 = m(x - x_1)$.

$$3y\sin t - 6\sin^2 t = -2x\cos t + 6\cos^2 t$$

Simplify.

$$3y\sin t + 2x\cos t = 6(\cos^2 t + \sin^2 t)$$

Use $\cos^2 t + \sin^2 t \equiv 1$.

$$3y\sin t + 2x\cos t = 6$$

Example 11

Show that the equation of the normal to the ellipse with equation $\frac{x^2}{a^2} + \frac{y^2}{b^2} = 1$ at the point $P(a\cos t, b\sin t)$ is $ax\sin t - by\cos t = (a^2 - b^2)\cos t\sin t$

$$\frac{dy}{dx} = \frac{b\cos t}{-a\sin t}$$

Find the gradient.

Gradient of normal is $\frac{a\sin t}{b\cos t}$

Use the perpendicular gradient rule.

Equation is $y - b\sin t = \frac{a\sin t}{b\cos t}(x - a\cos t)$

Use $y - y_1 = m(x - x_1)$ and simplify.

$$by\cos t - b^2\cos t\sin t = ax\sin t - a^2\cos t\sin t$$

$$ax\sin t - by\cos t = (a^2 - b^2)\cos t\sin t$$

- **An equation of the normal to the ellipse with equation $\frac{x^2}{a^2} + \frac{y^2}{b^2} = 1$ at the point $P(a\cos t, b\sin t)$ is $ax\sin t - by\cos t = (a^2 - b^2)\cos t\sin t$.**

You can use a similar method to find the general equation of a tangent to an ellipse.

- **An equation of the tangent to the ellipse with equation $\frac{x^2}{a^2} + \frac{y^2}{b^2} = 1$ at the point $P(a\cos t, b\sin t)$ is $bx\cos t + ay\sin t = ab$.**

Links The derivation of this result is left as an exercise. → **Exercise 3D Q3**

Example 12

The point $P\left(2, \frac{3\sqrt{3}}{2}\right)$ lies on the ellipse E with parametric equations $x = 4\cos\theta$, $y = 3\sin\theta$, $0 \leqslant \theta < 2\pi$.

a Find the value of θ at the point P.

The normal to the ellipse at P cuts the x-axis at the point A.

b Find the coordinates of the point A.

a $4\cos\theta = 2 \Rightarrow \cos\theta = \frac{1}{2}$ so $\theta = \frac{\pi}{3}, \frac{5\pi}{3}$

$3\sin\theta = 3\frac{\sqrt{3}}{2} \Rightarrow \sin\theta = \frac{\sqrt{3}}{2}$ so $\theta = \frac{\pi}{3}, \frac{2\pi}{3}$

So $\theta = \frac{\pi}{3}$

Set $a\cos\theta$ as the x-coordinate and $b\sin\theta$ as the y-coordinate and solve to find θ. Choose the value of θ in the given range that satisfies both equations.

b $\frac{dy}{dx} = \frac{3\cos\theta}{-4\sin\theta}$

Use the general point to find the gradient.

So gradient of normal is $\frac{4\sin\theta}{3\cos\theta}$

At P the gradient of the normal is

$\frac{4 \times \frac{\sqrt{3}}{2}}{3 \times \frac{1}{2}} = \frac{4\sqrt{3}}{3}$

Use the perpendicular gradient rule then substitute the value of θ.

This can be found by implicit differentiation on the Cartesian equation $\frac{x^2}{16} + \frac{y^2}{9} = 1$. Differentiating: $\frac{2}{16}x + \frac{2}{9}y\frac{dy}{dx} = 0$ so $\frac{dy}{dx} = -\frac{9x}{16y}$ and using the coordinates of P, $\frac{dy}{dx} = \frac{-18}{16 \times 3\frac{\sqrt{3}}{2}} = \frac{-3}{4\sqrt{3}}$

so normal gradient is $\frac{4\sqrt{3}}{3}$

Equation of normal at P is

$y - 3\frac{\sqrt{3}}{2} = \frac{4\sqrt{3}}{3}(x - 2)$

Cuts x-axis at $-9\sqrt{3} = 8\sqrt{3}(x - 2)$

So A is $\left(\frac{7}{8}, 0\right)$

Let $y = 0$ and solve to find x.

Example 13

Show that the condition for $y = mx + c$ to be a tangent to the ellipse $\frac{x^2}{a^2} + \frac{y^2}{b^2} = 1$ is $b^2 + a^2m^2 = c^2$.

The line meets the ellipse when $\frac{x^2}{a^2} + \frac{(mx + c)^2}{b^2} = 1$

Substitute $mx + c$ for y.

So $b^2x^2 + a^2m^2x^2 + 2a^2mxc + a^2c^2 = a^2b^2$

$x^2(b^2 + a^2m^2) + 2a^2mcx + a^2(c^2 - b^2) = 0$

Multiply out and rearrange as a quadratic equation in x.

A

To be a tangent there must be only one real root.
Therefore the discriminant of this quadratic is 0.

$(2a^2mc)^2 = 4(b^2 + a^2m^2)a^2(c^2 - b^2)$

So $4a^{42}m^2c^2 = 4a^2(b^2c^2 - b^4 + a^2m^2c^2 - a^2b^2m^2)$

$a^2m^2c^2 = b^2c^2 - b^4 + a^2m^2c^2 - a^2b^2m^2$

$b^4 + a^2b^2m^2 = b^2c^2$

$b^2 + a^2m^2 = c^2$

Use the properties of the discriminant.
← Pure Year 1, Chapter 2

Multiply out and simplify.

Cancel b^2.

Problem-solving

This is a general result about tangents to ellipses. Unless you are asked to prove it, you could quote it in your exam.

Example 14

The ellipse C has equation $\frac{x^2}{5^2} + \frac{y^2}{3^2} = 1$. The line l is normal to the ellipse at P and passes through the point Q, where C cuts the y-axis, as shown in the diagram.

Find the exact coordinates of the point R where l cuts the positive x-axis.

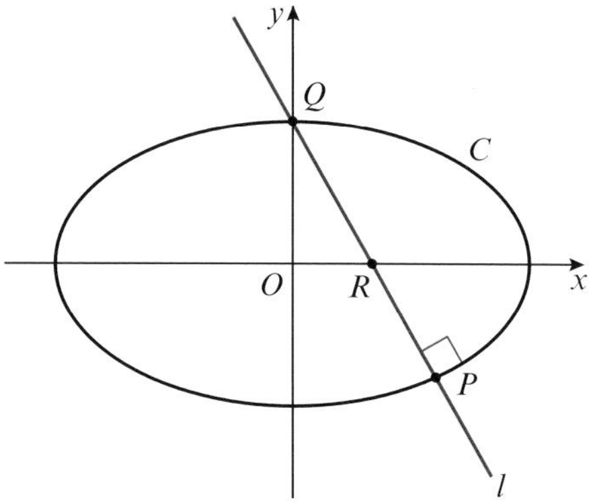

$\frac{x^2}{5^2} + \frac{y^2}{3^2} = 1$ so $a = 5$ and $b = 3$

$ax\sin\theta - by\cos\theta = (a^2 - b^2)\cos\theta\sin\theta$

$5x\sin\theta - 3y\cos\theta = 16\cos\theta\sin\theta$

Q cuts the y-axis at $(0, 3)$

$-9\cos\theta = 16\cos\theta\sin\theta$

$-9 = 16\sin\theta$

$\sin\theta = -\frac{9}{16}$

$\cos\theta = \sqrt{1 - \sin^2\theta}$

$= \frac{5\sqrt{7}}{16}$

So the equation of l is:

$5\left(-\frac{9}{16}\right)x - 3\left(\frac{5\sqrt{7}}{16}\right)y = 16\left(\frac{5\sqrt{7}}{16}\right)\left(-\frac{9}{16}\right)$

$-3x - \sqrt{7}y = -3\sqrt{7}$

When $y = 0$

$-3x = -3\sqrt{7}$

$x = \sqrt{7}$

So l cuts the x-axis at $(\sqrt{7}, 0)$.

Deduce the values for a and b from the general equation of an ellipse, $\frac{x^2}{a^2} + \frac{y^2}{b^2} = 1$

State the general equation for the normal of an ellipse and substitute $a = 5$ and $b = 3$.

The ellipse cuts the y-axis at $(0, \pm b)$ and $b = 3$.

Substitute $x = 0$, $y = 3$ into the general equation for the normal to an ellipse.

Use your value of $\sin\theta$ to find the value of $\cos\theta$

Problem-solving

The identity $\cos^2\theta + \sin^2\theta \equiv 1$ gives $\cos\theta = \pm\frac{5\sqrt{7}}{16}$
However, from the diagram you can see that P is in the fourth quadrant, so $\cos\theta$ must be positive.

Substitute your exact values for $\sin\theta$ and $\cos\theta$ to find the equation of l.

Substitute $y = 0$ to find the points where l cuts the x-axis.

Exercise 3D

A

1 Find the equations of tangents and normals to the following ellipses at the points given.

a $\frac{x^2}{4} + y^2 = 1$ at $(2\cos\theta, \sin\theta)$

b $\frac{x^2}{25} + \frac{y^2}{9} = 1$ at $(5\cos\theta, 3\sin\theta)$

2 Find equations of tangents and normals to the following ellipses at the points given.

a $\frac{x^2}{9} + \frac{y^2}{1} = 1$ at $\left(\sqrt{5}, \frac{2}{3}\right)$

b $\frac{x^2}{16} + \frac{y^2}{4} = 1$ at $(-2, \sqrt{3})$

P **3** Show that the equation of the tangent to the ellipse $\frac{x^2}{a^2} + \frac{y^2}{b^2} = 1$ at the point $(a\cos t, b\sin t)$ is $bx\cos t + ay\sin t = ab$.

4 **a** Show that the line $y = x + \sqrt{5}$ is a tangent to the ellipse with equation $\frac{x^2}{4} + \frac{y^2}{1} = 1$.

b Find the point of contact of this tangent.

5 **a** Find an equation of the normal to the ellipse with equation $\frac{x^2}{9} + \frac{y^2}{4} = 1$ at the point $P(3\cos\theta, 2\sin\theta)$.

This normal crosses the x-axis at the point $\left(-\frac{5}{6}, 0\right)$.

b Find the value of θ and the exact coordinates of the possible positions of P.

6 The line $y = 2x + c$ is a tangent to $x^2 + \frac{y^2}{4} = 1$.

Find the possible values of c.

7 The line with equation $y = mx + 3$ is a tangent to $x^2 + \frac{y^2}{5} = 1$.

Find the possible values of m.

E **8** The line $y = mx + 4$ $(m > 0)$ is a tangent to the ellipse E with equation $\frac{x^2}{3} + \frac{y^2}{4} = 1$ at the point P.

a Find the value of m. **(4 marks)**

b Find the coordinates of the point P. **(2 marks)**

The normal to E at P crosses the y-axis at the point A.

c Find the coordinates of A. **(5 marks)**

The tangent to E at P crosses the y-axis at the point B.

d Find the area of triangle APB. **(5 marks)**

E **9** The ellipse E has equation $\frac{x^2}{9} + \frac{y^2}{4} = 1$.

a Show that the gradient of the tangent to E at the point $P(3\cos\theta, 2\sin\theta)$ is $-\frac{2}{3}\cot\theta$. **(4 marks)**

b Show that the point $Q\left(\frac{9}{5}, -\frac{8}{5}\right)$ lies on E. **(2 marks)**

c Find the gradient of the tangent to E at Q. **(1 mark)**

The tangents to E at the points P and Q are perpendicular.

d Find the value of $\tan\theta$ and hence the exact coordinates of the two possible positions of P. **(4 marks)**

P **10** The line $y = mx + c$ is a tangent to both of the ellipses $\frac{x^2}{9} + \frac{y^2}{46} = 1$ and $\frac{x^2}{25} + \frac{y^2}{14} = 1$.

Find the possible values of m and c.

A E/P **11** The ellipse E has equation $\frac{x^2}{8^2} + \frac{y^2}{4^2} = 1$. The line l_1 is tangent to E at the point $P(8\cos\theta, 4\sin\theta)$ and the line l_2 is normal to E at the point $P(8\cos\theta, 4\sin\theta)$. Line l_1 cuts the x-axis at A and line l_2 cuts the y-axis at B. Find the equation of the line AB. **(6 marks)**

E/P **12** The ellipse E has equation $\frac{x^2}{5^2} + \frac{y^2}{3^2} = 1$. The line l_1 is tangent to E at the point $P(5\cos\theta, 3\sin\theta)$.

a Use calculus to show that an equation for l_1 is $3x\cos\theta + 5y\sin\theta = 15$. **(5 marks)**

The line l_1 cuts the y-axis at Q. The line l_2 passes through the point Q, perpendicular to l_1.

b Find the equation of the line l_2. **(3 marks)**

c Given that l_2 cuts the x-axis at $(-4, 0)$, show that $\cos\theta = \frac{4}{5}$ **(3 marks)**

E/P **13** The ellipse E has equation $\frac{x^2}{4} + \frac{y^2}{16} = 1$. The line l_1 is tangent to E at the point $P(2\cos t, 4\sin t)$.

a Use calculus to show that an equation for l_1 is $2x\cos t + y\sin t = 4$. **(5 marks)**

The line l_2 passes through the origin and is perpendicular to l_1. The lines l_1 and l_2 intersect at the point Q.

b Show that the coordinates of Q are $\left(\frac{8\cos t}{4\cos^2 t + \sin^2 t}, \frac{4\sin t}{4\cos^2 t + \sin^2 t}\right)$. **(4 marks)**

E/P **14** The line l_1 is tangent to the ellipse with equation $\frac{x^2}{a^2} + \frac{y^2}{b^2} = 1$ at the point $(a\cos t, b\sin t)$.

Show that the area of the shaded region is $ab\operatorname{cosec} 2t$. **(6 marks)**

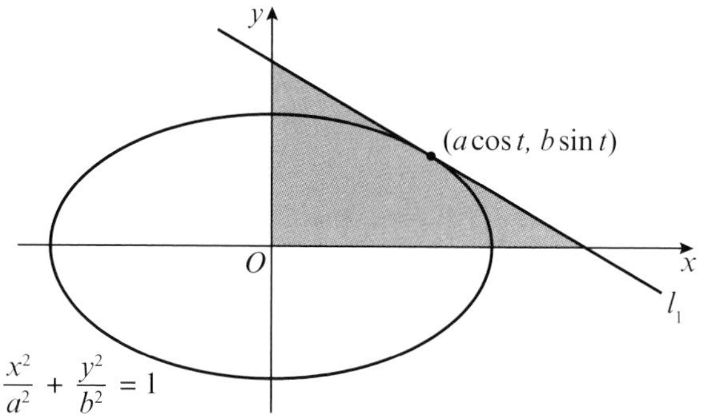

E/P **15** The diagram shows the ellipse with equation $\frac{x^2}{6^2} + \frac{y^2}{4^2} = 1$.

Show that the area of the shaded region is $8\pi - 6\sqrt{3}$.

(6 marks)

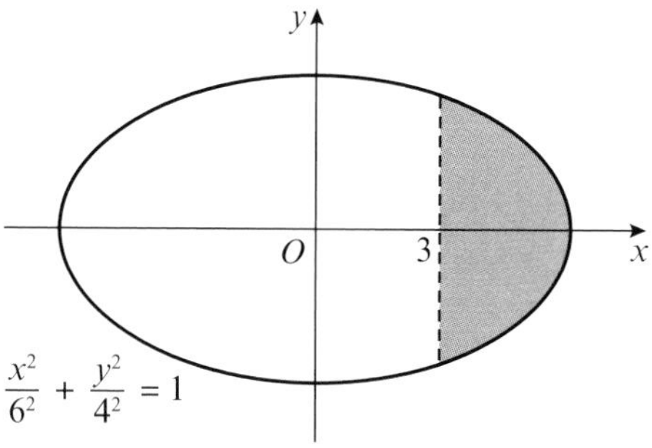

Problem-solving

Use the substitution $6\sin u = x$ and simplify the resulting integrand using an appropriate trigonometric identity.

Challenge

Prove that the area inside the ellipse with equation $\frac{x^2}{a^2} + \frac{y^2}{b^2} = 1$ is πab.

3.5 Tangents and normals to a hyperbola

A

You can find the equations of the tangent and normal to a hyperbola at a given point.

Example 15

Find the equation of the tangent to the hyperbola with equation $\frac{x^2}{9} - \frac{y^2}{4} = 1$ at the point $(6, 2\sqrt{3})$.

Differentiating, $\frac{2}{9}x - \frac{2}{4}y\frac{dy}{dx} = 0$ — Use implicit differentiation.

At $(6, 2\sqrt{3})$,

$\frac{12}{9} - \frac{4\sqrt{3}}{4}\frac{dy}{dx} = 0 \Rightarrow \frac{dy}{dx} = \frac{4\sqrt{3}}{9}$

Equation of tangent is

$y - 2\sqrt{3} = \frac{4\sqrt{3}}{9}(x - 6)$ — Use $y - y_1 = m(x - x_1)$.

or $y = \frac{4\sqrt{3}}{9}x - \frac{2\sqrt{3}}{3}$

Example 16

Show that the equation of the tangent to the hyperbola with equation $\frac{x^2}{a^2} - \frac{y^2}{b^2} = 1$ at the point $(a\cosh t, b\sinh t)$ can be written as $bx\cosh t - ay\sinh t = ab$.

$x = a\cosh t,\ y = b\sinh t$

$\frac{dy}{dx} = \frac{\frac{dy}{dt}}{\frac{dx}{dt}} = \frac{b\cosh t}{a\sinh t}$ — Use the chain rule to find $\frac{dy}{dx}$. Remember that $\frac{d}{dt}(\sinh t) = \cosh t$ and $\frac{d}{dt}(\cosh t) = \sinh t$ ← Core Pure Book 2, Chapter 6

Equation of tangent is

$y - b\sinh t = \frac{b\cosh t}{a\sinh t}(x - a\cosh t)$ — Use $y - y_1 = m(x - x_1)$.

$ay\sinh t - ab\sinh^2 t = bx\cosh t - ab\cosh^2 t$

$ay\sinh t + ab(\cosh^2 t - \sinh^2 t) = bx\cosh t$

$ay\sinh t + ab = bx\cosh t$ — Use $\cosh^2 t - \sinh^2 t \equiv 1$.

$bx\cosh t - ay\sinh t = ab$

- **An equation of the tangent to the hyperbola with equation $\frac{x^2}{a^2} - \frac{y^2}{b^2} = 1$ at the point $P(a\cosh t, b\sinh t)$ is $ay\sinh t + ab = bx\cosh t$.**

You can use the alternative form of a general point on a hyperbola to find a different general equation of a tangent to a hyperbola.

- **An equation of the tangent to the hyperbola with equation $\frac{x^2}{a^2} - \frac{y^2}{b^2} = 1$ at the point $P(a\sec\theta, b\tan\theta)$ is $bx\sec\theta - ay\tan\theta = ab$.**

Links The derivation of this result is left as an exercise. → Exercise 3E Q3

Example 17

Show that an equation of the normal to the hyperbola with equation $\frac{x^2}{a^2} - \frac{y^2}{b^2} = 1$ at $(a\sec\theta, b\tan\theta)$ is $by + ax\sin\theta = (a^2 + b^2)\tan\theta$.

$y = b\tan\theta,\ x = a\sec\theta$

$\frac{dy}{dx} = \frac{\frac{dy}{d\theta}}{\frac{dx}{d\theta}} = \frac{b\sec^2\theta}{a\sec\theta\tan\theta} = \frac{b}{a\sin\theta}$ — Use the chain rule to find $\frac{dy}{dx}$

So gradient of normal is $-\frac{a\sin\theta}{b}$ — Use the perpendicular gradient rule.

Equation of the normal is

$y - b\tan\theta = -\frac{a\sin\theta}{b}(x - a\sec\theta)$ — Use $y - y_1 = m(x - x_1)$.

$by - b^2\tan\theta = -ax\sin\theta + a^2\tan\theta$

So $by + ax\sin\theta = (a^2 + b^2)\tan\theta$

- **An equation of the normal to the hyperbola with equation $\frac{x^2}{a^2} - \frac{y^2}{b^2} = 1$ at the point $P(a\sec\theta, b\tan\theta)$ is $by + ax\sin\theta = (a^2 + b^2)\tan\theta$**

You can use the other form of a general point on a hyperbola to find a different general equation of a normal to a hyperbola.

- **An equation of the normal to the hyperbola with equation $\frac{x^2}{a^2} - \frac{y^2}{b^2} = 1$ at the point $P(a\cosh t, b\sinh t)$ is $ax\sinh t + by\cosh t = (a^2 + b^2)\sinh t\cosh t$**

Links The derivation of this result is left as an exercise → **Exercise 3E Q4**

Example 18

Show that the condition for the line $y = mx + c$ to be a tangent to the hyperbola $\frac{x^2}{a^2} - \frac{y^2}{b^2} = 1$ is that m and c satisfy $b^2 + c^2 = a^2m^2$.

$\frac{x^2}{a^2} - \frac{(mx + c)^2}{b^2} = 1$ — Substitute $mx + c$ for y in the equation of the hyperbola.

$b^2x^2 - a^2(m^2x^2 + 2mxc + c^2) = a^2b^2$

$(b^2 - a^2m^2)x^2 - 2mca^2x - a^2(c^2 + b^2) = 0$ — Multiply out and collect terms as a quadratic in x.

Since the line is a tangent the discriminant must be zero. — Use discriminant properties.

$4m^2c^2a^4 = -4(b^2 - a^2m^2)a^2(c^2 + b^2)$ — Cancel $4a^2$.

$m^2c^2a^2 = -b^4 - b^2c^2 + a^2m^2c^2 + a^2m^2b^2$

$b^2 + c^2 = a^2m^2$ — Cancel b^2.

Problem-solving

This is a general result about tangents to hyperbolas. Unless you are asked to prove it, you could quote it in your exam.

Example 19

A The tangent to the hyperbola with equation $\frac{x^2}{9} - \frac{y^2}{4} = 1$ at the point $(3\cosh t, 2\sinh t)$ crosses the y-axis at the point $(0, -1)$. Find the value of t.

Equation of tangent is

$3y\sinh t + 6 = 2x\cosh t$

Passes through $(0, -1)$

$-3\sinh t + 6 = 0$

so $\sinh t = 2$

Then $t = \text{arsinh}\, 2$

but $\text{arsinh}\, x = \ln(x + \sqrt{x^2 + 1})$

so $t = \ln(2 + \sqrt{5})$

Remember that for a hyperbola with equation $\frac{x^2}{a^2} - \frac{y^2}{b^2} = 1$, the equation of the tangent at point $(a\cosh t, b\sinh t)$ is $ay\sinh t + ab = bx\cosh t$. Here $a = 3$ and $b = 2$.

Substitute $x = 0$ and $y = -1$.

Use the formula for arsinh (x) from the formula booklet.

Example 20

The hyperbola H has equation $\frac{x^2}{36} - \frac{y^2}{9} = 1$

The line l_1 is the tangent to H at the point $P(6\cosh t, 3\sinh t)$. The line l_2 passes through the origin and is perpendicular to l_1. The lines l_1 and l_2 intersect at the point Q.

Show that the coordinates of the point Q are $\left(\frac{6\cosh t}{4\sinh^2 t + \cosh^2 t}, -\frac{12\sinh t}{4\sinh^2 t + \cosh^2 t}\right)$.

The general form of the equation of a tangent to a hyperbola is $ay\sinh t + ab = bx\cosh t$

So the equation of l_1 is

$6y\sinh t + 18 = 3x\cosh t$

$2y\sinh t + 6 = x\cosh t$

The gradient of l_1 is $\frac{\cosh t}{2\sinh t}$

The gradient of a perpendicular line is $-\frac{2\sinh t}{\cosh t}$

The equation of a perpendicular line through $(0, 0)$ is $y = -\frac{2x\sinh t}{\cosh t}$

l_1: $2y\sinh t + 6 = x\cosh t \Rightarrow y = \frac{x\cosh t - 6}{2\sinh t}$

At Q, $-\frac{2x\sinh t}{\cosh t} = \frac{x\cosh t - 6}{2\sinh t}$

$-4x\sinh^2 t = x\cosh^2 t - 6\cosh t$

$-x(4\sinh^2 t + \cosh^2 t) = -6\cosh t$

$x = \frac{6\cosh t}{4\sinh^2 t + \cosh^2 t}$

$y = \left(-\frac{2\sinh t}{\cosh t}\right)\left(\frac{6\cosh t}{4\sinh^2 t + \cosh^2 t}\right)$

$y = -\frac{12\sinh t}{4\sinh^2 t + \cosh^2 t}$

So the coordinates of Q are

$\left(\frac{6\cosh t}{4\sinh^2 t + \cosh^2 t}, -\frac{12\sinh t}{4\sinh^2 t + \cosh^2 t}\right)$

Here $a = 6$ and $b = 3$.

The gradients of perpendicular lines multiply to equal -1.

The line l_2 passes through $(0, 0)$, so its equation is $y = mx$.

Rearrange the equation for line l_1 into the form $y = \ldots$

The lines intersect at Q. Set the two equations equal to each other.

Simplify to obtain an expression for the x-coordinate.

Substitute the expression for the x-coordinate into $y = -\frac{2x\sinh t}{\cosh t}$

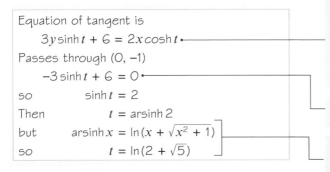

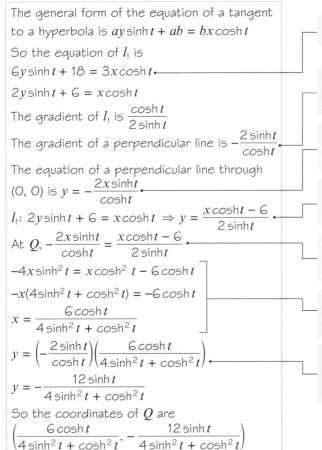

Exercise 3E

A

1 Find the equations of the tangents and normals to the hyperbolas with the following equations at the points indicated.

a $\frac{x^2}{16} - \frac{y^2}{2} = 1$ at the point (12, 4)

b $\frac{x^2}{36} - \frac{y^2}{12} = 1$ at the point (12, 6)

c $\frac{x^2}{25} - \frac{y^2}{3} = 1$ at the point (10, 3)

2 Find the equations of the tangents and normals to the hyperbolas with the following equations at the points indicated.

a $\frac{x^2}{25} - \frac{y^2}{4} = 1$ at the point $(5\cosh t, 2\sinh t)$

b $\frac{x^2}{1} - \frac{y^2}{9} = 1$ at the point $(\sec t, 3\tan t)$

P 3 Show that the equation of the tangent to the hyperbola $\frac{x^2}{a^2} - \frac{y^2}{b^2} = 1$ at the point $(a\sec t, b\tan t)$ is $bx\sec t - ay\tan t = ab$.

P 4 Show that the equation of the normal to the hyperbola $\frac{x^2}{a^2} - \frac{y^2}{b^2} = 1$ at the point $(a\cosh t, b\sinh t)$ is $ax\sinh t + by\cosh t = (a^2 + b^2)\sinh t\cosh t$.

5 The point $P(4\cosh t, 3\sinh t)$, $t \neq 0$, lies on the hyperbola with equation $\frac{x^2}{16} - \frac{y^2}{9} = 1$. The tangent at P crosses the y-axis at the point A.

a Find, in terms of t, the coordinates of A.

The normal to the hyperbola at P crosses the y-axis at B.

b Find, in terms of t, the coordinates of B.

c Find, in terms of t, the area of triangle APB.

6 The tangents from the points P and Q on the hyperbola with equation $\frac{x^2}{4} - \frac{y^2}{9} = 1$ meet at the point (1, 0). Find the exact coordinates of P and Q.

7 The line $y = 2x + c$ is a tangent to the hyperbola $\frac{x^2}{10} - \frac{y^2}{4} = 1$. Find the possible values of c.

8 The line $y = mx + 12$ is a tangent to the hyperbola $\frac{x^2}{49} - \frac{y^2}{25} = 1$ at the point P. Find the possible values of m.

9 The line with equation $y = mx + c$ is a tangent to both of the hyperbolas $\frac{x^2}{4} - \frac{y^2}{15} = 1$ and $\frac{x^2}{9} - \frac{y^2}{95} = 1$. Find the possible values of m and c.

10 The line $y = -x + c$, $c > 0$, touches the hyperbola $\frac{x^2}{25} - \frac{y^2}{16} = 1$ at the point P.

a Find the value of c.

b Find the exact coordinates of P.

E 11 The hyperbola H has equation $\frac{x^2}{a^2} - \frac{y^2}{b^2} = 1$.

a Use calculus to show that the equation of the normal to H at the point $(a\cosh t, b\sinh t)$, $t \neq 0$, may be written in the form $ax\sinh t + by\cosh t = (a^2 + b^2)\sinh t\cosh t$. **(4 marks)**

The line l_1 is the normal to H at the point $(a\cosh t, b\sinh t)$. Given that l_1 meets the x-axis at the point P.

b find, in terms of a, b and t, the coordinates of P. **(2 marks)**

A The line l_2 is the tangent to H at the point $(a, 0)$. Given that l_1 and l_2 meet at the point Q,

c find, in terms of a, b and t, the coordinates of Q. **(2 marks)**

P **12** The hyperbola H has equation $\frac{x^2}{49} - \frac{y^2}{25} = 1$.

The line l_1 is the tangent to H at the point $(7\sec\theta, 5\tan\theta)$.

a Use calculus to show that an equation of l_1 is $7y\sin\theta = 5x - 35\cos\theta$. **(5 marks)**

The line l_2 passes through the origin and is perpendicular to l_1. The lines l_1 and l_2 intersect at the point Q.

b Show that the coordinates of the point Q are $\left(\frac{175\cos\theta}{25 + 49\sin^2\theta}, \frac{-245\sin\theta\cos\theta}{25 + 49\sin^2\theta}\right)$. **(5 marks)**

P **13** P and Q are two distinct points on the hyperbola described by the equation $x^2 - 4y^2 = 16$. The line l passes through the point P and the point Q. The tangent to the hyperbola at P and the tangent to the hyperbola at Q intersect at the point (m, n). Show that an equation of the line l is $mx - 4ny = 16$. **(9 marks)**

P **14** Show that there are exactly two tangents to the hyperbola $\frac{x^2}{4^2} - \frac{y^2}{2^2} = 1$ passing through the point $(6, 4)$ and find each of their equations.

P **15** The hyperbola H has equation $x^2 - \frac{y^2}{4} = 1$.

The line l is a normal to the hyperbola at the point P with x-coordinate 2. The finite region R is bounded by the hyperbola H, the line l and the x-axis.

Show that the exact area of R is $10\sqrt{3} - \text{arcosh}\,2$. **(10 marks)**

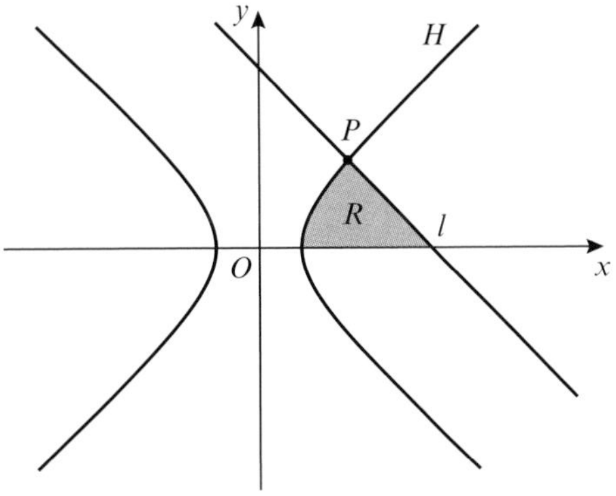

Problem-solving

You will need to use a substitution such as $x = \cosh u$ when integrating.

P **16** The point P lies on the hyperbola H with equation $x^2 - y^2 = 1$. The tangent to H at P cuts the asymptotes of H at the points A and B.

a Prove that P is the midpoint of the line segment AB. **(6 marks)**

b Prove that $OA \times OB$ remains constant as the position of P varies on H. **(3 marks)**

3.6 Loci

Each of the conic sections can be defined as a locus of points. For example, the parabola is the locus of points equidistant from a fixed point and a fixed straight line. You can use the properties of the conic sections, and the general points on each curve, to find other loci associated with these curves.

Example 21

The tangent to the ellipse with equation $\frac{x^2}{a^2} + \frac{y^2}{b^2} = 1$ at the point $P(a\cos t, b\sin t)$ crosses the x-axis at A and the y-axis at B.

Find an equation for the locus of the midpoint of AB as P moves round the ellipse.

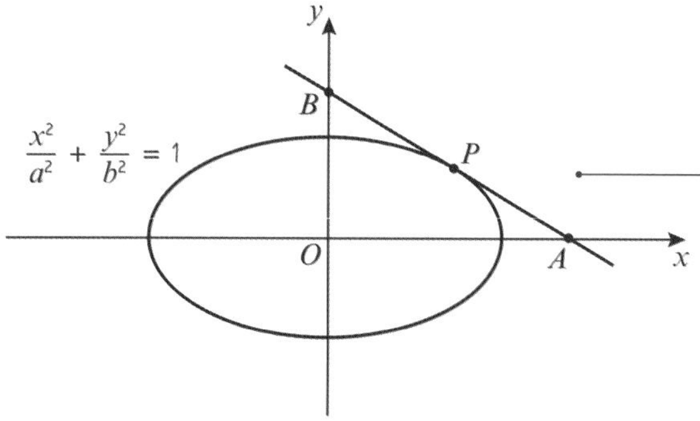

Parametric equations of the ellipse are
$x = a\cos t$ and $y = b\sin t$

Gradient: $\frac{dy}{dx} = \frac{\frac{dy}{dt}}{\frac{dx}{dt}} = \frac{b\cos t}{-a\sin t}$

Equation of tangent:
$y - b\sin t = \frac{b\cos t}{-a\sin t}(x - a\cos t)$
or $ay\sin t + bx\cos t = ab$
A is $(a\sec t, 0)$.
B is $(0, b\operatorname{cosec} t)$.

The midpoint of AB has coordinates (X, Y) where
$X = \frac{a\sec t}{2}$
$Y = \frac{b\operatorname{cosec} t}{2}$

Rearranging:
$\cos t = \frac{a}{2X}$ and $\sin t = \frac{b}{2Y}$

Using $\cos^2 t + \sin^2 t \equiv 1$ gives the locus
$\left(\frac{a}{2X}\right)^2 + \left(\frac{b}{2Y}\right)^2 = 1$

Online Explore the locus of the midpoint of AB using GeoGebra.

A diagram might help and it is always worth drawing a rough sketch.

Use the chain rule to find $\frac{dy}{dx}$

This result was found in Exercise 3D question **3**. You could quote it directly in a question such as this.

Use $x = 0$ to find B and $y = 0$ to find A.

To find the locus of the midpoint, let the coordinates of the midpoint be (X, Y) and then form parametric equations for X and Y.

Eliminate the parameter (t in this case) to find an equation in X and Y.

Problem-solving

In some questions, you may be asked to show that the locus has a particular shape, so you may need to rearrange the final equation into an appropriate form.

You might also need to use properties of the parabola and rectangular hyperbola when solving loci questions. This table summarises the results from the previous chapter.

	Parabola	**Rectangular hyperbola**
Standard Cartesian equation	$y^2 = 4ax$	$xy = c^2$
Parametric equations	$x = at^2, y = 2at$	$x = ct, y = \frac{c}{t}$
General point, P	$(at^2, 2at)$	$\left(ct, \frac{c}{t}\right)$
Equation of tangent at P	$ty = x + at^2$	$x + t^2y = 2ct$
Equation of normal at P	$y + tx = 2at + at^3$	$t^3x - ty = c(t^4 - 1)$

Example 22

A The normal at $P(ap^2, 2ap)$ and the normal at $Q(aq^2, 2aq)$ to the parabola with equation $y^2 = 4ax$ meet at R.

a Find the coordinates of R.

The chord PQ passes through the focus $(a, 0)$ of the parabola.

b Show that $pq = -1$.

c Show that the locus of R is a parabola with equation $y^2 = a(x - 3a)$.

a To find R, find the intersections of the normals.

Normal at P is $y + px = 2ap + ap^3$

Normal at Q is $y + qx = 2aq + aq^3$

Use the standard result for the equation of a normal to a parabola at $(at^2, 2at)$: $y + tx = 2at + at^3$

Subtracting,

$(p - q)x = 2a(p - q) + a(p^3 - q^3)$

$(p - q)x = 2a(p - q) + a(p - q)(p^2 + pq + q^2)$

$x = 2a + a(p^2 + pq + q^2)$

$y = 2ap + ap^3 - 2ap - ap^3 - ap^2q - apq^2$

$= -apq(p + q)$

So R is $(2a + a(p^2 + pq + q^2), -apq(p + q))$

Substitute for x to find y.

Problem-solving

The factorisations of $(p^3 \pm q^3) = (p \pm q)(p^2 \mp pq + q^2)$ are particularly useful in this type of problem and should be learned.

b Chord PQ has gradient

$$\frac{2a(p - q)}{a(p^2 - q^2)} = \frac{2(p - q)}{(p - q)(p + q)} = \frac{2}{p + q}$$

Use $\frac{y_1 - y_2}{x_1 - x_2}$

Equation of chord is

$$y - 2ap = \frac{2}{p + q}(x - ap^2)$$

$\Rightarrow y(p + q) = 2x + 2apq$

Since the chord passes through $(a, 0)$,

$0 = 2a + 2apq$

$\Rightarrow pq = -1$

Problem-solving

Notice that if you let $p = q$ in the equation of the chord you get the equation of the tangent at Q. This is sometimes a useful technique to use.

c Using $pq = -1$ the coordinates of R become

$(a + a(p^2 + q^2), a(p + q))$

Let R be (X, Y), then

$X = a + a(p^2 + q^2)$

$Y = a(p + q)$

So $X = a + a((p + q)^2 - 2pq)$

and using $pq = -1$

$X = 3a + a(p + q)^2$

The following technique is particularly useful when tackling questions of this sort.

Since $(p + q)^2 = p^2 + q^2 + 2pq$ then $p^2 + q^2 = (p + q)^2 - 2pq$.

Using $pq = -1$ gives $p^2 + q^2 = (p + q)^2 + 2$.

But $p + q = \frac{Y}{a}$

Now use Y to eliminate p and q.

So $X = 3a + a\left(\frac{Y}{a}\right)^2$

$\Rightarrow Y^2 = a(X - 3a)$

Rearrange to the specified form.

Exercise 3F

A P **1** The tangent at $P(ap^2, 2ap)$ and the tangent at $Q(aq^2, 2aq)$ to the parabola with equation $y^2 = 4ax$ meet at R.

a Find the coordinates of R.

The chord PQ passes through the focus $(a, 0)$ of the parabola.

b Show that the locus of R lies on the line $x = -a$.

Given instead that the chord PQ has gradient 2,

c find the locus of R.

E/P **2** The hyperbola H has equation $\frac{x^2}{a^2} - \frac{y^2}{b^2} = 1$. The line l_1 is tangent to H at the point $P(a \sec t, b \tan t)$.

a Use calculus to show that an equation for l_1 is $bx\sec t - ay\tan t = ab$. **(4 marks)**

The line l_1 cuts the x-axis at A and the y-axis at B.

b Show that the locus of the midpoint of AB is $\frac{a^2}{4x^2} - \frac{b^2}{4y^2} = 1$ **(5 marks)**

E/P **3** The hyperbola H has equation $\frac{x^2}{a^2} - \frac{y^2}{b^2} = 1$. The line l_1 is normal to H at the point $P(a \sec t, b \tan t)$.

a Use calculus to show that an equation for l_1 is $ax\sin t + by = (a^2 + b^2)\tan t$. **(4 marks)**

The line l_1 cuts the x-axis at A and the y-axis at B.

b Show that the locus of the midpoint of AB is $4a^2x^2 = (a^2 + b^2)^2 + 4b^2y^2$. **(5 marks)**

E/P **4** The ellipse E has equation $\frac{x^2}{25} + \frac{y^2}{9} = 1$. The line l_1 is normal to E at the point $P(5\cos\theta, 3\sin\theta)$.

a Use calculus to show that an equation for l_1 is $3y\cos\theta = 5x\sin\theta - 16\sin\theta\cos\theta$. **(4 marks)**

The line l_1 cuts the x-axis at M and the y-axis at N.

b Show that the locus of the midpoint of MN is $\frac{25x^2}{64} + \frac{9y^2}{64} = 1$ **(5 marks)**

E/P **5** The tangent at the point $P\left(cp, \frac{c}{p}\right)$ and the tangent at the point $Q\left(cq, \frac{c}{q}\right)$ to the rectangular hyperbola $xy = c^2$, intersect at the point R.

a Show that R is $\left(\frac{2cpq}{p+q}, \frac{2c}{p+q}\right)$. **(4 marks)**

b Show that the chord PQ has equation $ypq + x = c(p + q)$. **(3 marks)**

c Find the locus of R, given that:

i the chord PQ has gradient 2 **(2 marks)**

ii the chord PQ passes through the point $(1, 0)$ **(2 marks)**

iii the chord PQ passes through the point $(0, 1)$. **(2 marks)**

P **6** **a** Find the gradient of the parabola with equation $y^2 = 4ax$ at the point $P(at^2, 2at)$.

b Hence show that the equation of the tangent at this point is $x - ty + at^2 = 0$.

The tangent meets the y-axis at T, and O is the origin.

c Show that the coordinates of the centre of the circle through O, P and T are $\left(\frac{at^2}{2} + a, \frac{at}{2}\right)$.

d Deduce that, as t varies, the locus of the centre of this circle is another parabola.

A/P **7** The chord PQ to the rectangular hyperbola $xy = c^2$ passes through the point (0, 1).
Find the equation of the locus of the midpoint of PQ as P and Q vary. **(7 marks)**

P **8** The point P lies on the ellipse with equation $\frac{x^2}{4} + \frac{y^2}{16} = 1$. The point N is the foot of the perpendicular from point P to the line $y = 6$. M is the midpoint of PN.

a Find an equation for the locus of M as P moves around the ellipse. **(4 marks)**

b Show that this locus is a circle and state its centre and radius. **(3 marks)**

Challenge

The points A and B lie on an ellipse with equation $\frac{x^2}{a^2} + \frac{y^2}{b^2} = 1$, such that the chord AB has gradient k. Show that the locus of the midpoints of all possible such chords AB has equation $ka^2y + b^2x = 0$, and describe this locus.

Mixed exercise 3

1 The ellipse E has parametric equations $x = 4\cos\theta$, $y = 9\sin\theta$.

a Find a Cartesian equation of the ellipse.

b Sketch the ellipse, labelling any points of intersection with the coordinate axes.

c Find the equation of the normal to the ellipse at $P(4\cos\theta, 9\sin\theta)$.

2 The hyperbola H has parametric equations $x = \pm 2\cosh t$, $y = 5\sinh t$.

a Find a Cartesian equation of the hyperbola.

b Sketch the hyperbola, giving the equations of the asymptotes and show points of intersection of the hyperbola with the x-axis.

c Find the equation of the tangent to the hyperbola at $Q(2\cosh t, 5\sinh t)$.

E/P **3** A hyperbola of the form $\frac{x^2}{a^2} - \frac{y^2}{b^2} = 1$ has asymptotes with equations $y = \pm mx$ and passes through the point $(a, 0)$.

a Find an equation of the hyperbola in terms of x, y, a and m. **(4 marks)**

A point P on this hyperbola is equidistant from one of the hyperbola's asymptotes and the x-axis.

b Prove that, for all values of m, P lies on the curve with equation

$$(x^2 - y^2)^2 = 4x^2(x^2 - a^2)$$ **(3 marks)**

E/P **4** **a** Prove that the gradient of the chord joining the point $P\left(cp, \frac{c}{p}\right)$ and the point $Q\left(cq, \frac{c}{q}\right)$ on the rectangular hyperbola with equation $xy = c^2$ is $-\frac{1}{pq}$ **(5 marks)**

The points P, Q and R lie on a rectangular hyperbola, such that the angle QPR is a right angle.

b Prove that the angle between QR and the tangent at P is also a right angle. **(5 marks)**

A E/P **5** **a** Show that an equation of the tangent to the rectangular hyperbola with equation $xy = c^2$ (with $c > 0$) at the point $\left(ct, \frac{c}{t}\right)$ is

$$t^2y + x = 2ct$$ **(4 marks)**

Tangents are drawn from the point $(-3, 3)$ to the rectangular hyperbola with equation $xy = 16$.

b Find the coordinates of the points of contact of these tangents with the hyperbola. **(4 marks)**

E/P **6** The point P lies on the ellipse with equation $9x^2 + 25y^2 = 225$, and A and B are the points $(-4, 0)$ and $(4, 0)$ respectively.

a Prove that $PA + PB = 10$. **(4 marks)**

b Prove also that the normal at P bisects the angle APB. **(6 marks)**

E/P **7** A curve is given parametrically by $x = ct$, $y = \frac{c}{t}$

a Show that an equation of the tangent to the curve at the point $\left(ct, \frac{c}{t}\right)$ is $t^2y + x = 2ct$. **(4 marks)**

The point P is the foot of the perpendicular from the origin to this tangent.

b Show that the locus of P is the curve with equation $(x^2 + y^2)^2 = 4c^2xy$. **(6 marks)**

E/P **8** The points $P(ap^2, 2ap)$ and $Q(aq^2, 2aq)$ lie on the parabola with equation $y^2 = 4ax$. The angle $POQ = 90°$, where O is the origin.

a Prove that $pq = -4$. **(4 marks)**

Given that the normal at P to the parabola has equation

$$y + xp = ap^3 + 2ap$$

b write down an equation of the normal to the parabola at Q. **(1 mark)**

c Show that these two normals meet at the point R, with coordinates

$$(ap^2 + aq^2 - 2a, 4a(p + q))$$ **(3 marks)**

d Show that, as p and q vary, the locus of R has equation $y^2 = 16ax - 96a^2$. **(4 marks)**

E/P **9** Show that, for all values of m, the straight lines with equations $y = mx \pm \sqrt{b^2 + a^2m^2}$ are tangents to the ellipse with equation $\frac{x^2}{a^2} + \frac{y^2}{b^2} = 1$. **(6 marks)**

E/P **10** The chord PQ, where P and Q are points on $xy = c^2$, has gradient 1.

Show that the locus of the point of intersection of the tangents from P and Q is the line $y = -x$. **(6 marks)**

E/P **11** The ellipse E has equation $\frac{x^2}{36} + \frac{y^2}{16} = 1$. The line l_1 is tangent to E at the point $P(6\cos\theta, 4\sin\theta)$.

a Use calculus to show that an equation for l_1 is $2x\cos\theta + 3y\sin\theta = 12$. **(4 marks)**

The line l_1 cuts the x-axis at A and the y-axis at B.

b Show that the locus of the midpoint of AB is $\frac{9}{x^2} + \frac{4}{y^2} = 1$. **(5 marks)**

A P **12** The ellipse E has equation $\frac{x^2}{169} + \frac{y^2}{25} = 1$. The line l_1 is tangent to E at the point $P(13\cos\theta, 5\sin\theta)$.

a Use calculus to show that an equation for l_1 is $5x\cos\theta + 13y\sin\theta = 65$. **(5 marks)**

The line l_1 cuts the y-axis at A. The line l_2 passes through the point A, perpendicular to l_1.

b Find the equation of the line l_2. **(3 marks)**

c Given that l_2 cuts the x-axis at the focus of the ellipse $(-ae, 0)$, show that $\cos\theta = e$. **(3 marks)**

P **13** The hyperbola H has equation $\frac{x^2}{16} - \frac{y^2}{64} = 1$. The line l_1 is normal to H at the point $P(4\sec\theta, 8\tan\theta)$.

a Use calculus to show that an equation for l_1 is $x\sin\theta + 2y = 20\tan\theta$. **(4 marks)**

The line l_1 cuts the x-axis at A and the y-axis at B.

b Show that the locus of the midpoint of AB is also a hyperbola and find the equation of this hyperbola. **(6 marks)**

P **14** The ellipse E has equation $\frac{x^2}{a^2} + \frac{y^2}{b^2} = 1$. The line l_1 is normal to E at the point $P(a\cos t, b\sin t)$.

a Use calculus to show that an equation for l_1 is $ax\sin t - by\cos t = (a^2 - b^2)\cos t\sin t$. **(4 marks)**

The line l_1 cuts the x-axis at M and the y-axis at N.

b Show that the locus of the midpoint of MN is $4b^2y^2 + 4a^2x^2 = (a^2 - b^2)^2$. **(5 marks)**

P **15** The ellipse E with equation $\frac{x^2}{5^2} + \frac{y^2}{3^2} = 1$ has foci at S and S'. Prove that for any point P on the ellipse, $PS + PS' = 10$. **(5 marks)**

P **16** The line l_1 is tangent to the ellipse with equation $\frac{x^2}{a^2} + \frac{y^2}{b^2} = 1$. A line segment connects point P and the origin. Show that the area of the shaded region is $\frac{1}{2}ab\tan t$.

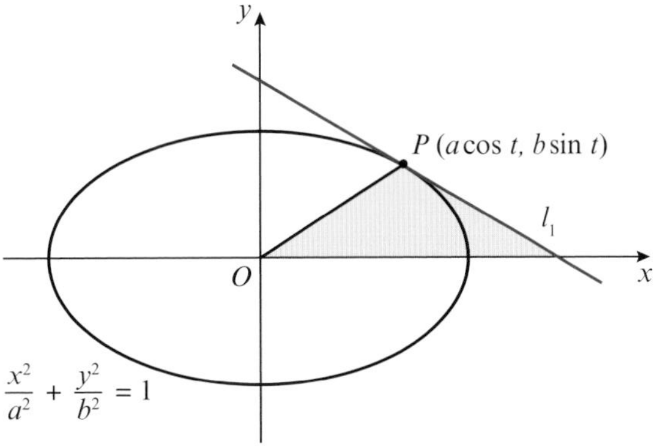

17 The line l_1 is tangent to the ellipse with equation $\frac{x^2}{6^2} + \frac{y^2}{3^2} = 1$ at the point $P\left(3, \frac{3\sqrt{3}}{2}\right)$. Show that the exact value for the area of the shaded region is $9\sqrt{3} - 3\pi$

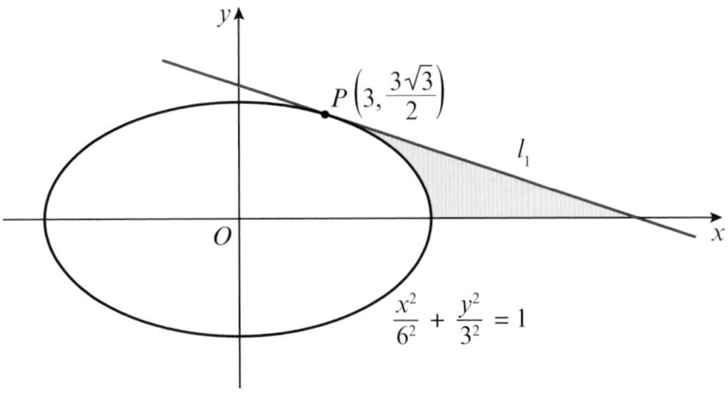

A P **18** The hyperbola H has equation $\frac{x^2}{9} - \frac{y^2}{16} = 1$. The tangents to the hyperbola at points P and Q both meet one directrix of H at a single point A with y-coordinate 0, and the other directrix of H at points B and C. Find the area of triangle ABC.

E/P **19** The hyperbola H has equation $x^2 - y^2 = 1$. The tangents to the hyperbola at points P and Q meet at the point $\left(\frac{1}{3}, 0\right)$.

a Find the exact coordinates of P and Q. **(3 marks)**

b Show that the exact area of the region R enclosed by the tangents at P and Q and the hyperbola H is arcosh $3 - k\sqrt{2}$, where k is a rational constant to be found. **(7 marks)**

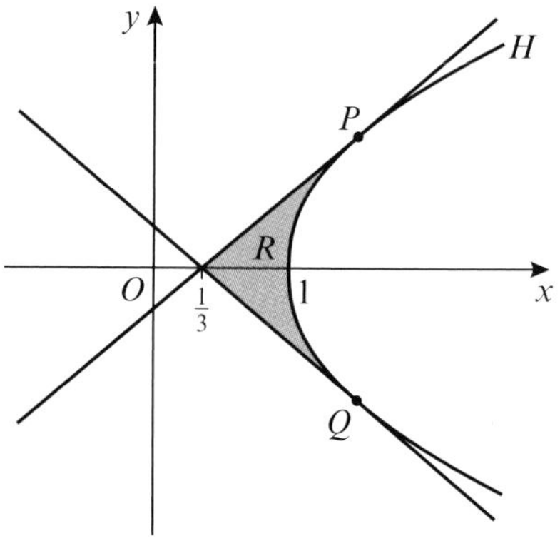

Challenge

Let P be a point on an ellipse with eccentricity e. The normal to the ellipse at P meets the major axis at Q. Prove that $QS = ePS$, where S is a focus.

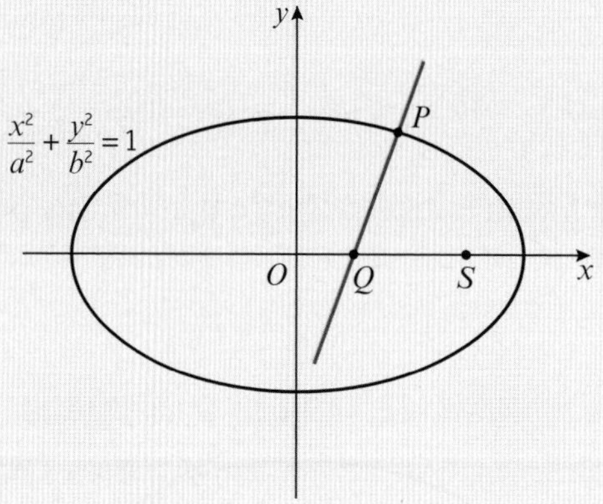

Summary of key points

1 A standard **ellipse** has Cartesian equation $\frac{x^2}{a^2} + \frac{y^2}{b^2} = 1$

- The standard ellipse has parametric equations $x = a\cos t$, $y = b\sin t$, $0 \leqslant t < 2\pi$
- A general point P on an ellipse has coordinates $(a\cos t, b\sin t)$.

2 A standard **hyperbola** has Cartesian equation $\frac{x^2}{a^2} - \frac{y^2}{b^2} = 1$

- The standard hyperbola has parametric equations $x = \pm a\cosh t$, $y = b\sinh t$, $t \in \mathbb{R}$
- The standard hyperbola has alternative parametric equations
  $x = a\sec\theta$, $y = b\tan\theta$, $-\pi \leqslant \theta < \pi$, $\theta \neq \pm\frac{\pi}{2}$
- A general point P on a hyperbola has coordinates $(\pm a\cosh t, b\sinh t)$ or $(a\sec\theta, b\tan\theta)$.

A

3 For all points, P, on a conic section, the ratio of the distance of P from a fixed point (called the **focus**) and a fixed straight line (called the **directrix**) is constant. This ratio, e, is known as the **eccentricity** of the curve.

- If $0 < e < 1$, the point P describes an ellipse.
- If $e = 1$, the point P describes a parabola.
- If $e > 1$, the point P describes a hyperbola.

4 For an ellipse with equation $\frac{x^2}{a^2} + \frac{y^2}{b^2} = 1$, and $a > b$,

- the eccentricity, $0 < e < 1$, is given by $b^2 = a^2(1 - e^2)$
- the foci are at $(\pm ae, 0)$
- the directrices are $x = \pm\frac{a}{e}$

5 For a hyperbola with equation $\frac{x^2}{a^2} - \frac{y^2}{b^2} = 1$,

- the eccentricity, $e > 1$, is given by $b^2 = a^2(e^2 - 1)$
- the foci are at $(\pm ae, 0)$
- the directrices are $x = \pm\frac{a}{e}$

6 An equation of the tangent to the ellipse with equation $\frac{x^2}{a^2} + \frac{y^2}{b^2} = 1$ at the point $P(a\cos t, b\sin t)$ is $bx\cos t + ay\sin t = ab$.

7 An equation of the normal to the ellipse with equation $\frac{x^2}{a^2} + \frac{y^2}{b^2} = 1$ at the point $P(a\cos t, b\sin t)$ is $ax\sin t - by\cos t = (a^2 - b^2)\cos t\sin t$.

8
- An equation of the tangent to the hyperbola with equation $\frac{x^2}{a^2} - \frac{y^2}{b^2} = 1$ at the point $P(a\cosh t, b\sinh t)$ is $ay\sinh t + ab = bx\cosh t$.
- An equation of the tangent to the hyperbola with equation $\frac{x^2}{a^2} - \frac{y^2}{b^2} = 1$ at the point $P(a\sec\theta, b\tan\theta)$ is $bx\sec\theta - ay\tan\theta = ab$.

9
- An equation of the normal to the hyperbola with equation $\frac{x^2}{a^2} - \frac{y^2}{b^2} = 1$ at the point $P(a\cosh t, b\sinh t)$ is $ax\sinh t + by\cosh t = (a^2 + b^2)\sinh t\cosh t$.
- An equation of the normal to the hyperbola with equation $\frac{x^2}{a^2} - \frac{y^2}{b^2} = 1$ at the point $P(a\sec\theta, b\tan\theta)$ is $by + ax\sin\theta = (a^2 + b^2)\tan\theta$.

4 Inequalities

Objectives

After completing this chapter you should be able to:

- Manipulate inequalities involving algebraic fractions → pages 93–96
- Use graphs to find solutions to inequalities → pages 96–99
- Solve inequalities involving modulus signs → pages 99–102

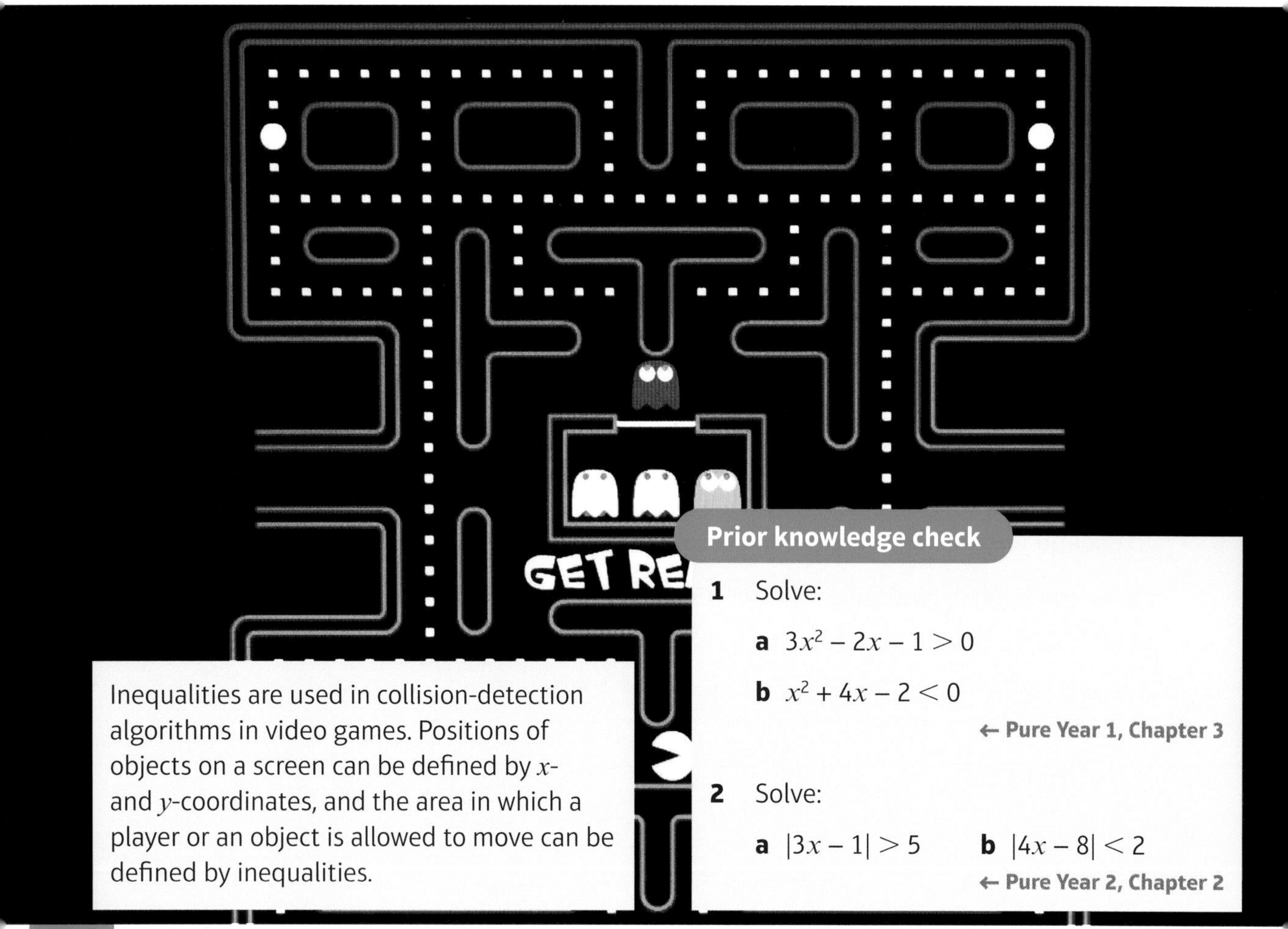

Inequalities are used in collision-detection algorithms in video games. Positions of objects on a screen can be defined by x- and y-coordinates, and the area in which a player or an object is allowed to move can be defined by inequalities.

Prior knowledge check

1 Solve:

a $3x^2 - 2x - 1 > 0$

b $x^2 + 4x - 2 < 0$

← Pure Year 1, Chapter 3

2 Solve:

a $|3x - 1| > 5$ **b** $|4x - 8| < 2$

← Pure Year 2, Chapter 2

4.1 Algebraic methods

If you multiply both sides of an inequality by a negative number you reverse the direction of the inequality sign.

You need to be more careful if you multiply or divide both sides of an inequality by a variable or expression. If the variable or expression could take either a positive or a negative value then you don't know which direction is correct for the inequality sign. You can overcome this problem by multiplying by an expression squared.

Suppose you want to solve the inequality $\frac{1}{x} > x$, $x \neq 0$.

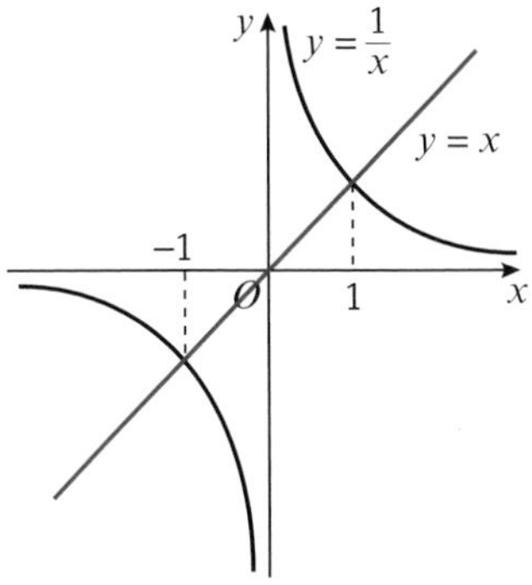

The values of x where the graph of $y = \frac{1}{x}$ is above the graph of $y = x$ give you the solution: $x < -1$ or $0 < x < 1$.

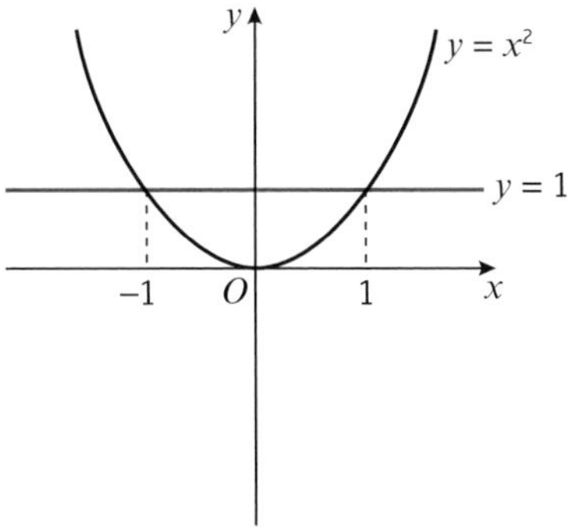

If you multiply both sides of the inequality by x you get $1 > x^2$. The solution to this inequality is $-1 < x < 1$, which is not the required solution.

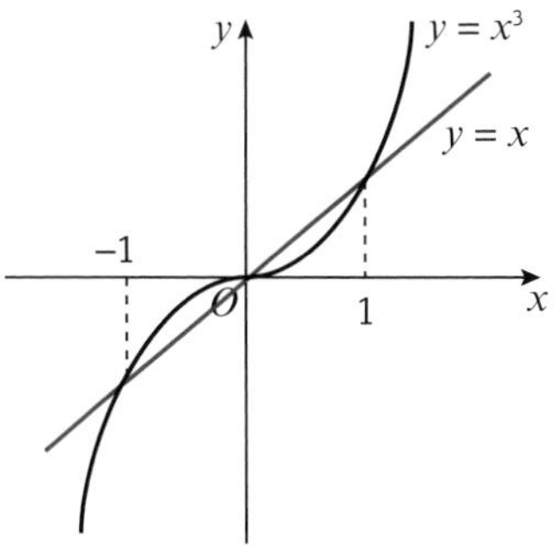

If you multiply both sides of the inequality by x^2 you get $x > x^3$. The graph of $y = x$ is above the graph of $y = x^3$ for $x < -1$ and $0 < x < 1$, which is the solution to the original inequality.

In the third example above, you can solve the inequality $x > x^3$ by algebraically rearranging and factorising.

$$x^3 - x < 0$$

You can add or subtract any term from both sides of an inequality.

$$x(x^2 - 1) < 0$$

$$x(x - 1)(x + 1) < 0$$

The **critical values** are $x = 0$, $x = 1$ and $x = -1$. You can consider a sketch of the graph of $y = x(x - 1)(x + 1)$ to work out which intervals satisfy the inequality.

- **To solve an inequality involving algebraic fractions:**
  - **Step 1: multiply by an expression squared to remove fractions**
  - **Step 2: rearrange the inequality to get 0 on one side**
  - **Step 3: find critical values**
  - **Step 4: use a sketch to identify the correct intervals**

Example 1

Use algebra to solve the inequality $\frac{x^2}{x-2} < x + 1$, $x \neq 2$.

Multiply both sides by $(x - 2)^2$

$(x-2)^2 \times \frac{x^2}{x-2} < (x-2)^2 \times (x+1)$

$(x-2)^{\not 2} \times \frac{x^2}{\cancel{(x-2)}} < (x-2)^2 \times (x+1)$

$(x-2)x^2 - (x+1)(x-2)^2 < 0$

$(x-2)(x^2 - (x+1)(x-2)) < 0$

$(x-2)(x^2 - x^2 + x + 2) < 0$

or $(x-2)(x+2) < 0$

Critical values are $x = \pm 2$

The sketch of $y = (x-2)(x+2)$ is

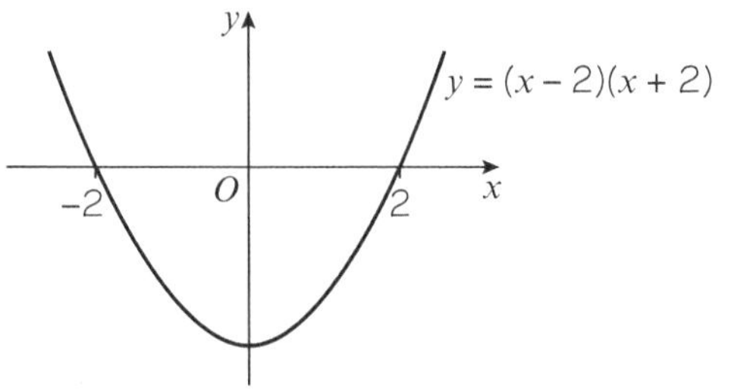

The solution to $(x-2)(x+2) < 0$ is $-2 < x < 2$.

Problem-solving A natural first step would be to multiply both sides by $(x - 2)$ but we cannot be sure that this is positive. A simple solution is to multiply both sides of the inequality by $(x - 2)^2$ as this will always be positive.

Do **not** aim to multiply out but cancel, collect terms on one side and **factorise**.

This is a quadratic inequality so you can solve it in the usual way. ← **Pure Year 1, Chapter 3**

Watch out When a question says 'Use algebra…' you can still use a sketch to identify which intervals to include in your solution set. However, you should make sure you show algebraic working to find the critical values.

When the inequality is not strict you have to be a bit more careful. In the above example, the left-hand side of the inequality is undefined when $x = 2$, so you cannot include $x = 2$ in your solution set.

Hint Values for which one side of the inequality is undefined will usually be explicitly excluded. In the above example you are given $x \neq 2$.

- **When solving an inequality involving $\leqslant$ or $\geqslant$, check whether or not each of your critical values should be included in the solution set.**

Example 2

Find all values of x such that $\frac{x}{x+1} \leqslant \frac{2}{x+3}$, where $x \neq -1$ and $x \neq -3$, and express your answer using set notation.

Multiply both sides by $(x + 1)^2(x + 3)^2$ — In order to remove the fractions and guarantee that you are not multiplying by a negative quantity, use $(x + 1)^2(x + 3)^2$.

So

$$\frac{x(x + 1)^2(x + 3)^2}{x + 1} \leqslant \frac{2(x + 1)^2(x + 3)^2}{x + 3}$$ — Cancel terms on each side.

$x(x + 1)(x + 3)^2 - 2(x + 1)^2(x + 3) \leqslant 0$ — Collect terms on LHS.

$(x + 1)(x + 3)(x(x + 3) - 2(x + 1)) \leqslant 0$ — Factorise as much as possible.

$(x + 1)(x + 3)(x^2 + x - 2) \leqslant 0$

$(x + 1)(x + 3)(x + 2)(x - 1) \leqslant 0$ — To find the critical values you need to solve $(x + 1)(x + 3)(x + 2)(x - 1) = 0$.

So the critical values are:

$x = -1, -3, -2$ or 1

A sketch of $y = (x + 1)(x + 3)(x + 2)(x - 1)$ is

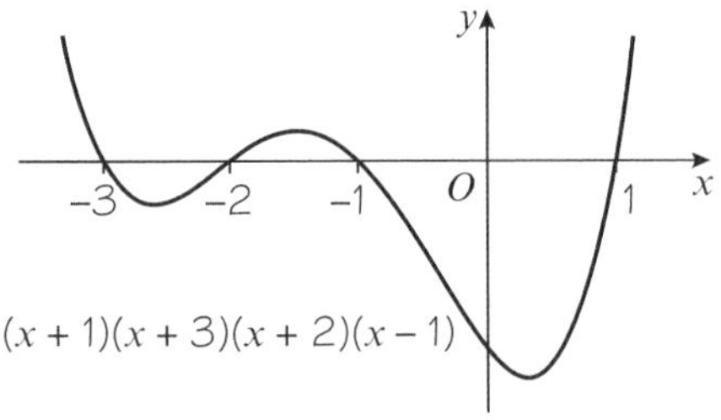

The curve $y = (x + 1)(x + 3)(x + 2)(x - 1)$ is a quartic graph with positive x^4 coefficient, so it starts in top left and ends in top right and passes through $(-3, 0)$, $(-2, 0)$, $(-1, 0)$ and $(1, 0)$.

The solution to $(x + 1)(x + 3)(x + 2)(x - 1) \leqslant 0$ corresponds to the sections of this graph that are on or below the x-axis.

So the solution is

$\{x : -3 < x \leqslant -2\} \cup \{x : -1 < x \leqslant 1\}$ — The inequality is non-strict so you need to check whether the critical values should be included in the solution. The conditions $x \neq -1$ and $x \neq -3$ are given in the question, so use strict inequalities to exclude these values.

Exercise 4A

1 Solve the following inequalities.

a $x^2 < 5x + 6$

b $x(x + 1) \geqslant 6$

c $\frac{2}{x^2 + 1} > 1$

d $\frac{2}{x^2 - 1} > 1$

e $\frac{x}{x - 1} \leqslant 2x \quad x \neq 1$

f $\frac{3}{x + 1} < \frac{2}{x}$

g $\frac{3}{(x + 1)(x - 1)} < 1$

h $\frac{2}{x^2} \geqslant \frac{3}{(x + 1)(x - 2)}$

i $\frac{2}{x - 4} < 3$

j $\frac{3}{x + 2} > \frac{1}{x - 5}$

2 Solve the following inequalities, giving your answers using set notation.

a $\frac{3x^2 + 5}{x + 5} > 1$

b $\frac{3x}{x - 2} > x$

c $\frac{1 + x}{1 - x} > \frac{2 - x}{2 + x}$

d $\frac{x^2 + 7x + 10}{x + 1} > 2x + 7$

e $\frac{x + 1}{x^2} > 6$

f $\frac{x^2}{x + 1} > \frac{1}{6}$

(E) **3** **a** Use algebra to find the set of values for which $\frac{2x+1}{x+5} < \frac{x+2}{x+4}$ **(6 marks)**

(E) **4** **a** Use algebra to find the set of values for which $\frac{x}{2x+1} < \frac{1}{x-3}$, giving your answer in set notation. **(6 marks)**

(E/P) **5** A teacher asks a student to solve the inequality $\frac{x}{3x+4} < \frac{1}{x}$

The student's attempt was as follows:

$\frac{x}{3x+4} < \frac{1}{x}$

$x^2 < 3x + 4$

$x^2 - 3x - 4 < 0$

$(x - 4)(x + 1) < 0$

$-1 < x < 4$

a Identify the mistake made by the student and explain why it will produce an incorrect answer. **(2 marks)**

b Solve the inequality correctly. **(6 marks)**

(E/P) **6** Use algebra to solve $\frac{4}{x} < x < \frac{1}{2x+1}$, giving your answer using set notation. **(6 marks)**

Challenge

Solve $\frac{1}{1-e^x} < \frac{1}{e^x}$

Hint You probably won't be able to sketch the graph in this question. Find the critical values, then test values within each interval to determine the solution set.

4.2 Using graphs to solve inequalities

- **If you can sketch the graphs of $y = f(x)$ and $y = g(x)$ then you can solve an inequality such as $f(x) < g(x)$ by observing when one curve is above the other. The critical values will be the solutions to the equation $f(x) = g(x)$.**

Watch out If you are asked to solve an inequality **algebraically** you should not start by sketching graphs.

Example 3

a On the same set of axes, sketch the graphs of the curves with equations $y = \frac{7x}{3x+1}$ and $y = 4 - x$.

b Find the points of intersection of $y = \frac{7x}{3x+1}$ and $y = 4 - x$.

c Solve $\frac{7x}{3x+1} < 4 - x$.

a Sketch $y = 4 - x$ and $y = \frac{7x}{3x + 1}$:

$y = 4 - x$ is a straight line crossing the axes at (4, 0) and (0, 4).

$y = \frac{7x}{3x + 1}$ crosses the coordinate axes at (0, 0).

There is a vertical asymptote at $x = -\frac{1}{3}$

There is a horizontal asymptote at $y = \frac{7}{3}$

So the sketch looks like this

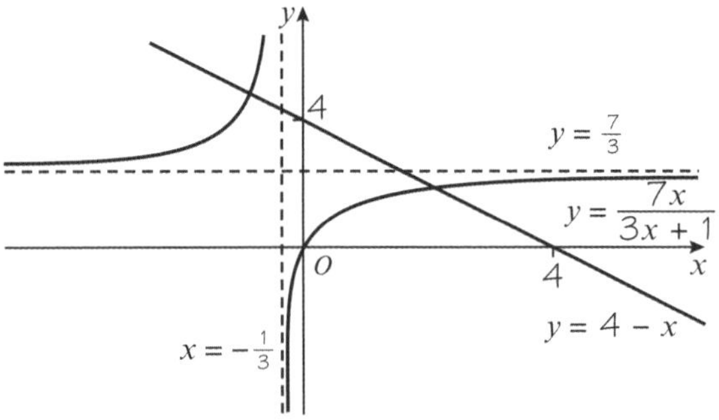

b Using algebra to find critical values:

$\frac{7x}{3x + 1} = 4 - x$

$7x = 12x + 4 - 3x^2 - x$

$3x^2 - 4x - 4 = 0$

$(3x + 2)(x - 2) = 0$

So $x = -\frac{2}{3}$ or 2

c Marking these points on the graph:

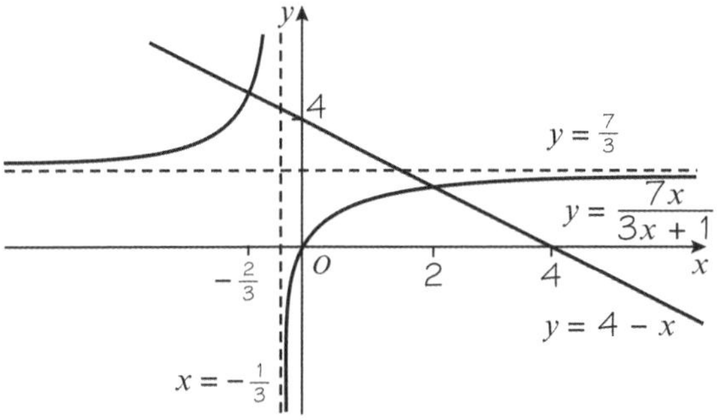

So the solution is

$x < -\frac{2}{3}$ or $-\frac{1}{3} < x < 2$

Problem-solving

To sketch unfamiliar curves, look for:

- points where the curve meets or crosses the axes
- vertical asymptotes (where the denominators of fractions equal 0)
- behaviour on either side of vertical asymptotes
- behaviour as x gets very large or very small

You can find **horizontal asymptotes** by rearranging the fraction to see how it behaves as $x \to \infty$.

$$\frac{7x}{3x+1} = \frac{7}{3}\left(\frac{3x}{3x+1}\right) = \frac{7}{3}\left(1 - \frac{1}{3x+1}\right)$$

As $x \to \infty$, $\frac{1}{3x+1} \to 0$ so the curve has a horizontal asymptote at $y = \frac{7}{3}$

Multiply both sides by $3x + 1$. This is an equation, not an inequality, so you don't need to multiply by an expression squared.

Multiply out and collect terms to form a quadratic equation.

Solve the equation in the usual way: this one factorises.

Look on the sketch for the places where the line is above the curve.
These places will give the solution.

Watch out Any vertical asymptotes will also be critical values when you are finding your solution set.

Online Explore the solution to the inequality using GeoGebra.

Exercise 4B

1 Sketch the graphs of the following functions.

a $y = x^2 - 5x + 6$

b $y = x^3 + 2x^2 - 3x$

c $y = \dfrac{1}{x + 1}$

d $y = \dfrac{4x}{1 - 2x}$

2 Sketch each of the following pairs of functions on the same sets of axes.

a $y = x^2 - 2x + 1$ and $y = 4 - 4x^2$

b $y = x$ and $y = \dfrac{1}{x}$

c $y = 2x - 1$ and $y = \dfrac{3}{x - 2}$

d $y = 4 - 3x$ and $y = \dfrac{x}{4x - 2}$

3 Find the points of intersection of the following pairs of functions.

a $y = \dfrac{2}{x + 1}$ and $y = \dfrac{1}{x - 3}$

b $y = x - 2$ and $y = \dfrac{3x}{x + 2}$

c $y = x^2 - 4$ and $y = \dfrac{4(x + 2)}{x - 2}$

(E) **4** **a** On the same set of axes, sketch the graphs of $y = x - 1$ and $y = \dfrac{4}{x - 1}$ **(3 marks)**

b Find the points of intersection of $y = x - 1$ and $y = \dfrac{4}{x - 1}$ **(2 marks)**

c Write down the solution to the inequality $x - 1 > \dfrac{4}{x - 1}$ **(2 marks)**

(E/P) **5** $f(x) = \dfrac{3}{x^2}, x \neq 0$ and $g(x) = \dfrac{2}{3 - x}, x \neq 3$

a Sketch $y = f(x)$ and $y = g(x)$ on the same set of axes. **(3 marks)**

b Solve $f(x) = g(x)$ **(2 marks)**

c Hence write down the solution to the inequality $f(x) > g(x)$. Give your answer using set notation. **(3 marks)**

(E/P) **6** **a** On the same set of axes, sketch the graphs of $y = \dfrac{3x}{2 - x}$ and $y = \dfrac{4x}{(x - 1)^2}$ **(4 marks)**

b Find the points of intersection of $y = \dfrac{3x}{2 - x}$ and $y = \dfrac{4x}{(x - 1)^2}$ **(2 marks)**

c Hence, or otherwise, solve the inequality $\dfrac{3x}{2 - x} \leqslant \dfrac{4x}{(x - 1)^2}$ **(2 marks)**

(E/P) **7** **a** On the same set of axes, sketch the graphs of $y = x - 2$ and $y = \dfrac{6(2 - x)}{(x + 2)(x - 3)}$ **(4 marks)**

b Find the points of intersection of $y = x - 2$ and $y = \dfrac{6(2 - x)}{(x + 2)(x - 3)}$ **(3 marks)**

c Write down the solution to the inequality $x - 2 \leqslant \dfrac{6(2 - x)}{(x + 2)(x - 3)}$ **(2 marks)**

(E) **8** **a** On the same set of axes, sketch the graphs of $y = \dfrac{1}{x}$ and $y = \dfrac{x}{x + 2}$ **(3 marks)**

b Find the points of intersection of $y = \dfrac{1}{x}$ and $y = \dfrac{x}{x + 2}$ **(2 marks)**

c Solve $\dfrac{1}{x} > \dfrac{x}{x + 2}$ **(2 marks)**

Challenge

a Sketch the circle with equation $(x-2)^2 + (y-4)^2 = 10$.

b Determine the coordinates of all points of intersection between this circle and the curve with equation $y = \frac{4x-5}{x-2}$

c Sketch this curve on the same set of axes as your answer to part **a**.

d Hence, or otherwise, find the solutions to the inequality

$$(x-2)^2 + \left(\frac{4x-5}{x-2} - 4\right)^2 < 10$$

4.3 Modulus inequalities

A You need to be able to solve inequalities that include modulus signs. It is often useful to sketch the relevant modulus graph when solving inequalities like this.

Example 4

Solve $|x^2 - 4x| < 3$

Sketch $y = |x^2 - 4x|$ and $y = 3$:

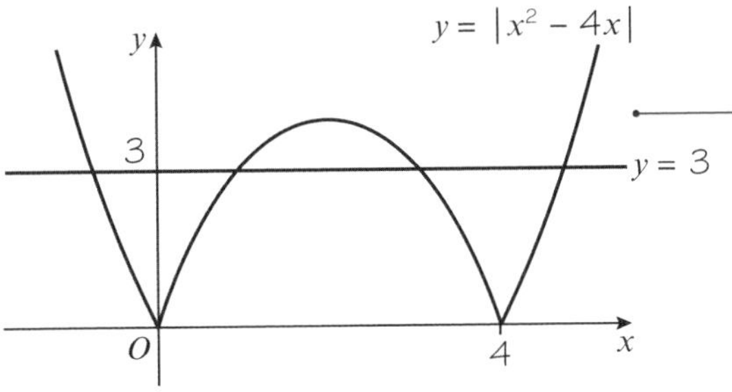

To find the critical values, solve $|x^2 - 4x| = 3$

$x^2 - 4x = 3 \Rightarrow x^2 - 4x - 3 = 0$

$(x-2)^2 - 4 - 3 = 0$

$(x-2)^2 = 7$

$x = 2 \pm \sqrt{7}$

$-(x^2 - 4x) = 3 \Rightarrow x^2 - 4x + 3 = 0$

$(x-3)(x-1) = 0$

$x = 1$ or 3

Sketch $y = |x^2 - 4x|$ and $y = 3$ on the same set of axes. To sketch $y = |x^2 - 4x|$ consider the graph of $y = x^2 - 4x$, and reflect any sections of the graph that are below the x-axis in the x-axis.

← **Pure Year 2, Section 2.5**

Watch out Solve $|x^2 - 4x| = 3$ to find the critical values. You need to consider the two separate cases: when the argument of $|x^2 - 4x|$ is positive and when it is negative. Use your sketch to determine whether these critical values all correspond to points of intersection.

Complete the square or use the quadratic formula.

The line $y = 3$ intersects the graph of $y = |x^2 - 4x|$ at four places, so all of these values of x correspond to points of intersection. Look at example 6 for a situation where this is not the case.

A

Marking these values on the sketch:

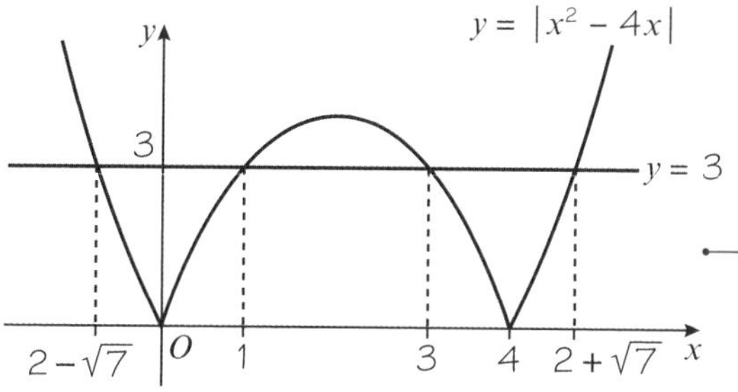

You need to identify where the points of intersection are on the sketch.

So the solution is:

$2 - \sqrt{7} < x < 1$ or $3 < x < 2 + \sqrt{7}$

Finally write down the solution to the inequality: the points where the line $y = 3$ is above the curve.

Sometimes a little simple rearranging first can make the sketching much simpler.

Example 5

Solve $|3x| + x \leqslant 2$

Rearranging gives:

$|3x| \leqslant 2 - x$

Sketching $y = |3x|$ and $y = 2 - x$ gives

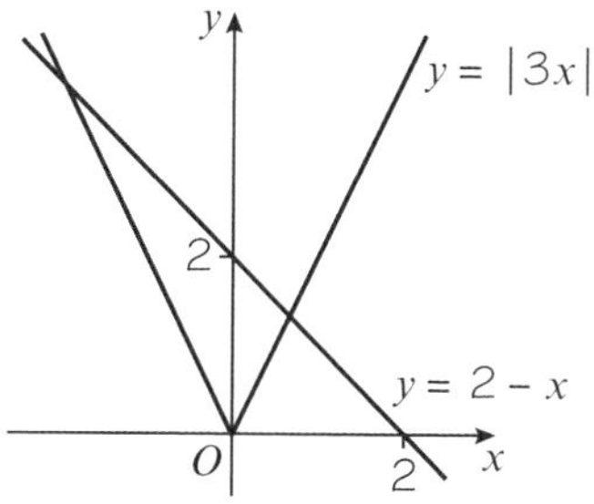

Problem-solving

Sketching $y = |3x| + x$ is quite difficult so it is usually simpler to rearrange and isolate the modulus function.

Critical values are given by:

$3x = 2 - x$

$4x = 2$

$x = \frac{1}{2}$

or

$-3x = 2 - x$

$-2 = 2x$

$x = -1$

Find the critical values in the usual way. Remember the two cases.

So the line is above $|3x|$ for

$-1 \leqslant x \leqslant \frac{1}{2}$

By considering the positions of the critical values, identify the places where the line is above the V-shaped graph.

Sometimes care must be taken to identify the correct roots when solving modulus equations.

Example 6

Find all values of x such that $|x^2 - 19| \leqslant 5(x - 1)$, expressing your answer in set notation.

Sketching both graphs:

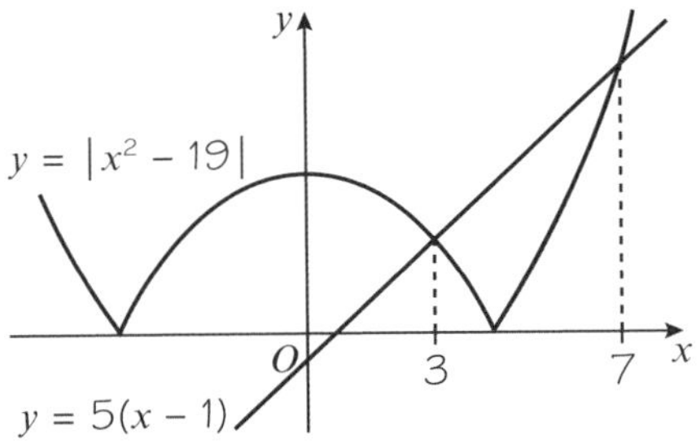

Sketch the graphs.

Online Explore the solution to the inequality using GeoGebra.

$x^2 - 19 = 5x - 5 \Rightarrow x^2 - 5x - 14 = 0$

$(x - 7)(x + 2) = 0$

$x = 7$ or -2

$-(x^2 - 19) = 5x - 5 \Rightarrow x^2 + 5x - 24 = 0$

$(x + 8)(x - 3) = 0$

$x = -8$ or 3

Find the critical values.

Watch out Solving the equations gives four values but the graphs only have two crossing points. The valid critical values are $x = 3$ and $x = 7$.

The set of points for which the line is above the curve can be written as $\{x : 3 \leqslant x \leqslant 7\}$.

Write down the solution.

Exercise 4C

1 Solve the following inequalities.

a $|x - 6| > 6x$

b $|x - 3| > x^2$

c $|(x - 2)(x + 6)| < 9$

d $|2x + 1| \geqslant 3$

e $|2x| + x > 3$

f $\dfrac{x + 3}{|x| + 1} < 2$

2 a On the same set of axes, sketch the graphs of $y = |3x - 2|$ and $y = 2x + 4$.

b Solve, giving your answer in set notation, $|3x - 2| \leqslant 2x + 4$.

3 a On the same set of axes, sketch the graphs of $y = |x^2 - 4|$ and $y = \dfrac{4}{x^2 - 1}$

b Solve $|x^2 - 4| \leqslant \dfrac{4}{x^2 - 1}$

E/P 4 Solve the inequality $\dfrac{3 - x}{|x| + 1} > 2$, giving your answer in set notation. **(5 marks)**

E/P 5 Solve the inequality $\left|\dfrac{x}{x + 2}\right| < 1 - x$, giving your answer in set notation. **(5 marks)**

Problem-solving To sketch $y = \dfrac{x}{x + 2}$ rearrange it into the form $y = A + \dfrac{B}{x + 2}$ for constants A and B.

A E/P **6** **a** On the same set of axes, sketch the graphs of $y = \frac{1}{x - a}$ and $y = 4|x - a|$. **(5 marks)**

b Solve, giving your answer in terms of the constant a, $\frac{1}{x - a} < 4|x - a|$. **(3 marks)**

E/P **7** Solve $\frac{4x}{|x| + 2} < x$ **(6 marks)**

E/P **8** A student attempts to solve the inequality $|x^2 + x - 8| < 4x + 2$.

The working is shown below:

> $x^2 + x - 8 = 4x + 2 \Rightarrow x^2 - 3x - 10 = 0$
>
> and
>
> $-x^2 - x + 8 = 4x + 2 \Rightarrow x^2 + 5x - 6 = 0$
>
> So critical values are $x = -6, -2, 1, 5$.
>
> Solution is:
>
> $-6 < x < -2$ and $1 < x < 5$

a Identify the mistake in the student's answer. **(1 mark)**

b Find the correct values of x for which the inequality is satisfied. **(3 marks)**

Challenge

$f(x) = x^3 + 3x^2 - 13x - 15$

a Show that $(x + 1)$ is a factor of $f(x)$.

b Find the other factors and hence sketch the graph of $y = f(x)$.

c Hence or otherwise, solve the inequality $|x^3 + 3x^2 - 13x - 15| \leqslant x + 5$.

Mixed exercise 4

E **1** Use algebra to solve $\frac{1}{x - 2} \leqslant \frac{2}{x}$ **(6 marks)**

E **2** Use algebra to solve $\frac{2x^2 - 2}{x + 2} > 4$. **(4 marks)**

E **3** Use algebra to solve $\frac{2x^2 - 3x + 4}{x - 2} < 4x - 2$. **(4 marks)**

E **4** Use algebra to find the set of values of x for which $\frac{x + 1}{2x - 3} < \frac{1}{x - 3}$, giving your answer in set notation. **(6 marks)**

E **5** Use algebra to find the set of values of x for which $\frac{(x + 3)(x + 9)}{x - 1} > 3x - 5$, giving your answer in set notation. **(4 marks)**

6 **a** Sketch, on the same axes, the line with equation $y = 2x + 2$ and the graph with equation $y = \frac{2x + 4}{x - 2}$

b Solve the inequality $2x + 2 > \frac{2x + 4}{x - 2}$

7 **a** Sketch, on the same set of axes, the graph with equation $y = \frac{2x - 4}{x^2 - 2}$ and the line with equation $y = 2 - 4x$.

b Solve the inequality $2 - 4x < \frac{2x - 4}{x^2 - 2}$

P 8 **a** Sketch, on the same set of axes, the graphs with equations $y = \frac{x - 2}{3x - 1}$ and $y = \frac{2}{x + 2}$ **(4 marks)**

b Solve the inequality $\frac{x - 2}{3x - 1} < \frac{2}{x + 2}$ **(3 marks)**

P 9 **a** Sketch, on the same set of axes, the graphs with equations $y = \frac{x + 1}{x - 2}$ and $y = \frac{2x - 1}{x + 4}$ **(4 marks)**

b Solve the inequality $\frac{x + 1}{x - 2} < \frac{2x - 1}{x + 4}$ **(3 marks)**

A 10 Solve the inequality $|x^2 - 7| < 3(x + 1)$

11 Solve the inequality $\frac{x^2}{|x| + 6} < 1$

E 12 Find the set of values of x for which $|x - 1| > 6x - 1$ **(3 marks)**

E 13 Find the complete set of values of x for which $|x^2 - 2| > 2x$ **(3 marks)**

E 14 **a** Sketch, on the same set of axes, the graph with equation $y = |2x - 3|$, and the line with equation $y = 5x - 1$ **(3 marks)**

b Solve the inequality $|2x - 3| < 5x - 1$ **(3 marks)**

E 15 **a** Use algebra to find the exact solution of $|2x^2 + x - 6| = 6 - 3x$ **(4 marks)**

b On the same diagram, sketch the curve with equation $y = |2x^2 + x - 6|$ and the line with equation $y = 6 - 3x$ **(3 marks)**

c Find the set of values of x for which $|2x^2 + x - 6| > 6 - 3x$ **(1 mark)**

E/P 16 **a** On the same diagram, sketch the graphs of $y = |x^2 - 4|$ and $y = |2x - 1|$, showing the coordinates of the points where the graphs meet the x-axis. **(4 marks)**

b Solve $|x^2 - 4| = |2x - 1|$, giving your answers in surd form where appropriate. **(4 marks)**

c Hence, or otherwise, find the set of values of x for which $|x^2 - 4| > |2x - 1|$ **(1 mark)**

A E/P **17** A teacher asks a student to solve the inequality $|x^2 + 3x + 1| > 3x + 2$, expressing their answer in set notation. The student's work is shown below.

> We find critical values
>
> $x^2 + 3x + 1 = 3x + 2 \Rightarrow x^2 - 1 \Rightarrow x = \pm 1$
>
> and
>
> $x^2 + 3x + 1 = -2 - 3x \Rightarrow x^2 + 6x + 3 = 0 \Rightarrow x = -3 \pm \sqrt{6}$
>
> Hence inequality is satisfied when x is in the set
>
> $\{x : x < -3 - \sqrt{6}\} \cup \{x : -1 < x < -3 + \sqrt{6}\} \cup \{x : x > 1\}$

a Identify the mistake in the student's working. **(1 mark)**

b Write down the correct solution to the problem. **(3 marks)**

Challenge

Solve the inequality $|x^2 - 5x + 2| > |x - 3|$

Give your answer in set notation, expressing any critical values as surds where appropriate.

Summary of key points

1 To solve an inequality involving algebraic fractions:
- Step 1: multiply by an expression squared to remove fractions
- Step 2: rearrange the inequality to get 0 on one side
- Step 3: find critical values
- Step 4: use a sketch to identify the correct intervals

2 When solving an inequality involving $\leqslant$ or $\geqslant$, check whether or not each of your critical values should be included in the solution set.

3 If you can sketch the graphs of $y = \text{f}(x)$ and $y = \text{g}(x)$ then you can solve an inequality such as $\text{f}(x) < \text{g}(x)$ by observing when one curve is above the other. The critical values will be the solutions to the equation $\text{f}(x) = \text{g}(x)$.

Review exercise 1

In this exercise, AS students may use, without proof, the result that, for the general parabola $y^2 = 4ax, \frac{dy}{dx} = \frac{2a}{y}$

E **1** Find the magnitude of the vector $(-\mathbf{i} - \mathbf{j} + \mathbf{k}) \times (-\mathbf{i} + \mathbf{j} - \mathbf{k})$. **(3)**

← Section 1.1

E/P **2** $\mathbf{p} = \begin{pmatrix} 2 \\ -1 \\ 3 \end{pmatrix}$ and $\mathbf{q} = \begin{pmatrix} k \\ 1 \\ 0 \end{pmatrix}$, where k is a real constant.

a Find $\mathbf{p} \times \mathbf{q}$, giving your answer as a column vector in terms of k. **(3)**

b Hence find the least possible value of $|\mathbf{p} \times \mathbf{q}|$, and state the value of k for which it occurs. **(3)**

← Section 1.1

E/P **3** Referred to a fixed origin O, the position vectors of three non-linear points A, B and C are $\mathbf{a}$, $\mathbf{b}$ and $\mathbf{c}$ respectively. By considering $\overrightarrow{AB} \times \overrightarrow{AC}$, prove that the area of triangle ABC can be expressed in the form $\frac{1}{2}|\mathbf{a} \times \mathbf{b} + \mathbf{b} \times \mathbf{c} + \mathbf{c} \times \mathbf{a}|$. **(5)**

← Section 1.2

E **4** The figure shows a right prism with triangular ends ABC and DEF, and parallel edges AD, BE, CF.

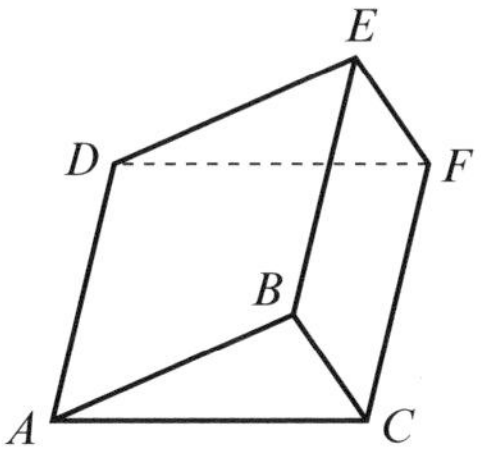

Given that

A is (2, 7, −1), B is (5, 8, 2), C is (6, 7, 4) and D is (12, 1, −9),

a find $\overrightarrow{AB} \times \overrightarrow{AC}$ **(3)**

b find $\overrightarrow{AD}.(\overrightarrow{AB} \times \overrightarrow{AC})$ **(3)**

c Calculate the volume of the prism. **(2)**

← Sections 1.1, 1.3

E/P **5** The points A, B, C and D have coordinates (3, 1, 2), (5, 2, −1), (6, 4, 5) and (−7, 6, −3) respectively.

a Find $\overrightarrow{AC} \times \overrightarrow{AD}$. **(3)**

b Find a vector equation of the line through A which is perpendicular to $\overrightarrow{AC}$ and $\overrightarrow{AD}$. **(3)**

c Verify that B lies on this line. **(2)**

d Find the volume of the tetrahedron $ABCD$. **(2)**

← Sections 1.1, 1.3

A E/P **6** The points A, B and C have position vectors, relative to a fixed origin O,

$\mathbf{a} = 2\mathbf{i} - \mathbf{j}$

$\mathbf{b} = \mathbf{i} + 2\mathbf{j} + 3\mathbf{k}$

$\mathbf{c} = 2\mathbf{i} + 3\mathbf{j} + 2\mathbf{k}$

respectively. The plane Π passes through A, B and C.

a Find $\overrightarrow{AB} \times \overrightarrow{AC}$. **(3)**

b Show that a Cartesian equation of Π is $3x - y + 2z = 7$. **(3)**

The line l has equation

$(\mathbf{r} - 5\mathbf{i} - 5\mathbf{j} - 3\mathbf{k}) \times (2\mathbf{i} - \mathbf{j} - 2\mathbf{k}) = \mathbf{0}$

The line l and the plane Π intersect at the point T.

c Find the coordinates of T. **(4)**

d Show that A, B and T lie on the same straight line. **(4)**

← Sections 1.1, 1.4

E/P **7** Vector equations of the two straight lines l and m are respectively

$\mathbf{r} = \mathbf{j} + 3\mathbf{k} + t(2\mathbf{i} + \mathbf{j} - \mathbf{k})$

$\mathbf{r} = \mathbf{i} + \mathbf{j} - \mathbf{k} + u(-2\mathbf{i} + \mathbf{j} + \mathbf{k})$

a Show that these lines do not intersect. **(4)**

A The point A with parameter t_1 lies on l and the point B with parameter u_1 lies on m.

b Write down the vector $\overrightarrow{AB}$ in terms of **i**, **j**, **k**, t_1 and u_1. **(1)**

Given that the line AB is perpendicular to both l and m,

c find the values of t_1 and u_1 and show that, in this case, the length of AB is $\frac{7}{\sqrt{5}}$ **(4)**

← Section 1.4

E **8** A line L passes through the points with position vectors $\begin{pmatrix}2\\5\\0\end{pmatrix}$ and $\begin{pmatrix}-1\\3\\2\end{pmatrix}$.

a Find the direction cosines of L. **(3)**

b Hence or otherwise, write a Cartesian equation of L. **(2)**

← Section 1.4

E **9** The points A, B and C lie on the plane Π and, relative to a fixed origin O, they have position vectors

$\mathbf{a} = 3\mathbf{i} - \mathbf{j} + 4\mathbf{k}$

$\mathbf{b} = -\mathbf{i} + 2\mathbf{j}$

$\mathbf{c} = 5\mathbf{i} - 3\mathbf{j} + 7\mathbf{k}$

respectively.

a Find $\overrightarrow{AB} \times \overrightarrow{AC}$. **(3)**

b Obtain the equation of Π in the form $\mathbf{r}.\mathbf{n} = p$. **(3)**

The point D has position vector $5\mathbf{i} + 2\mathbf{j} + 3\mathbf{k}$.

c Calculate the volume of the tetrahedron $ABCD$. **(2)**

← Sections 1.1, 1.3, 1.5

E **10** The plane Π_1 has vector equation

$\mathbf{r} = 5\mathbf{i} + \mathbf{j} + u(-4\mathbf{i} + \mathbf{j} + 3\mathbf{k}) + v(\mathbf{j} + 2\mathbf{k})$

where u and v are parameters.

a Find a vector $\mathbf{n}_1$ normal to Π_1. **(3)**

The plane Π_2 has equation $3x + y - z = 3$.

b Write down a vector $\mathbf{n}_2$ normal to Π_2. **(1)**

A **c** Show that $4\mathbf{i} + 13\mathbf{j} + 25\mathbf{k}$ is perpendicular to both $\mathbf{n}_1$ and $\mathbf{n}_2$. **(2)**

Given that the point (1, 1, 1) lies on both Π_1 and Π_2,

d write down an equation of the line of intersection of Π_1 and Π_2 in the form $\mathbf{r} = \mathbf{a} + t\mathbf{b}$, where t is a parameter. **(4)**

← Section 1.5

E/P **11** Relative to a fixed origin O, the point A has position vector $a(4\mathbf{i} + \mathbf{j} + 2\mathbf{k})$ and the plane Π has equation

$\mathbf{r}.(\mathbf{i} - 5\mathbf{j} + 3\mathbf{k}) = 5a$,

where a is a scalar constant.

a Show that A lies in the plane Π. **(3)**

The point B has position vector $a(2\mathbf{i} + 11\mathbf{j} - 4\mathbf{k})$.

b Show that $\overrightarrow{BA}$ is perpendicular to the plane Π. **(3)**

c Calculate, to the nearest one tenth of a degree, $\angle OBA$. **(3)**

← Section 1.5

E/P **12** The line l_1 has equation

$\mathbf{r} = \mathbf{i} + 6\mathbf{j} - \mathbf{k} + \lambda(2\mathbf{i} + 3\mathbf{k})$

and the line l_2 has equation

$\mathbf{r} = 3\mathbf{i} + p\mathbf{j} + \mu(\mathbf{i} - 2\mathbf{j} + \mathbf{k})$

where p is a constant.

The plane Π_1 contains l_1 and l_2.

a Find a vector which is normal to Π_1. **(3)**

b Show that an equation for Π_1 is $6x + y - 4z = 16$. **(3)**

c Find the value of p. **(2)**

The plane Π_2 has equation

$\mathbf{r}.(\mathbf{i} + 2\mathbf{j} + \mathbf{k}) = 2$

d Find an equation for the line of intersection of Π_1 and Π_2, giving your answer in the form $(\mathbf{r} - \mathbf{a}) \times \mathbf{b} = \mathbf{0}$. **(4)**

← Section 1.5

E/P **13** The plane Π passes through the points $P(-1, 3, -2)$, $Q(4, -1, -1)$ and $R(3, 0, c)$, where c is a constant.

A **a** Find, in terms of c, $\overrightarrow{RP} \times \overrightarrow{RQ}$. **(3)**

Given that $\overrightarrow{RP} \times \overrightarrow{RQ} = 3\mathbf{i} + d\mathbf{j} + \mathbf{k}$, where d is a constant,

b find the value of c and show that $d = 4$. **(2)**

c Find an equation of Π in the form $\mathbf{r.n} = p$, where p is a constant. **(3)**

The point S has position vector $\mathbf{i} + 5\mathbf{j} + 10\mathbf{k}$. The point S' is the image of S under reflection in Π.

d Find the position vector of S'. **(4)**

← Sections 1.1, 1.5

E **14** The points A, B and C lie on the plane Π_1 and, relative to a fixed origin O, they have position vectors

$\mathbf{a} = \mathbf{i} + 3\mathbf{j} - \mathbf{k}$

$\mathbf{b} = 3\mathbf{i} + 3\mathbf{j} - 4\mathbf{k}$

$\mathbf{c} = 5\mathbf{i} - 2\mathbf{j} - 2\mathbf{k}$

respectively.

a Find $(\mathbf{b} - \mathbf{a}) \times (\mathbf{c} - \mathbf{a})$. **(2)**

b Find an equation of Π_1, giving your answer in the form $\mathbf{r.n} = p$. **(2)**

The plane Π_2 has Cartesian equation $x + z = 3$ and Π_1 and Π_2 intersect in the line l.

c Find an equation of l in the form $(\mathbf{r} - \mathbf{p}) \times \mathbf{q} = \mathbf{0}$. **(3)**

The point P is the point on l that is nearest to the origin O.

d Find the coordinates of P. **(3)**

← Section 1.1, 1.5

E/P **15** The points $A(2, 0, -1)$ and $B(4, 3, 1)$ have position vectors $\mathbf{a}$ and $\mathbf{b}$ respectively with respect to a fixed origin O.

a Find $\mathbf{a} \times \mathbf{b}$. **(2)**

The plane Π_1 contains the points O, A and B.

b Verify that an equation of Π_1 is $x - 2y + 2z = 0$. **(3)**

The plane Π_2 has equation $\mathbf{r.n} = d$ where $\mathbf{n} = 3\mathbf{i} + \mathbf{j} - \mathbf{k}$ and d is a constant.

A Given that B lies on Π_2,

c find the value of d. **(3)**

The planes Π_1 and Π_2 intersect in the line L.

d Find an equation of L in the form $\mathbf{r} = \mathbf{p} + t\mathbf{q}$, where t is a parameter. **(3)**

e Find the position vector of the point X on L where OX is perpendicular to L. **(4)**

← Sections 1.1, 1.5

E/P **16** The points A, B and C have position vectors $\mathbf{j} + 2\mathbf{k}$, $2\mathbf{i} + 3\mathbf{j} + \mathbf{k}$ and $\mathbf{i} + \mathbf{j} + 3\mathbf{k}$, respectively, relative to the origin O. The plane Π contains the points A, B and C.

a Find a vector which is perpendicular to Π. **(4)**

b Find the area of triangle ABC. **(3)**

c Find a vector equation of Π in the form $\mathbf{r.n} = p$. **(3)**

d Hence, or otherwise, obtain a Cartesian equation of Π. **(2)**

e Find the distance of the origin O from Π. **(2)**

The point D has position vector $3\mathbf{i} + 4\mathbf{j} + \mathbf{k}$. The distance of D from Π is $\frac{1}{\sqrt{17}}$

f Using this distance, or otherwise, calculate the acute angle between the line AD and Π, giving your answer in degrees to one decimal place. **(3)**

← Sections 1.2, 1.5

E/P **17** The plane Π passes through the points $A(-1, -1, 1)$, $B(4, 2, 1)$ and $C(2, 1, 0)$.

a Find a vector equation of the line perpendicular to Π which passes through the point $D(1, 2, 3)$. **(3)**

b Find the volume of the tetrahedron $ABCD$. **(3)**

c Obtain the equation of Π in the form $\mathbf{r.n} = p$. **(3)**

A

The perpendicular from D to the plane Π meets Π at the point E.

d Find the coordinates of E. **(3)**

e Show that $DE = \frac{11\sqrt{35}}{35}$ **(2)**

The point D' is the reflection of D in Π.

f Find the coordinates of D'. **(4)**

← Sections 1.3, 1.5

E/P **18** Relative to a fixed origin O the lines l_1 and l_2 have equations

$l_1 : \mathbf{r} = -\mathbf{i} + 2\mathbf{j} - 4\mathbf{k} + s(-2\mathbf{i} + \mathbf{j} + 3\mathbf{k})$

$l_2 : \mathbf{r} = -\mathbf{j} + 7\mathbf{k} + t(-\mathbf{i} + \mathbf{j} - \mathbf{k})$

where s and t are variable parameters.

a Show that the lines intersect and are perpendicular to each other. **(4)**

b Find a vector equation of the straight line l_3 which passes through the point of intersection of l_1 and l_2 and the point with position vector $4\mathbf{i} + \lambda\mathbf{j} - 3\mathbf{k}$, where λ is a real number. **(4)**

The line l_3 makes an angle θ with the plane containing l_1 and l_2.

c Find $\sin\theta$ in terms of λ. **(4)**

Given that l_1, l_2 and l_3 are coplanar,

d find the value of λ. **(3)**

← Sections 1.4, 1.5

E **19** Referred to a fixed origin O, the planes Π_1 and Π_2 have equations $\mathbf{r}.(2\mathbf{i} - \mathbf{j} + 2\mathbf{k}) = 9$ and $\mathbf{r}.(4\mathbf{i} + 3\mathbf{j} - \mathbf{k}) = 8$ respectively.

a Determine the shortest distance from O to the line of intersection of Π_1 and Π_2. **(3)**

b Find, in vector form, an equation of the plane Π_3 which is perpendicular to Π_1 and Π_2 and passes through the point with position vector $2\mathbf{j} + \mathbf{k}$. **(3)**

c Find the position vector of the point that lies in Π_1, Π_2 and Π_3. **(3)**

← Sections 1.4, 1.5

E **20** The rectangular hyperbola, H, with equation $x = 8t, y = \frac{8}{t}$ intersects the line with equation $y = \frac{1}{4}x + 4$ at the points A and B. The midpoint of AB is M. Find the coordinates of M. **(4)**

← Section 2.3

E **21** The curve C has equations $x = 3t^2$, $y = 6t$.

a Sketch the graph of the curve C. **(3)**

The curve C intersects the line with equation $y = x - 72$ at the points A and B.

b Find the length AB, giving your answer as a surd in its simplest form. **(4)**

← Section 2.1

E/P **22** The points $P(1, a)$, where $a > 0$, and $Q(b, 6)$ lie on the parabola C with equation $y^2 = 4x$. The perpendicular bisector of PQ meets the parabola at the points M and N. Show that the x-coordinates of M and N can be written in the form $x = \lambda \pm \mu\sqrt{29}$, where λ and μ are rational numbers to be found. **(6)**

← Section 2.2

E **23** A parabola C has equation $y^2 = 16x$. The point S is the focus of the parabola.

a Write down the coordinates of S. **(1)**

The point P with coordinates (16, 16) lies on C.

b Find an equation of the line SP, giving your answer in the form $ax + by + c = 0$, where a, b and c are integers. **(2)**

The line SP intersects C at the point Q, where P and Q are distinct points.

c Find the coordinates of Q. **(4)**

← Section 2.2

E/P **24** The diagram shows the parabola C with equation $y^2 = 20x$. The straight line l with gradient $\frac{4}{3}$ passes through the focus, S, of the parabola and intersects C at the point P with positive y-coordinate.

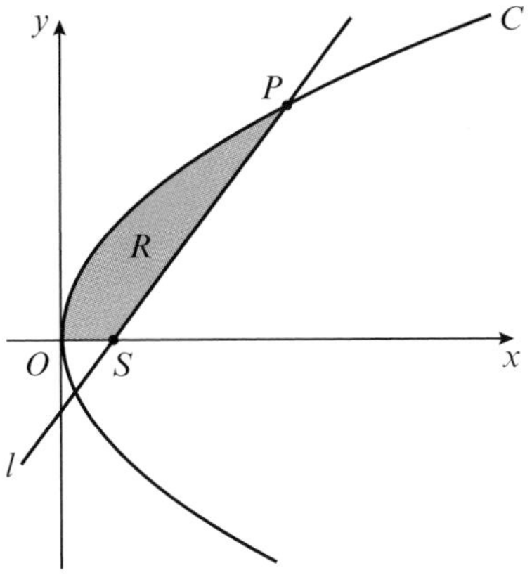

Find the area of the shaded region R bounded by C, l and the x-axis. **(6)**

← Section 2.2

E **25** A rectangular hyperbola H has parametric equations $x = 4t$ and $y = \frac{4}{t}$, $t \neq 0$. The straight line l with equation $2x - y = -4$ intersects H at the points P and Q. Find the coordinates of P and Q. **(4)**

← Section 2.3

E **26** The curve H with equation $x = 8t$, $y = \frac{16}{t}$ intersects the line with equation $y = \frac{1}{4}x + 4$ at the points A and B. The midpoint of AB is M. Find the coordinates of M. **(5)**

← Section 2.3

P **27** The diagram shows the straight line $x + 2y = 12$ that intersects the rectangular hyperbola $xy = 10$ at the points P and Q.

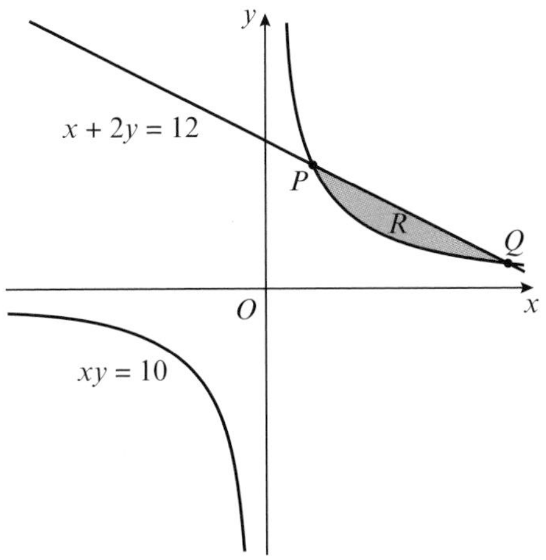

a Find the coordinates of P and Q. **(2)**

b Find the exact area of the shaded region. Leave your answer in the form $a + b\ln c$, where a, b and c are rational numbers to be found. **(5)**

← Section 2.3

E **28** The point $P(24t^2, 48t)$ lies on the parabola with equation $y^2 = 96x$. The point P also lies on the rectangular hyperbola with equation $xy = 144$.

a Find the value of t and, hence, the coordinates of P. **(3)**

b Find an equation of the tangent to the parabola at P, giving your answer in the form $y = mx + c$, where m and c are real constants. **(3)**

c Find an equation of the tangent to the rectangular hyperbola at P, giving your answer in the form $y = mx + c$, where m and c are real constants. **(4)**

← Section 2.4

E/P **29** The point $P(at^2, 2at)$, where $t > 0$, lies on the parabola with equation $y^2 = 4ax$. The tangent and normal to the parabola at P cut the x-axis at the points T and N respectively. Prove that $\frac{PT}{PN} = t$. **(6)**

← Section 2.4

E/P **30** A rectangular hyperbola H has cartesian equation $xy = 9$. The point $\left(3t, \frac{3}{t}\right)$ is a general point on H.

a Show that an equation of the tangent to H at $\left(3t, \frac{3}{t}\right)$ is $x + t^2y = 6t$. **(2)**

The tangent to H at $\left(3t, \frac{3}{t}\right)$ cuts the x-axis at A and the y-axis at B. The point O is the origin of the coordinate system.

b Prove that, as t varies, the area of the triangle OAB is constant. **(3)**

← Section 2.4

E/P **31** The point $P\left(ct, \frac{c}{t}\right)$ lies on the hyperbola with equation $xy = c^2$, where c is a positive constant.

a Show that an equation of the normal to the hyperbola at P is
$t^3x - ty - c(t^4 - 1) = 0$. **(4)**

The normal to the hyperbola at P meets the line $y = x$ at G. Given that $t \neq \pm 1$,

b show that $PG^2 = c^2\left(t^2 + \frac{1}{t^2}\right)$. **(5)**

← Section 2.4

E/P **32** The parabola C has equation $y^2 = 32x$.

a Write down the coordinates of the focus S of C. **(1)**

b Write down the equation of the directrix of C. **(1)**

The points P (2, 8) and Q (32, −32) lie on C.

c Prove that the line joining P and Q goes through S. **(3)**

The tangent to C at P and the tangent to C at Q intersect at the point D.

d Prove that D lies on the directrix of C. **(5)**

← Sections 2.2, 2.4

E/P **33** The point $P(at^2, 2at)$, $t \neq 0$, lies on the parabola with equation $y^2 = 4ax$, where a is a positive constant.

a Show that an equation of the normal to the parabola at P is
$y + xt = 2at + at^3$. **(3)**

The normal to the parabola at P meets the parabola again at Q.

b Find, in terms of t, the coordinates of Q. **(5)**

← Section 2.4

E **34** The point $P(2, 8)$ lies on the parabola C with equation $y^2 = 4ax$. Find:

a the value of a **(1)**

b an equation of the tangent to C at P **(3)**

The tangent to C at P cuts the x-axis at the point X and the y-axis at the point Y.

c Find the exact area of the triangle OXY. **(4)**

← Section 2.4

E **35** **a** Show that the normal to the rectangular hyperbola $xy = c^2$, at the point $P\left(ct, \frac{c}{t}\right)$, $t \neq 0$, has equation
$y = t^2x + \frac{c}{t} - ct^3$. **(3)**

The normal to the hyperbola at P meets the hyperbola again at the point Q.

b Find, in terms of t, the coordinates of the point Q. **(4)**

Given that the midpoint of PQ is (X, Y) and that $t \neq \pm 1$,

c show that $\frac{X}{Y} = -\frac{1}{t^2}$ **(4)**

← Section 2.4

E **36** The rectangular hyperbola C has equation $xy = c^2$, where c is a positive constant.

a Show that the tangent to C at the point $P\left(cp, \frac{c}{p}\right)$ has equation $p^2y = -x + 2cp$. **(3)**

The point Q has coordinates $Q\left(cq, \frac{c}{q}\right)$, $q \neq p$.

The tangents to C at P and Q meet at N. Given that $p + q \neq 0$,

b show that the y-coordinate of N is $\frac{2c}{p + q}$ **(4)**

The line joining N to the origin O is perpendicular to the chord PQ.

c Find the value of p^2q^2. **(4)**

← Section 2.4

E/P **37** The point P lies on the rectangular hyperbola $xy = c^2$, where c is a positive constant.

a Show that an equation of the tangent to the hyperbola at the point $P\left(cp, \frac{c}{p}\right)$, $p > 0$, is $yp^2 + x = 2cp$. **(3)**

This tangent at P cuts the x-axis at the point S.

b Write down the coordinates of S. **(1)**

c Find an expression, in terms of p, for the length of PS. **(2)**

The normal at P cuts the x-axis at the point R. Given that the area of triangle RPS is $41c^2$,

d find, in terms of c, the coordinates of the point P. **(5)**

← Section 2.4

P **38** A point P lies on hyperbola H with equation $xy = c^2$. Prove that the locus of the midpoints of OP, where O is the origin, form a hyperbola and state its equation. **(3)**

← Section 2.5

P **39** A point P with coordinates (x, y) moves so that its distance from the point $(5, 0)$ is equal to its distance from the line with equation $x = -5$.

Prove that the locus of P has an equation of the form $y^2 = 4ax$, stating the value of a. **(5)**

← Section 2.5

A E **40** An ellipse has equation $\frac{x^2}{16} + \frac{y^2}{9} = 1$.

a Sketch the ellipse. **(2)**

b Find the value of the eccentricity e. **(2)**

c State the coordinates of the foci of the ellipse. **(2)**

← Sections 3.1, 3.3

E **41** The hyperbola H has equation $\frac{x^2}{16} - \frac{y^2}{4} = 1$.

Find:

a the value of the eccentricity of H **(2)**

b the distance between the foci of H. **(2)**

The ellipse E has equation $\frac{x^2}{16} + \frac{y^2}{4} = 1$.

c Sketch H and E on the same diagram, showing the coordinates of the points where each curve crosses the axes. **(4)**

← Sections 3.1, 3.2, 3.3

A E/P **42** An ellipse, with equation $\frac{x^2}{9} + \frac{y^2}{4} = 1$, has foci S and S'.

a Find the coordinates of the foci of the ellipse. **(2)**

b Using the focus–directrix property of the ellipse, prove that, for any point P on the ellipse, $SP + S'P = 6$. **(5)**

← Sections 3.1, 3.3

E **43** **a** Find the eccentricity of the ellipse with equation $3x^2 + 4y^2 = 12$. **(3)**

b Find an equation of the tangent to the ellipse with equation $3x^2 + 4y^2 = 12$ at the point with coordinates $\left(1, \frac{3}{2}\right)$. **(4)**

This tangent meets the y-axis at G. Given that S and S' are the foci of the ellipse,

c find the area of triangle $SS'G$. **(5)**

← Sections 3.3, 3.4

E **44** The points S_1 and S_2 have Cartesian coordinates $\left(-\frac{a}{2}\sqrt{3}, 0\right)$ and $\left(\frac{a}{2}\sqrt{3}, 0\right)$ respectively.

a Find a Cartesian equation of the ellipse which has S_1 and S_2 as its two foci, and a major axis of length $2a$. **(4)**

b Write down the equations of the directrices of this ellipse. **(1)**

Given that parametric equations of this ellipse are

$$x = a\cos\phi,\ y = b\sin\phi$$

c express b in terms of a. **(4)**

The point P is such that $\phi = \frac{\pi}{4}$ and the point Q such that $\phi = \frac{\pi}{2}$.

d Show that an equation of the chord PQ is $(\sqrt{2} - 1)x + 2y - a = 0$. **(3)**

← Section 3.3

E/P **45** **a** Find the eccentricity of the ellipse

$$\frac{x^2}{9} + \frac{y^2}{4} = 1$$ **(2)**

b Find also the coordinates of both foci and equations of both directrices of this ellipse. **(2)**

A

c Show that an equation for the tangent to this ellipse at the point $P(3\cos\theta, 2\sin\theta)$ is

$$\frac{x\cos\theta}{3} + \frac{y\sin\theta}{2} = 1 \qquad \textbf{(4)}$$

d Show that, as θ varies, the foot of the perpendicular from the origin to the tangent at P lies on the curve

$$(x^2 + y^2)^2 = 9x^2 + 4y^2 \qquad \textbf{(6)}$$

← Sections 3.3, 3.4

E/P **46** **a** Show that an equation of the normal to the ellipse $\frac{x^2}{a^2} + \frac{y^2}{b^2} = 1$ at the point $P(a\cos\theta, b\sin\theta)$ is

$$ax\sec\theta - by\operatorname{cosec}\theta = a^2 - b^2 \qquad \textbf{(3)}$$

The normal at P cuts the x-axis at G.

b Show that the coordinates of M, the midpoint of PG are

$$\left(\frac{2a^2 - b^2}{2a}\cos\theta, \frac{b}{2}\sin\theta\right) \qquad \textbf{(3)}$$

c Prove that, as θ varies, the locus of M is an ellipse and determine the equation of this ellipse. **(4)**

Given that the normal at P meets the y-axis at H and that O is the origin,

d prove that, if $a > b$, then the ratio of the area of $\triangle OMG$ to the area of $\triangle OGH$ is $b^2 : 2(a^2 - b^2)$. **(4)**

← Sections 3.4, 3.5

E/P **47** The diagram shows the ellipse with equation $\frac{x^2}{8^2} + \frac{y^2}{4^2} = 1$. The point P has coordinates $(4, 2\sqrt{3})$

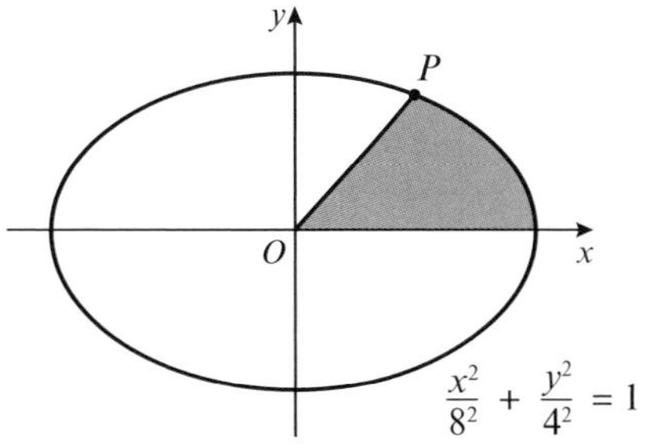

A

Show that the exact value for the area of the shaded region is $a\pi$, where a is a rational number to be found. **(6)**

← Section 3.2

E/P **48** The line with equation $y = mx + c$ is a tangent to the ellipse with equation

$$\frac{x^2}{a^2} + \frac{y^2}{b^2} = 1$$

a Show that $c^2 = a^2m^2 + b^2$. **(4)**

b Hence, or otherwise, find the equations of the tangents from the point (3, 4) to the ellipse with equation $\frac{x^2}{16} + \frac{y^2}{25} = 1$. **(4)**

← Section 3.4

E/P **49** The ellipse E has equation $\frac{x^2}{a^2} + \frac{y^2}{b^2} = 1$ and the line L has equation $y = mx + c$, where $m > 0$ and $c > 0$.

a Show that, if L and E have any points of intersection, the x-coordinates of these points are the roots of the equation $(b^2 + a^2m^2)x^2 + 2a^2mcx + a^2(c^2 - b^2) = 0$. **(4)**

Hence, given that L is a tangent to E,

b show that $c^2 = b^2 + a^2m^2$. **(2)**

The tangent L meets the negative x-axis at the point A and the positive y-axis at the point B, and O is the origin.

c Find, in terms of m, a and b, the area of the triangle OAB. **(3)**

d Prove that, as m varies, the minimum area of the triangle OAB is ab. **(3)**

e Find, in terms of a, the x-coordinate of the point of contact of L and E when the area of the triangle is a minimum. **(2)**

← Section 3.4

E/P **50** **a** Find equations for the tangent and normal to the rectangular hyperbola $x^2 - y^2 = 1$, at the point P with coordinates $(\cosh t, \sinh t)$, $t > 0$. **(5)**

A The tangent and normal cut the x-axis at T and G respectively. The perpendicular from P to the x-axis meets an asymptote in the first quadrant at Q.

b Show that GQ is perpendicular to this asymptote. **(4)**

The normal cuts the y-axis at R.

c Show that R lies on the circle with centre at T and radius TG. **(4)**

← Section 3.5

P **51** The point P lies on the hyperbola $\frac{x^2}{a^2} - \frac{y^2}{b^2} = 1$, and N is the foot of the perpendicular from P onto the x-axis. The tangent to the hyperbola at P meets the x-axis at T.

Show that $OT \times ON = a^2$, where O is the origin. **(6)**

← Section 3.5

P **52** The hyperbola C has equation $\frac{x^2}{a^2} - \frac{y^2}{b^2} = 1$.

a Show that an equation of the normal to C at the point $P(a\sec t, b\tan t)$ is

$$ax\sin t + by = (a^2 + b^2)\tan t$$ **(6)**

The normal to C at P cuts the x-axis at the point A and S is a focus of C. Given that the eccentricity of C is $\frac{3}{2}$, and that $OA = 3OS$, where O is the origin,

b determine the possible values of t, for $0 \leqslant t \leqslant 2\pi$. **(3)**

← Section 3.5

E **53 a** Show that the hyperbola $x^2 - y^2 = a^2$, $a > 0$, has eccentricity equal to $\sqrt{2}$. **(3)**

b Hence state the coordinates of the focus S and an equation of the corresponding directrix L, where both S and L lie in the region $x > 0$. **(2)**

The perpendicular from S to the line $y = x$ meets the line $y = x$ at P and the perpendicular from S to the line $y = -x$ meets the line $y = -x$ at Q.

A **c** Show that both P and Q lie on the directrix L and give the coordinates of P and Q. **(3)**

Given that the line SP meets the hyperbola at the point R,

d prove that the tangent at R passes through the point Q. **(4)**

← Sections 3.3, 3.5

E/P **54** Show that the equations of the tangents with gradient m to the hyperbola with equation $x^2 - 4y^2 = 4$ are

$$y = mx \pm \sqrt{4m^2 - 1}, \text{ where } |m| > \tfrac{1}{2}$$ **(6)**

← Section 3.5

E **55** An ellipse has equation $\frac{x^2}{a^2} + \frac{y^2}{b^2} = 1$, where a and b are constants and $a > b$.

a Find an equation of the tangent at the point $P(a\cos t, b\sin t)$. **(3)**

b Find an equation of the normal at the point $P(a\cos t, b\sin t)$. **(3)**

The normal at P meets the x-axis at the point Q. The tangent at P meets the y-axis at the point R.

c Find, in terms of a, b and t, the coordinates of M, the midpoint of QR. **(4)**

Given that $0 < t < \frac{\pi}{2}$,

d prove that, as t varies, the locus of M has equation $\left(\frac{2ax}{a^2 - b^2}\right)^2 + \left(\frac{b}{2y}\right)^2 = 1$. **(4)**

← Sections 3.5, 3.6

E/P **56 a** Find the equations for the tangent and normal to the hyperbola

$$\frac{x^2}{a^2} - \frac{y^2}{b^2} = 1$$

at the point $(a\sec\theta, b\tan\theta)$. **(6)**

b If these lines meet the y-axis at P and Q respectively, prove that the circle with PQ as a diameter passes through the foci of the hyperbola. **(5)**

← Sections 3.5, 3.6

E/P **57** Use algebra to solve $\frac{2}{x-2} < \frac{1}{x+1}$ **(6)**

← Section 4.1

E **58** Find the set of values of x for which

$$\frac{x^2}{x-2} > 2x$$ **(5)**

← Section 4.1

E **59** Find the set of values of x for which

$$\frac{x^2 - 12}{x} > 1$$ **(5)**

← Section 4.1

E **60** Find the set of values of x for which

$$2x - 5 > \frac{3}{x}$$

giving your answer using set notation. **(5)**

← Section 4.1

E/P **61** Given that k is a constant and that $k > 0$, find, in terms of k, the set of values of x for which $\frac{x+k}{x+4k} > \frac{k}{x}$ **(7)**

← Section 4.1

E **62** **a** On the same set of axes, sketch the graphs of

$y = 2 - x$ and $y = -\frac{2}{x-1}$ **(3)**

b Find the points of intersection of

$y = 2 - x$ and $y = -\frac{2}{x-1}$ **(2)**

c Write down the solution to the inequality

$$2 - x > -\frac{2}{x-1}$$ **(2)**

← Section 4.2

E **63** **a** On the same set of axes sketch the graphs of $y = \frac{4x}{2-x}$ and $y = \frac{2x}{(x+1)^2}$ **(4)**

b Find the points of intersection of

$y = \frac{4x}{2-x}$ and $y = \frac{2x}{(x+1)^2}$ **(2)**

c Hence, or otherwise, solve the inequality

$$\frac{4x}{2-x} \leqslant \frac{2x}{(x+1)^2}$$

giving your answer using set notation. **(2)**

← Section 4.2

A E **64** **a** On the same set of axes, sketch the graphs of

$y = |x - 5|$ and $y = |3x - 2|$. **(3)**

b Finds the coordinates of the points of intersection of $y = |x - 5|$ and $y = |3x - 2|$. **(3)**

c Write down the solution to the inequality

$$|x - 5| < |3x - 2|$$ **(2)**

← Section 4.3

E **65** **a** Sketch the graph of $y = |x + 2|$. **(2)**

b Use algebra to solve the inequality $2x > |x + 2|$. **(4)**

← Section 4.3

E **66** **a** Sketch the graph of $y = |x - 2a|$, given that $a > 0$. **(2)**

b Solve $|x - 2a| > 2x + a$, where $a > 0$. **(4)**

← Section 4.3

E/P **67** Solve the inequality $\left|\frac{x}{x-3}\right| < 8 - x$, giving your answer in set notation. **(6)**

← Section 4.3

E **68** **a** On the same set of axes, sketch the graphs of $y = x$ and $y = |2x - 1|$. **(3)**

b Use algebra to find the coordinates of the points of intersection of the two graphs. **(2)**

c Hence, or otherwise, find the set of values of x for which $|2x - 1| > x$. **(4)**

← Section 4.3

E/P **69** Use algebra to find the set of real values of x for which $|x - 3| > 2|x + 1|$. **(5)**

← Section 4.3

E/P **70** Solve, for x, the inequality $|5x + a| \leqslant |2x|$, where $a > 0$. **(6)**

← Section 4.3

E/P **71** **a** Using the same set of axes, sketch the curve with equation $y = |x^2 - 6x + 8|$ and the line with equation $2y = 3x - 9$.

A State the coordinates of the points where the curve and the line meet the x-axis. **(4)**

b Use algebra to find the coordinates of the points where the curve and the line intersect and, hence, solve the inequality $2|x^2 - 6x + 8| > 3x - 9$. **(5)**

← Section 4.3

E **72** **a** Sketch, on the same set of axes, the graph of $y = |(x - 2)(x - 4)|$, and the line with equation $y = 6 - 2x$. **(3)**

b Find the exact values of x for which $|(x - 2)(x - 4)| = 6 - 2x$. **(3)**

c Hence solve the inequality $|(x - 2)(x - 4)| < 6 - 2x$. **(2)**

← Section 4.3

E/P **73**

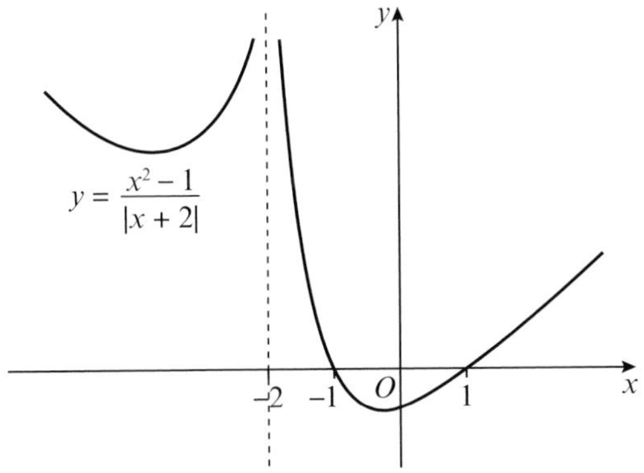

The diagram above shows a sketch of the curve with equation

$$y = \frac{x^2 - 1}{|x + 2|}, \quad x \neq -2$$

The curve crosses the x-axis at $x = 1$ and $x = -1$ and the line $x = -2$ is an asymptote of the curve.

a Use algebra to solve the equation

$$\frac{x^2 - 1}{|x + 2|} = 3(1 - x)$$ **(6)**

b Hence, or otherwise, find the set of values of x for which

$$\frac{x^2 - 1}{|x + 2|} < 3(1 - x)$$

Give your answer using set notation. **(2)**

← Section 4.3

Challenge

1 The hyperbola with equation $xy = c^2$ is rotated through 135° anticlockwise about the origin. Show that the resulting curve can be written in the form $x^2 - y^2 = k^2$, giving k in terms of c. ← Section 2.3

2 Solve in the range $0 < x < 2\pi$,

$$\frac{1}{1 - \sin x} < \frac{1}{\sin x}$$

← Chapter 4

A **3** The lines L_1 and L_2 intersect and have direction cosines l_1, m_1, n_1 and l_2, m_2, n_2 respectively.

a By means of a diagram, show that there are two lines that bisect the angles between L_1 and L_2.

b Show that these lines have direction ratios $l_1 + l_2, m_1 + m_2, n_1 + n_2$ and $l_1 - l_2, m_1 - m_2, n_1 - n_2$ respectively, and explain why these are not, in general, direction cosines.

← Section 1.4

4 **a** Prove that for two lines $y = m_1x + c_1$ and $y = m_2x + c_2$, $m_1 \neq m_2$, the acute angle α between the two lines satisfies

$$\tan\alpha = \frac{m_2 - m_1}{1 + m_1m_2}$$

b Hence, or otherwise, prove that the normal to an ellipse at any point P bisects the angle SPS', where S and S' are the foci of the ellipse. ← Section 3.3

5 The t-formulae

Objectives

After completing this chapter you should be able to:

- State the t-formulae → pages 117–120
- Apply the t-formulae to trigonometric identities → pages 120–122
- Use the t-formulae to solve trigonometric equations → pages 122–124
- Use the t-formulae for modelling with trigonometry → pages 124–126

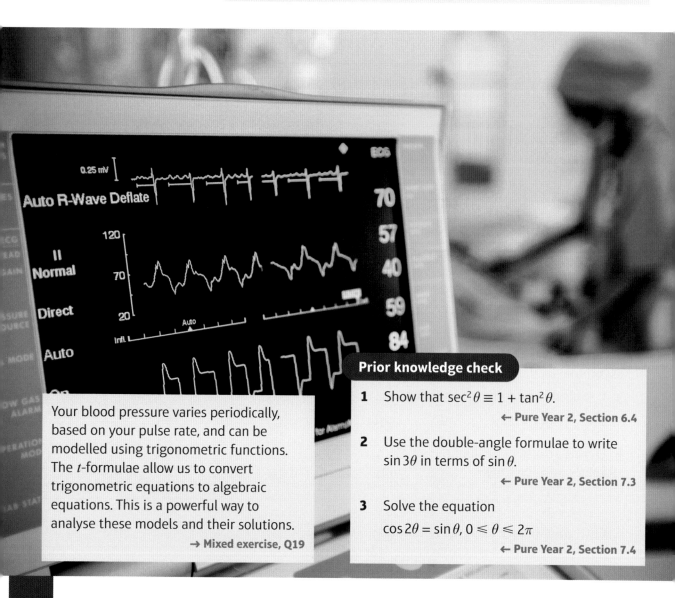

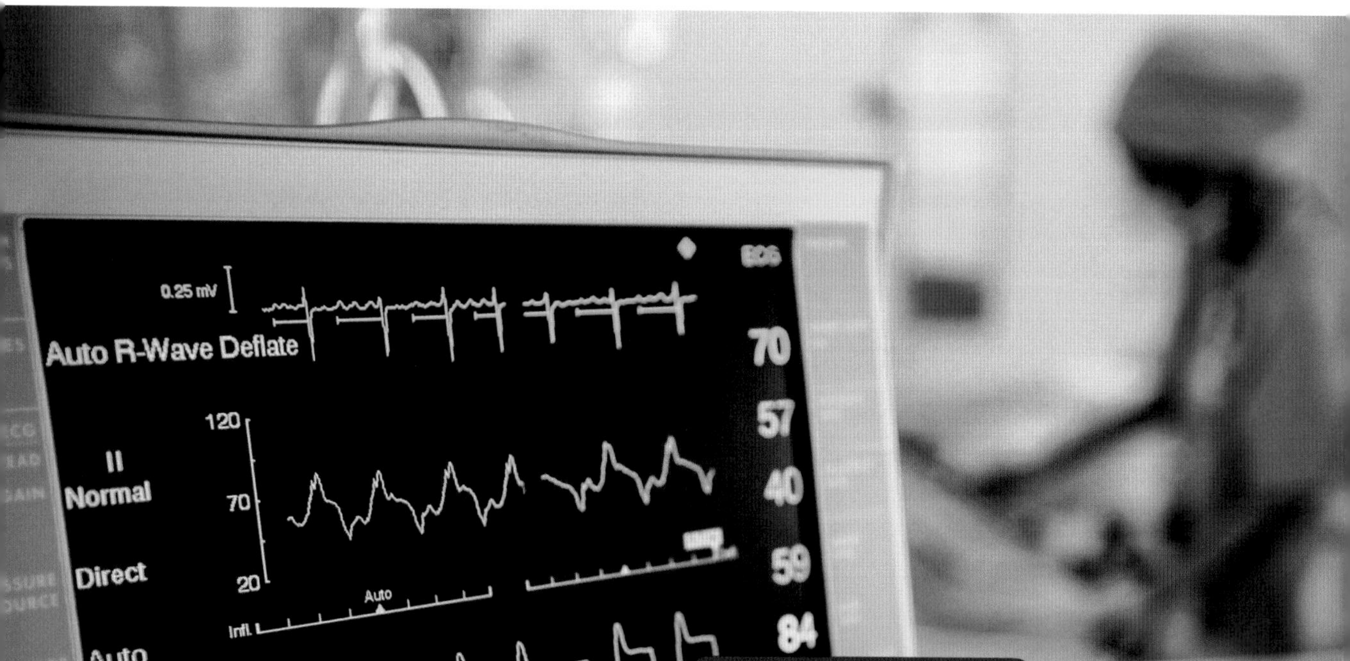

Your blood pressure varies periodically, based on your pulse rate, and can be modelled using trigonometric functions. The t-formulae allow us to convert trigonometric equations to algebraic equations. This is a powerful way to analyse these models and their solutions.

→ Mixed exercise, Q19

Prior knowledge check

1 Show that $\sec^2\theta \equiv 1 + \tan^2\theta$.

← Pure Year 2, Section 6.4

2 Use the double-angle formulae to write $\sin 3\theta$ in terms of $\sin\theta$.

← Pure Year 2, Section 7.3

3 Solve the equation $\cos 2\theta = \sin\theta$, $0 \leqslant \theta \leqslant 2\pi$

← Pure Year 2, Section 7.4

5.1 The t-formulae

The t-formulae are a set of formulae that allow you to express $\sin\theta$, $\cos\theta$ and $\tan\theta$ in terms of $t = \tan\frac{\theta}{2}$. They can be very useful for **solving trigonometric equations**, and **proving trigonometric identities**, as they allow you to write expressions involving $\sin\theta$, $\cos\theta$ and $\tan\theta$ in terms of a single variable, t.

- **When $t = \tan\frac{\theta}{2}$:**
  - $\sin\theta = \frac{2t}{1+t^2}$
  - $\cos\theta = \frac{1-t^2}{1+t^2}$
  - $\tan\theta = \frac{2t}{1-t^2}$

Watch out You should learn these formulae. They are not given in the formulae booklet, and provided you are not asked to prove or derive them, you may quote them in your exam.

You need to know how to derive the t-formulae using the definitions of the trigonometric ratios and the **double-angle formulae**. You can do this by constructing a right-angled triangle with acute angle $\frac{\theta}{2}$

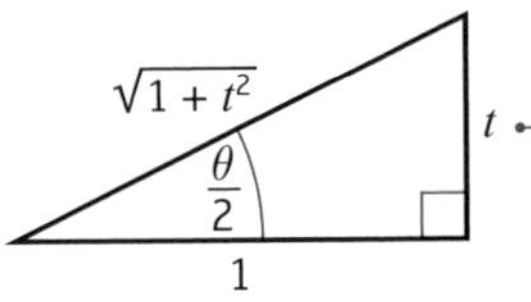

Make the side opposite the angle t, and the side adjacent 1. Then by Pythagoras' theorem, the hypotenuse has length $\sqrt{1+t^2}$.

Applying the definitions of the trigonometric ratios to this triangle gives:

$$\tan\frac{\theta}{2} = t$$

$$\sin\frac{\theta}{2} = \frac{t}{\sqrt{1+t^2}}$$

$$\cos\frac{\theta}{2} = \frac{1}{\sqrt{1+t^2}}$$

Therefore, using double-angle formulae gives

$$\sin\theta \equiv 2\sin\frac{\theta}{2}\cos\frac{\theta}{2} = 2\frac{t}{\sqrt{1+t^2}} \times \frac{1}{\sqrt{1+t^2}} = \frac{2t}{1+t^2}$$

$$\cos\theta \equiv \cos^2\frac{\theta}{2} - \sin^2\frac{\theta}{2} = \left(\frac{1}{\sqrt{1+t^2}}\right)^2 - \left(\frac{t}{\sqrt{1+t^2}}\right)^2 = \frac{1-t^2}{1+t^2}$$

$$\tan\theta \equiv \frac{\sin\theta}{\cos\theta} = \frac{2t}{1+t^2} \times \frac{1+t^2}{1-t^2} = \frac{2t}{1-t^2}$$

Links The double-angle formulae are:

$\sin 2\theta \equiv 2\sin\theta\cos\theta$

$\cos 2\theta \equiv \cos^2\theta - \sin^2\theta$

$\equiv 2\cos^2\theta - 1 \equiv 1 - 2\sin^2\theta$

$\tan 2\theta \equiv \frac{2\tan\theta}{1-\tan^2\theta}$

← Pure Year 2, Section 7.3

The above proof assumes that the angle $\frac{\theta}{2}$ is acute, but the formulae hold in general.

To see this, you can also derive the formulae purely algebraically.

$$\sin\theta \equiv 2\sin\frac{\theta}{2}\cos\frac{\theta}{2} \equiv 2\tan\frac{\theta}{2}\cos^2\frac{\theta}{2} \equiv \frac{2\tan\frac{\theta}{2}}{\sec^2\frac{\theta}{2}} \equiv \frac{2\tan\frac{\theta}{2}}{1+\tan^2\frac{\theta}{2}} \equiv \frac{2t}{1+t^2}$$

Similarly,

$$\cos\theta \equiv \cos^2\frac{\theta}{2} - \sin^2\frac{\theta}{2} \equiv \cos^2\frac{\theta}{2}\left(1 - \tan^2\frac{\theta}{2}\right)$$

$$\equiv \frac{1 - \tan^2\frac{\theta}{2}}{\sec^2\frac{\theta}{2}} \equiv \frac{1 - \tan^2\frac{\theta}{2}}{1 + \tan^2\frac{\theta}{2}} \equiv \frac{1 - t^2}{1 + t^2}$$

Finally, prove the identity for $\tan\theta$ exactly as before.

$$\tan\theta \equiv \frac{\sin\theta}{\cos\theta} \equiv \frac{2t}{1 + t^2} \times \frac{1 + t^2}{1 - t^2} \equiv \frac{2t}{1 - t^2}$$

Links These algebraic proofs make use of the identity $\sec^2\theta \equiv 1 + \tan^2\theta$.

$\sec\theta = \frac{1}{\cos\theta}$

$\operatorname{cosec}\theta = \frac{1}{\sin\theta}$

$\cot\theta = \frac{1}{\tan\theta}$ ← **Pure Year 2, Chapter 6**

Example 1

Given that $\tan\frac{\theta}{2} = \frac{3}{4}$, find the exact values of:

a $\sin\theta$ **b** $\cos\theta$

a $t = \frac{3}{4}$

So $\sin\theta = \frac{2t}{1 + t^2} = \frac{2\left(\frac{3}{4}\right)}{1 + \left(\frac{3}{4}\right)^2} = \frac{24}{25}$ — Use the t-formula for sin.

b $\cos\theta = \frac{1 - t^2}{1 + t^2} = \frac{1 - \left(\frac{3}{4}\right)^2}{1 + \left(\frac{3}{4}\right)^2} = \frac{7}{25}$ — Use the t-formula for cos.

Example 2

Given that $\frac{\pi}{2} \leqslant \frac{\theta}{2} < \pi$ and $\sin\frac{\theta}{2} = \frac{8}{17}$, find:

a the exact value of $\cot\theta$

b the value of $\sec\theta + \operatorname{cosec}\theta$, correct to 3 significant figures.

a $\cos\frac{\theta}{2} = -\sqrt{1 - \sin^2\frac{\theta}{2}} = -\sqrt{1 - \left(\frac{8}{17}\right)^2} = -\frac{15}{17}$ — Use $\sin^2\frac{\theta}{2} + \cos^2\frac{\theta}{2} \equiv 1$ to find the exact value of $\cos\frac{\theta}{2}$

So $\tan\frac{\theta}{2} = \frac{\sin\frac{\theta}{2}}{\cos\frac{\theta}{2}} = \frac{8}{17} \div \left(-\frac{15}{17}\right) = -\frac{8}{15}$

Watch out Make sure you choose the correct sign when taking square roots. Since $\frac{\pi}{2} \leqslant \frac{\theta}{2} < \pi$ you know that $\cos\frac{\theta}{2}$ must be negative.

Set $t = \tan\frac{\theta}{2} = -\frac{8}{15}$

$\cot\theta \equiv \frac{1}{\tan\theta} = \frac{1 - t^2}{2t} = \frac{1 - \left(-\frac{8}{15}\right)^2}{2\left(-\frac{8}{15}\right)} = -\frac{161}{240}$ — Use the definition of cot and the t-formula for tan.

$$\text{b} \quad \sec\theta + \operatorname{cosec}\theta \equiv \frac{1}{\cos\theta} + \frac{1}{\sin\theta}$$

$$= \frac{1+t^2}{1-t^2} + \frac{1+t^2}{2t}$$

$$= \frac{1+\left(-\frac{8}{15}\right)^2}{1-\left(-\frac{8}{15}\right)^2} + \frac{1+\left(-\frac{8}{15}\right)^2}{2\left(-\frac{8}{15}\right)}$$

$$= \frac{22831}{38640} = 0.591 \text{ (3 s.f.)}$$

Use the definitions of sec and cosec and the t-formulae for cos and sin.

Exercise 5A

1 Given that $\tan\frac{\theta}{2} = \frac{2}{3}$, use the t-formulae to find the exact values of:

a $\sin\theta$ **b** $\cos\theta$

c $\tan\theta$

2 Given that $\tan\frac{\theta}{2} = 2$, use the t-formulae to find the exact values of:

a $\sin\theta$ **b** $\cos\theta$

c $\tan\theta$ **d** $\sec\theta + \cot\theta$

3 Given that $\sin\frac{\theta}{2} = \frac{4}{5}$ and that $0 \leqslant \frac{\theta}{2} < \frac{\pi}{2}$, use the t-formulae to find the values of:

a $\sin\theta$ **b** $\cos\theta$

c $\sec\theta$ **d** $\dfrac{\cos\theta}{\sin\theta(1+\cot\theta)}$

4 Given that $\cos\frac{\theta}{2} = -\frac{5}{13}$ and that $\frac{\pi}{2} \leqslant \frac{\theta}{2} < \pi$, use the t-formulae to find the values of:

a $\cos\theta$ **b** $\tan^2\theta$

c $\sec\theta + \operatorname{cosec}\theta$ **d** $\dfrac{\sec\theta}{\operatorname{cosec}\theta + \cos\theta}$

5 Suppose that $\operatorname{cosec}\frac{\theta}{2} = \frac{25}{24}$ where $\frac{\pi}{2} \leqslant \frac{\theta}{2} < \pi$. Use the t-formulae to find the values of:

a $\tan\theta$ **b** $\sin 2\theta$

c $\cos 2\theta$ **d** $\cot 2\theta$

E **6** Suppose that $0 \leqslant \frac{\theta}{2} < \frac{\pi}{2}$ and that $\sin\theta = \dfrac{\sqrt{3}-1}{2\sqrt{2}}$

a Show that $\tan\theta = \dfrac{\sqrt{3}-1}{\sqrt{3}+1}$ **(2 marks)**

b Using the t-formulae, find $\sin 2\theta$ and $\cos 2\theta$. **(3 marks)**

c Hence deduce the value of θ. **(1 mark)**

(E) **7** Suppose that $\frac{\pi}{2} \leqslant x < \pi$ and that $\cos x = -\frac{\sqrt{2+\sqrt{2}}}{2}$

a Show that $\tan x = -\frac{\sqrt{2-\sqrt{2}}}{\sqrt{2+\sqrt{2}}}$ **(2 marks)**

b Using the t-formulae, find $\tan 2x$. **(2 marks)**

c Hence deduce the value of x. **(1 mark)**

(E/P) **8** Given that $t = \tan\frac{5\pi}{12}$,

a show that $t^2 - 4t + 1 = 0$ **(3 marks)**

b show further that $t^2 = \frac{2+\sqrt{3}}{2-\sqrt{3}}$ **(3 marks)**

c deduce the exact value of t. **(1 mark)**

(P) **9** Consider the following diagram, where θ is an acute angle.

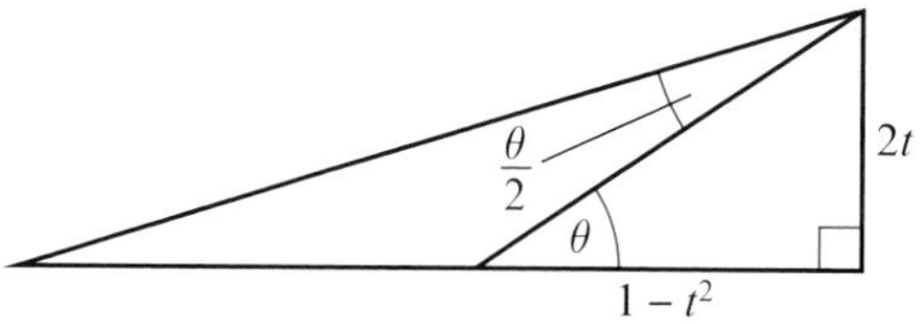

Show that $t = \tan\frac{\theta}{2}$ and hence derive the t-formulae for $\sin\theta$, $\cos\theta$ and $\tan\theta$.

5.2 Applying the t-formulae to trigonometric identities

You can use the t-formulae to prove trigonometric identities.

Example 3

Prove that $\frac{1 + \operatorname{cosec}\theta}{\cot\theta} \equiv \frac{1 + \tan\frac{\theta}{2}}{1 - \tan\frac{\theta}{2}}$, $\theta \neq (2n+1)\frac{\pi}{2}$, $n \in \mathbb{Z}$.

Let $t = \tan\frac{\theta}{2}$. Then $\operatorname{cosec}\theta = \frac{1+t^2}{2t}$ and $\cot\theta = \frac{1-t^2}{2t}$.

Use the t-formulae with the identities $\operatorname{cosec}\theta \equiv \frac{1}{\sin\theta}$ and $\cot\theta \equiv \frac{1}{\tan\theta}$

So $\frac{1+\operatorname{cosec}\theta}{\cot\theta} = \frac{2t + 1 + t^2}{1 - t^2}$

The denominator is the difference of two squares.

$$= \frac{(1+t)^2}{(1-t)(1+t)}$$

$$= \frac{1+t}{1-t}$$

$$= \frac{1 + \tan\frac{\theta}{2}}{1 - \tan\frac{\theta}{2}}$$

Example 4

Prove that $\tan 2\theta \cot\theta \equiv 1 + \sec 2\theta$, $\theta \neq (2n+1)\frac{\pi}{4}$, $n \in \mathbb{Z}$.

Let $t = \tan\theta$

We have $\cot\theta = \frac{1}{t}$, $\tan 2\theta = \frac{2t}{1-t^2}$ and $\sec 2\theta = \frac{1+t^2}{1-t^2}$

$$\tan 2\theta \cot\theta = \frac{2}{1-t^2}$$
$$= \frac{1-t^2+1+t^2}{1-t^2}$$
$$= \frac{1-t^2}{1-t^2} + \frac{1+t^2}{1-t^2}$$
$$= 1 + \frac{1+t^2}{1-t^2}$$
$$= 1 + \sec 2\theta$$

Hence $\tan 2\theta \cot\theta \equiv 1 + \sec 2\theta$

Since this equation uses θ and 2θ it makes sense to use $t = \tan\theta$ rather than $t = \tan\frac{\theta}{2}$

From the t-formulae and the definition of $\cot\theta$.

Problem-solving

When proving identities, you should always start from one side and work towards the other side. For example, if you start with the left-hand side, look at the right-hand side to give you an idea of what form you need your expression for t to be in.

Here, you need to find $\sec 2\theta = \frac{1+t^2}{1-t^2}$ on the right-hand side, so you need to try to isolate this term in your expression.

Exercise 5B

P 1 Using the t-formulae, prove the following trigonometric identities.

a $\sin^2\theta + \cos^2\theta \equiv 1$

b $\frac{\tan^2\theta}{\tan^2\theta + 1} \equiv \sin^2\theta$

c $\frac{\operatorname{cosec}\theta}{\sin\theta} - \frac{\cot\theta}{\tan\theta} \equiv 1$, $\theta \neq n\pi$, $n \in \mathbb{Z}$

d $\cot 2\theta + \tan\theta \equiv \operatorname{cosec} 2\theta$, $\theta \neq \frac{n\pi}{2}$, $n \in \mathbb{Z}$

P 2 Using the t-formulae, prove the following trigonometric identities.

a $\tan\theta + \cot\theta \equiv \sec\theta\operatorname{cosec}\theta$, $\theta \neq \frac{n\pi}{2}$, $n \in \mathbb{Z}$

b $\frac{1+\cos\theta}{\sin\theta} \equiv \frac{\sin\theta}{1-\cos\theta}$, $\theta \neq n\pi$, $n \in \mathbb{Z}$

c $\frac{1-\sin\theta}{\cos\theta} \equiv \frac{\cos\theta}{1+\sin\theta}$, $\theta \neq \frac{(2n+1)\pi}{2}$, $n \in \mathbb{Z}$

d $\tan\theta\sin\theta + \cos\theta \equiv \sec\theta$, $\theta \neq \frac{(2n+1)\pi}{2}$, $n \in \mathbb{Z}$

E/P 3 Using the substitution $t = \tan\frac{\theta}{2}$, prove that $\sin\theta + \sin\theta\cot^2\theta \equiv \operatorname{cosec}\theta$ for $\theta \neq n\pi$, $n \in \mathbb{Z}$ **(4 marks)**

E/P 4 Using the substitution $t = \tan\frac{\theta}{2}$, prove that

$$\frac{\cos\theta}{1-\sin\theta} - \frac{\cos\theta}{1+\sin\theta} \equiv 2\tan\theta \text{ for } \theta \neq \frac{(2n+1)\pi}{2}, n \in \mathbb{Z}$$ **(4 marks)**

E/P 5 Using the substitution $t = \tan\frac{x}{2}$, prove that

$$\cos^2 x \equiv \frac{\operatorname{cosec} x \cos x}{\tan x + \cot x}$$ **(4 marks)**

E/P 6 Using the substitution $t = \tan\frac{\theta}{2}$, prove that

$$\frac{\cos\theta}{1+\sin\theta} + \frac{1+\sin\theta}{\cos\theta} \equiv 2\sec\theta \text{ for } \theta \neq \frac{(2n+1)\pi}{2}, n \in \mathbb{Z}$$ **(4 marks)**

E/P 7 Using the substitution $t = \tan\frac{\theta}{2}$, prove that

$$\sec\theta + \tan\theta \equiv \frac{\cos\theta}{1-\sin\theta} \text{ for } \theta \neq \frac{(2n+1)\pi}{2}, n \in \mathbb{Z}$$ **(4 marks)**

E/P 8 Using the substitution $t = \tan x$, prove that

$$\frac{1+\sin 2x - \cos 2x}{\sin 2x + \cos 2x - 1} \equiv \frac{1+\tan x}{1-\tan x} \text{ for } x \neq n\pi, \frac{(2n+1)\pi}{4}, n \in \mathbb{Z}$$ **(4 marks)**

E/P 9 Using the substitution $t = \tan\frac{\theta}{2}$, prove that

$$\frac{\cos\theta}{1-\sin\theta} - \tan\theta \equiv \sec\theta, \text{ for } \theta \neq \frac{(2n+1)\pi}{2}, n \in \mathbb{Z}$$ **(4 marks)**

E/P 10 Using the substitution $t = \tan\frac{\theta}{2}$, prove that

$$\tan^2\theta + \tan\theta\sec\theta + 1 \equiv \frac{1+\sin\theta}{\cos^2\theta} \text{ for } \theta \neq \frac{(2n+1)\pi}{2}, n \in \mathbb{Z}$$ **(4 marks)**

E/P 11 Use the substitution $t = \tan x$ to prove the identity

$$\frac{\cos 2x}{1-\sin 2x} \equiv \frac{\cot x + 1}{\cot x - 1}, x \neq (4n+1)\frac{\pi}{4}, n \in \mathbb{Z}$$ **(5 marks)**

Challenge

Using the t-formulae, prove the identity

$$\frac{\sin^3\theta + \cos^3\theta}{\sin\theta + \cos\theta} \equiv 1 - \sin\theta\cos\theta$$

5.3 Solving trigonometric equations

The t-formulae can be used to convert equations given in terms of different trigonometric functions of θ into equations in t.

- **To solve trigonometric equations using the t-formulae:**
  - **use the substitution $t = \tan\frac{\theta}{2}$**
  - **write any trigonometric functions in the equation in terms of t**
  - **solve the resulting equation algebraically to find the value(s) of t**
  - **find corresponding values of θ which satisfy the original equation.**

Watch out This substitution is only valid when $\tan\frac{\theta}{2}$ is defined. If the original equation has solutions of the form $\theta = (2n + 1)$, $n \in \mathbb{Z}$ this method will not find those solutions.

Example 5

Solve $2\sin\theta - 3\cos\theta = 1$ for $0 \leqslant \theta \leqslant 2\pi$. Give your answers to 2 decimal places.

Using the substitution $t = \tan\frac{\theta}{2}$

$\frac{4t}{1+t^2} - \frac{3(1-t^2)}{1+t^2} = 1$

$4t - 3 + 3t^2 = 1 + t^2$

$2t^2 + 4t - 4 = 0$

$(t+1)^2 - 3 = 0$

$t = -1 \pm\sqrt{3}$

So $\tan\frac{\theta}{2} = -1 \pm \sqrt{3}$, $0 \leqslant \frac{\theta}{2} \leqslant \pi$.

$\tan\frac{\theta}{2} = -1 + \sqrt{3}$

$\frac{\theta}{2} = 0.6319\ldots$ so $\theta = 1.26$ (2 d.p.)

$\tan\frac{\theta}{2} = -1 - \sqrt{3}$

$\frac{\theta}{2} = 1.9216\ldots$ so $\theta = 3.84$ (2 d.p.)

Apply the t-formulae so that everything is in terms of t.

Multiply both sides by $1 + t^2$.

Solve the resulting quadratic equation by completing the square, or using the quadratic formula.

Problem-solving

Use the substitution to find the corresponding values of θ that lie within the given range. The range of values of θ is $0 \leqslant \theta \leqslant 2\pi$, so the range of values for $\frac{\theta}{2}$ will be $0 \leqslant \frac{\theta}{2} \leqslant \pi$. Make sure you solve separately for each value of t.

You could also solve the original equation by writing $2\sin\theta - 3\cos\theta$ in the form $R\cos(\theta + \alpha)$.

← **Pure Year 2, Section 7.5**

Exercise 5C

1 Using the t-formulae, solve the following trigonometric equations for θ in the range $0 \leqslant \theta \leqslant 2\pi$, giving your answers to 2 decimal places in each case.

a $2\sin\theta - \cos\theta = 2$

b $\sin\theta + 5\cos\theta = -1$

c $\tan\theta - 5\sec\theta = 7$

d $7\cot\theta + 3\operatorname{cosec}\theta = 9$

e $2\cot\theta - \operatorname{cosec}\theta = 0$

E) 2 **a** Using the substitution $t = \tan\theta$, show that the equation $\sin 2\theta - 2\cos 2\theta = 1 - \sqrt{3}\cos 2\theta$ can be written as $(\sqrt{3}-1)t^2 - 2t - (\sqrt{3}-3) = 0$. **(3 marks)**

b Hence find the exact solutions of $\sin 2\theta - 2\cos 2\theta = 1 - \sqrt{3}\cos 2\theta$ in the range $0 \leqslant \theta \leqslant 2\pi$. **(3 marks)**

/P) 3 **a** Using the substitution $t = \tan\frac{x}{2}$, show that the equation $16\cot x - 9\tan x = 0$, $x \neq \frac{n\pi}{2}$, $n \in \mathbb{Z}$, can be written as $4t^4 - 17t^2 + 4 = 0$. **(3 marks)**

b Hence find all solutions of $16\cot x - 9\tan x = 0$ in the range $0 \leqslant x \leqslant 2\pi$ to two decimal places. **(4 marks)**

Problem-solving

This quartic equation is a quadratic in t^2. Solve it using the substitution $u = t^2$.

E **4 a** Using the substitution $t = \tan\frac{\theta}{2}$, show that the equation $10\sin\theta\cos\theta - 3\cos\theta = -3$ can be written as $t(t-2)(3t^2 - 4t - 5) = 0$. **(3 marks)**

b Hence find all solutions of $10\sin\theta\cos\theta - 3\cos\theta = -3$ in the range $0 \leqslant \theta \leqslant 2\pi$ to 2 decimal places. **(4 marks)**

E/P **5 a** Using the substitution $t = \tan\theta$, show that the equation $3\sin 2\theta + \cos 2\theta + 3\tan 2\theta = 1$, $\theta \neq \frac{(2n+1)\pi}{2}$, $n \in \mathbb{Z}$, can be written as $t^4 - t^2 + 6t = 0$. **(3 marks)**

b Given that $(t + 2)$ is a factor of $t^4 - t^2 + 6t$, find all solutions of $3\sin 2\theta + \cos 2\theta + 3\tan 2\theta = 1$ in the range $0 \leqslant \theta \leqslant 2\pi$ to 2 decimal places. **(4 marks)**

E/P **6 a** Using the substitution $t = \tan\theta$, show that the equation $\tan\theta + \cos 2\theta = 1$ can be written as $t^3 - 2t^2 + t = 0$. **(3 marks)**

b Hence find all solutions of $\tan\theta + \cos 2\theta = 1$ in the range $0 \leqslant \theta \leqslant 2\pi$. **(4 marks)**

E/P **7 a** Using the substitution $t = \tan\theta$, show that the equation $2\sin 2\theta - \cos 4\theta - 4\tan\theta = -1$ can be written as $t^5 + t^3 - 2t^2 = 0$. **(4 marks)**

b Hence find all solutions of $2\sin 2\theta - \cos 2\theta - 4\tan\theta = -1$ in the range $0 \leqslant \theta \leqslant 2\pi$. **(3 marks)**

E/P **8** Solve $5\cos\theta - 12\operatorname{cosec}\theta = 12$ for $0 \leqslant \theta \leqslant 2\pi$. **(7 marks)**

Challenge

Show that the equation $5\sin 2\theta + 12\cos\theta = -12$ has exactly two solutions in the range $0 \leqslant \theta \leqslant 2\pi$, and state their values.

5.4 Modelling with trigonometry

A Trigonometric functions appear frequently in mathematical models describing quantities that vary periodically. By adding or subtracting multiples of different trigonometric functions, more complicated situations can be modelled.

Models involving different trigonometric functions can be simplified and analysed using the t-formulae.

Example 6

The displacement of a particle moving in a straight line, s m, at time x seconds is given by

$$s = \sin 4x + 2\sin 2x + 2$$

a Show that the velocity of the particle at time x seconds is given by $v = \frac{8}{(1+t^2)^2}(1 - 3t^2)$, m s^{-1}, where $t = \tan x$.

b Hence find the value of x where $0 \leqslant x \leqslant \pi$ for which the displacement is maximised.

a Differentiating, we have

$v = \frac{ds}{dx} = 4\cos 4x + 4\cos 2x$

Substituting the t-formulae and using a double-angle formula:

$$v = 4\left(\frac{(1-t^2)^2}{(1+t^2)^2} - \frac{4t^2}{(1+t^2)^2}\right) + \frac{4(1-t^2)}{1+t^2}$$

$$= \frac{4}{(1+t^2)^2}(1 - 2t^2 + t^4 - 4t^2 + 1 - t^4)$$

$$= \frac{8}{(1+t^2)^2}(1 - 3t^2)$$

b Solving for $\frac{ds}{dx} = 0$ we have

$t = \pm\frac{1}{\sqrt{3}}$ which implies that $x = \frac{\pi}{6}, \frac{5\pi}{6}$

To check which point is a maximum we differentiate again.

$$\frac{d^2s}{dx^2} = \frac{dv}{dx}$$

$$= -16\sin 4x - 8\sin 2x$$

$$= -\frac{16}{(1+t^2)^2}(4t(1-t^2) + t(1+t^2))$$

$$= -\frac{16t}{(1+t^2)^2}(5 - 3t^2)$$

When $t = \frac{1}{\sqrt{3}}$, $\frac{d^2s}{dx^2} < 0$ and when $t = -\frac{1}{\sqrt{3}}$, $\frac{d^2s}{dx^2} > 0$

so the maximum is at $t = \frac{1}{\sqrt{3}}$

Converting back to x, we get

$\tan x = \frac{1}{\sqrt{3}} \Rightarrow x = \frac{\pi}{6}$

To find the velocity you differentiate the displacement with respect to time.

← Statistics and Mechanics Year 1, Chapter 11

$\cos 2x \equiv \cos^2 x - \sin^2 x$, so $\cos 4x = \cos^2 2x - \sin^2 2x$

Problem-solving

You could also substitute using the t-formulae **before** you differentiate:

$$\sin 4x + 2\sin 2x + 2 = \frac{4t(1-t^2)}{(1+t^2)^2} + \frac{4t}{1+t^2} + 2$$

You still need to differentiate with respect to x, so use $\frac{ds}{dx} = \frac{ds}{dt} \times \frac{dt}{dx}$, together with $\frac{dt}{dx} = \sec^2 x = 1 + t^2$.

tan is periodic with period π, so we get infinitely many solutions, but only one of them is in the correct range.

Exercise 5D

(P) **1** The displacement of a particle moving in a straight line, s m, at time x seconds is given by $s = 10 - 5\sin x - 12\cos x$, $0 \leqslant x \leqslant 2\pi$.

a Show that $\frac{ds}{dx} = \frac{1}{1+t^2}(5t^2 + 24t - 5)$ where $t = \tan\frac{x}{2}$ **(6 marks)**

b Hence find all values of x for which displacement is minimised. **(3 marks)**

(P) **2** The displacement of a particle moving in a straight line, s m, at time x seconds is given by $s = 1 + 2\sin x - \cos 2x$, $0 \leqslant x \leqslant 2\pi$.

a Show that $\frac{ds}{dx} = \frac{2}{(1+t^2)^2}(1 - t^2)(t^2 + 4t + 1)$ where $t = \tan\frac{x}{2}$ **(6 marks)**

b Hence find all values of x for which the particle is stationary. **(3 marks)**

A E/P **3** The height in cm of a car chassis above the road x seconds after it drives over a speed bump is modelled by the function $\text{h}(x) = 3\sin 2x - 4\cos 2x + 25,\ 0 \leqslant x \leqslant \pi$.

a Show that the vertical velocity of the chassis at time x is given by $\text{v}(x) = \frac{-2}{1+t^2}(3t^2 - 8t - 3)$ where $t = \tan x$. **(6 marks)**

b Find the time between oscillations according to the model. **(2 marks)**

c Using part **a** find the value of x for which the displacement is minimised. **(3 marks)**

E/P **4** The figure below shows the graph of the function $y(x) = \frac{1}{2}\sin\frac{x}{5} + \sin\frac{2x}{5} + \frac{1}{2}\cos\frac{x}{5} + 2$.

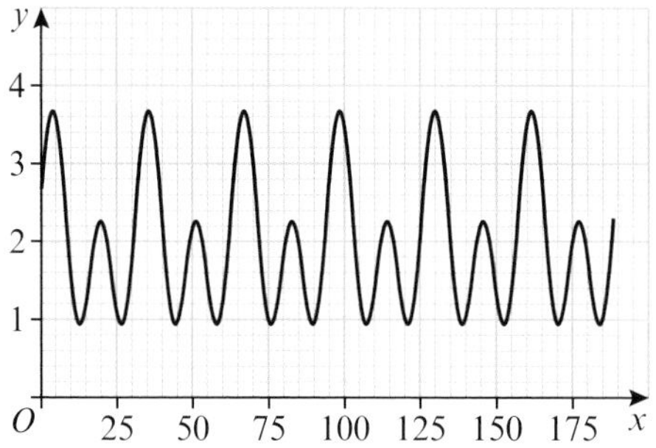

a Show that

$$\frac{\text{d}y}{\text{d}x} = \frac{(3t^2 - 8t - 5)(t^2 + 2t - 1)}{10(1 + t^2)^2} \text{ where } t = \tan\frac{x}{10}.$$ **(6 marks)**

Below is a graph showing the intensity of x-rays emitted over time by a pulsar, a type of rotating neutron star that emits a beam of x-rays in a specific direction.

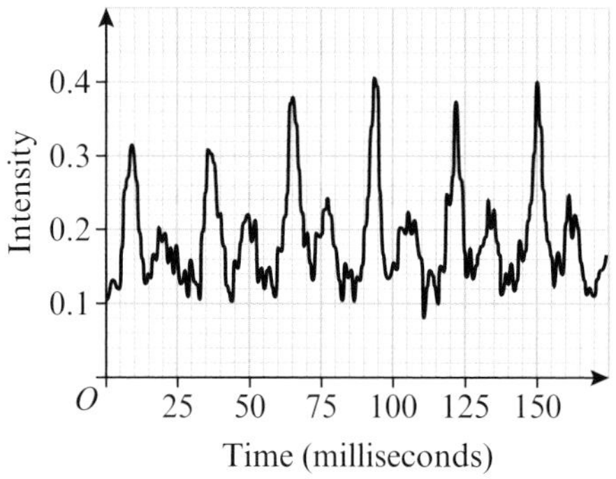

The graph of $ky(x)$, where k is a constant and x is measured in milliseconds can be used to model the predicted intensity of x-ray radiation observed on Earth.

b **i** Suggest a value of k that could be used to approximate the observed data with the graph of $ky(x)$.

ii Why might such a model be suitable for predicting the times of the peaks, but not the intensity of those peaks? **(3 marks)**

c Use the second graph and the result from part **a** to estimate, to the nearest millisecond, the time of the most intense peak in the observed data. **(6 marks)**

Mixed exercise 5

1 Consider the following diagram.

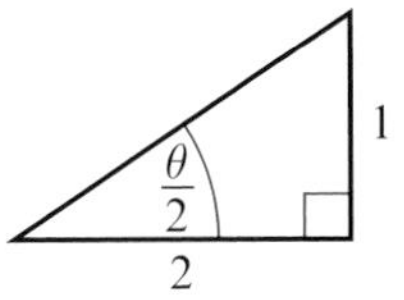

Using the t-formulae with $t = \tan\frac{\theta}{2}$, find the values of:

a $\cos\theta$ **b** $\sin\theta$ **c** $\sec\theta + \tan\theta$ **d** $\sec\theta\operatorname{cosec}\theta$

2 Consider the following diagram.

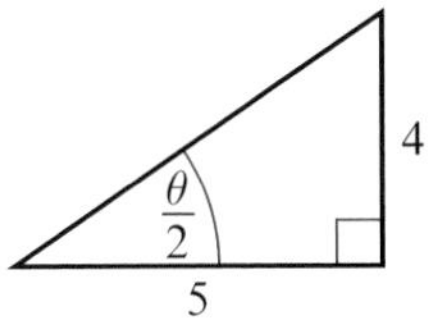

Using the t-formulae with $t = \tan\frac{\theta}{2}$, find the values of:

a $\tan\theta$ **b** $\sec\theta$ **c** $\sin\theta$ **d** $\cot\theta + \operatorname{cosec}\theta$

3 Given that $\tan\theta = 3$, use the substitution $t = \tan\theta$ to find:

a $\sin 2\theta$ **b** $\cos 2\theta$ **c** $\tan^2 2\theta$ **d** $\dfrac{\sec 2\theta}{\operatorname{cosec} 2\theta + \cot 2\theta}$

4 Given that $t = \tan\theta = -2$, use the t-formulae find:

a $\tan 2\theta$ **b** $\sec 2\theta\operatorname{cosec} 2\theta$ **c** $\sec^2 2\theta$ **d** $\cot 2\theta + \tan 2\theta$

5 **a** Using the t-formulae, show that $\tan^2\theta \equiv \sec^2\theta - 1$.

Suppose that $\pi \leqslant \theta \leqslant \frac{3\pi}{2}$, and that $\sec\theta = \dfrac{-2\sqrt{2}}{1+\sqrt{3}}$

b Using the result from part **a** or otherwise, find $\tan\theta$.

c Using the t-formulae, compute $\sin 2\theta$ and $\cos 2\theta$.

d Hence deduce the value of θ.

E **6** Let $t = \tan\frac{\pi}{8}$

a By writing down expressions for $\cos\frac{\pi}{4}$ and $\sin\frac{\pi}{4}$ in terms of t, find the exact value of t. **(4 marks)**

b Using the identity $\sec^2\theta \equiv \tan^2\theta + 1$, find $\sec\frac{\pi}{8}$, and hence deduce the values of $\sin\frac{\pi}{8}$ and $\cos\frac{\pi}{8}$ **(2 marks)**

E/P **7** Using the substitution $t = \tan\frac{x}{2}$, show that $\dfrac{1 + \sin x - \cos x}{\sin x + \cos x - 1} \equiv \dfrac{1 + \sin x}{\cos x}$

for $x \neq \dfrac{(2n+1)\pi}{2}$, $n \in \mathbb{Z}$. **(4 marks)**

(E/P) **8** Using the substitution $t = \tan\frac{\theta}{2}$, show that $\tan^2\theta - \sin^2\theta \equiv \tan^2\theta\sin^2\theta$ for $\theta \neq \frac{(2n+1)\pi}{2}$, $n \in \mathbb{Z}$. **(4 marks)**

(E/P) **9** Using the substitution $t = \tan\frac{\theta}{2}$, show that $\sin\theta\cos\theta\tan\theta \equiv 1 - \cos^2\theta$. **(4 marks)**

(E/P) **10** Using the substitution $t = \tan\frac{\theta}{2}$, show that $\frac{1+\sin\theta}{1-\sin\theta} - \frac{1-\sin\theta}{1+\sin\theta} \equiv 4\tan\theta\sec\theta$ for $\theta \neq \frac{(2n+1)\pi}{2}$, $n \in \mathbb{Z}$. **(4 marks)**

(E/P) **11** Using the substitution $t = \tan\frac{x}{2}$, show that $\frac{1+\tan^2 x}{1-\tan^2 x} \equiv \frac{1}{\cos^2 x - \sin^2 x}$ for $x \neq \frac{(2n+1)\pi}{4}$, $n \in \mathbb{Z}$. **(4 marks)**

(E/P) **12** Using the substitution $t = \tan\frac{\theta}{2}$, show that $\frac{1}{1-\sin\theta} - \frac{1}{1+\sin\theta} \equiv 2\tan\theta\sec\theta$ for $\theta \neq \frac{(2n+1)\pi}{2}$, $n \in \mathbb{Z}$. **(4 marks)**

(E/P) **13** Using the substitution $t = \tan\frac{\theta}{2}$, show that $\tan\theta + \frac{\cos\theta}{1+\sin\theta} \equiv \sec\theta$ for $\theta \neq \frac{(2n+1)\pi}{2}$, $n \in \mathbb{Z}$. **(4 marks)**

(E/P) **14** Using the substitution $t = \tan\frac{\theta}{2}$, show that $(\sin\theta + \cos\theta)(\tan\theta + \cot\theta) \equiv \sec\theta + \operatorname{cosec}\theta$ for $\theta \neq \frac{n\pi}{2}$, $n \in \mathbb{Z}$. **(4 marks)**

(E) **15** **a** Using the substitution $t = \tan\frac{x}{2}$, show that the equation $3\cos x - \sin x = -1$ can be written as $t^2 + t - 2 = 0$. **(3 marks)**

b Hence find all solutions of $3\cos x - \sin x = -1$ in the range $0 \leqslant x < 2\pi$ to 2 decimal places. **(3 marks)**

(E) **16** **a** Using the substitution $t = \tan\frac{\theta}{2}$, show that the equation $\sin\theta + \cos\theta = -\frac{1}{5}$ can be written as $2t^2 - 5t - 3 = 0$. **(3 marks)**

b Hence find all solutions of $\sin\theta + \cos\theta = -\frac{1}{5}$ in the range $0 \leqslant \theta \leqslant 2\pi$ to 2 decimal places. **(3 marks)**

(E) **17** **a** Using the substitution $t = \tan\frac{\theta}{2}$, show that the equation $6\tan\theta + 12\sin\theta + \cos\theta = 1$ can be written as $t(t-2)(t^2 - 4t - 9) = 0$. **(4 marks)**

b Hence find all solutions of $6\tan\theta + 12\sin\theta + \cos\theta = 1$ in the range $0 \leqslant \theta \leqslant 2\pi$ to 2 decimal places. **(2 marks)**

(E) **18** **a** Using the substitution $t = \tan\frac{x}{2}$, show that the equation $5\cot x + 4\operatorname{cosec} x = \frac{9}{4}$ can be written as $2t^2 + 9t - 18 = 0$. **(4 marks)**

b Hence find all solutions of $5\cot x + 4\operatorname{cosec} x = \frac{9}{4}$ in the range $0 \leqslant x \leqslant 2\pi$ to 2 decimal places. **(2 marks)**

19 The graph below shows how arterial blood pressure varies over time in humans.

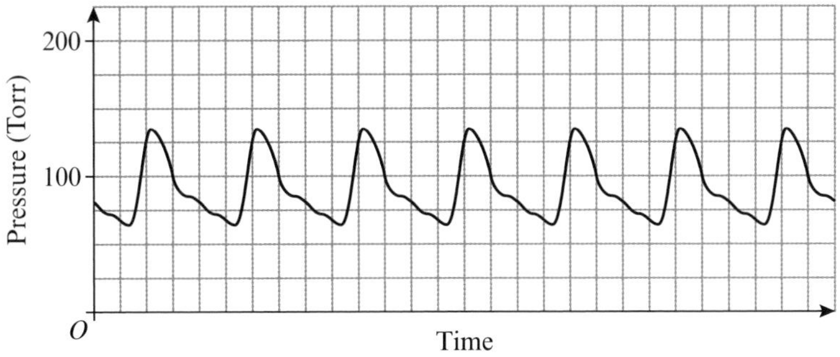

Ursula is trying to model blood pressure mathematically, and uses the following function to describe blood pressure at time x seconds.

$$\text{p}(x) = 8\sin 5x + 16\cos 5x - 4\sin 10x + \frac{16}{3}\cos 10x + 100.$$

The graph of $y = \text{p}(x)$ is shown below.

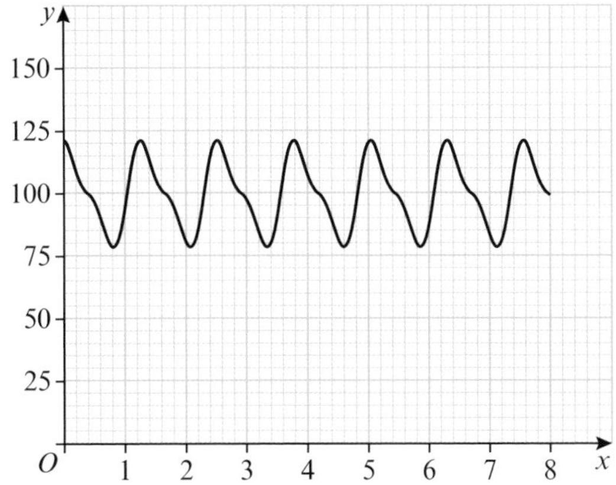

a Using the t-formulae, show that $\dfrac{\text{d}p}{\text{d}x} = \dfrac{-80t(t+2)(3t^2 - 8t + 7)}{3(1+t^2)^2}$ **(5 marks)**

b This model is very simple. What might it fail to take into account? **(1 mark)**

c Using the figure and the result from part **a**, find the time in seconds of the first pressure low-point in the model. **(3 marks)**

Challenge

a Given that $\tan\frac{\theta}{2} = \frac{1}{4}$, find the values of $\tan\theta$, $\sin\theta$ and $\cos\theta$ as fractions in their lowest terms.

b Hence construct a right-angled triangle with integer sides and acute angle θ.

c Given that $\tan\frac{\theta}{2}$ is a rational number between 0 and 1, show that it is possible to construct a right-angled triangle with integer sides and acute angle θ.

A **Pythagorean triple** is a set of three positive integers a, b and c, such that $a^2 + b^2 = c^2$. A Pythagorean triple is **primitive** if a, b and c have no common factors.

d Prove that there are infinitely many primitive Pythagorean triples.

Summary of key points

1 The t-formulae are a set of formulae that allow you to express $\sin\theta$, $\cos\theta$ and $\tan\theta$ in terms of $t = \tan\frac{\theta}{2}$

- $\sin\theta = \frac{2t}{1+t^2}$
- $\cos\theta = \frac{1-t^2}{1+t^2}$
- $\tan\theta = \frac{2t}{1-t^2}$

2 You can use the t-formulae to prove trigonometric identities.

3 To solve trigonometric equations using the t-formulae:

- use the substitution $t = \tan\frac{\theta}{2}$
- write any trigonometric functions in the equation in terms of t
- solve the resulting equation algebraically to find the value(s) of t
- find corresponding values of θ which satisfy the original equation.

4 Models involving different trigonometric functions can be simplified and analysed using the t-formulae.

Taylor series

6

Objectives

After completing this chapter you should be able to:

- Derive and use Taylor series for simple functions → pages 132–135
- Use series expansions to evaluate limits → pages 135–139
- Use the Taylor series method to find a series solution to a differential equation → pages 139–143

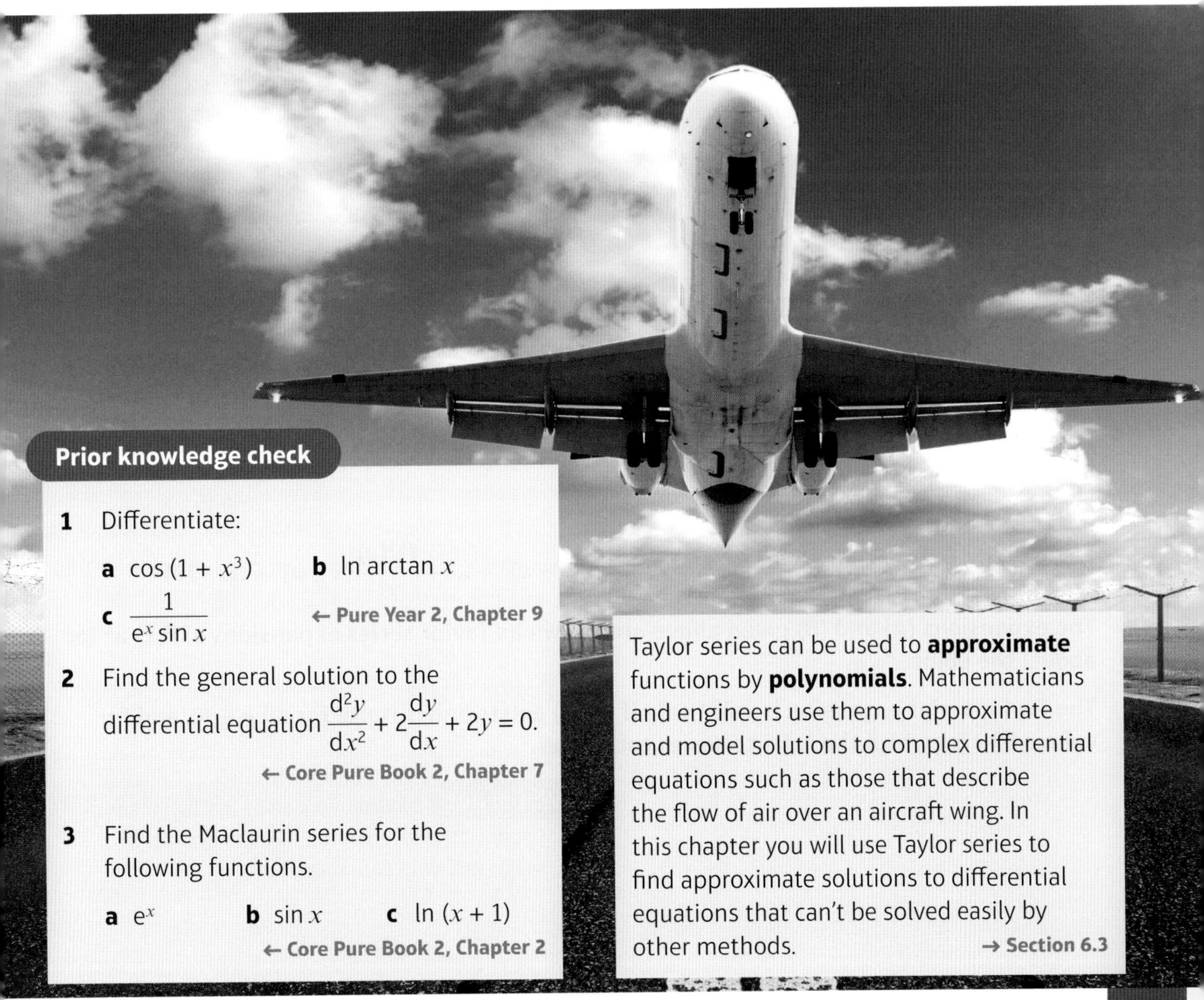

Prior knowledge check

1 Differentiate:

a $\cos(1 + x^3)$ **b** $\ln \arctan x$

c $\dfrac{1}{e^x \sin x}$ ← Pure Year 2, Chapter 9

2 Find the general solution to the differential equation $\dfrac{d^2y}{dx^2} + 2\dfrac{dy}{dx} + 2y = 0$.

← Core Pure Book 2, Chapter 7

3 Find the Maclaurin series for the following functions.

a e^x **b** $\sin x$ **c** $\ln(x + 1)$

← Core Pure Book 2, Chapter 2

Taylor series can be used to **approximate** functions by **polynomials**. Mathematicians and engineers use them to approximate and model solutions to complex differential equations such as those that describe the flow of air over an aircraft wing. In this chapter you will use Taylor series to find approximate solutions to differential equations that can't be solved easily by other methods. → Section 6.3

6.1 Taylor series

A In Core Pure Book 2 you used Maclaurin series expansions to write a function of x as an infinite series in ascending powers of x. However, the conditions of the Maclaurin series expansion mean that some functions, such as $\ln x$, cannot be expanded in this way.

Links The Maclaurin series expansion requires that $f^{(n)}(0)$ exists and is finite for all $n \in \mathbb{N}$.
If $f(x) = \ln x$ then $f'(x) = \frac{1}{x}$ so $f'(0)$ is undefined.
← **Core Pure Book 2, Section 2.3**

The construction of the Maclaurin series expansion focuses on $x = 0$ and, for a value of x very close to 0, a few terms of the series may well give a good approximation of the function.

For values of x further away from 0, even if they are in the interval of validity, more and more terms of the series are required to give a good degree of accuracy.

Note An extreme example of this is in using $x = 1$ in the series for $\ln(1 + x)$ to find $\ln 2$; thousands of terms of the series are required to reach 4 significant figure accuracy.

To overcome these problems, a series expansion focusing on $x = a$ can be derived.

This series expansion, called a **Taylor series**, is a more general form of the Maclaurin series.

Consider the functions f and g, where $f(x + a) \equiv g(x)$.

Note For example, $f(x) = \ln x$, $g(x) = \ln(x + 1)$

Then $f^{(r)}(x + a) = g^{(r)}(x)$, $r = 1, 2, 3\dots$.

In particular, $f^{(r)}(a) = g^{(r)}(0)$, $r = 1, 2, 3\dots$

So the Maclaurin series expansion for g,

$$g(x) = g(0) + g'(0)x + \frac{g''(0)}{2!}x^2 + \frac{g'''(0)}{3!}x^3 + \dots + \frac{g^{(r)}(0)}{r!}x^r + \dots$$

becomes

- $$\mathbf{f(x + a) = f(a) + f'(a)x + \frac{f''(a)}{2!}x^2 + \frac{f'''(a)}{3!}x^3 + \dots + \frac{f^{(r)}(a)}{r!}x^r + \dots \quad (A)}$$

Note The Taylor series allows you to approximate the value of $f(x)$ close to $x = a$.

Replacing x by $x - a$, gives a second useful form:

- $$\mathbf{f(x) = f(a) + f'(a)(x - a) + \frac{f''(a)}{2!}(x - a)^2 + \frac{f'''(a)}{3!}(x - a)^3 + \dots + \frac{f^{(r)}(a)}{r!}(x - a)^r + \dots \quad (B)}$$

The expansions (A) and (B) given above are known as Taylor series expansions of $f(x)$ at (or about) the point $x = a$.

The Taylor series expansion is valid only if $f^{(n)}(a)$ exists and is finite for all $n \in \mathbb{N}$, and for values of x for which the infinite series converges.

Watch out Neither version of the Taylor series expansion is given in the formula booklet so make sure you learn them both.

Example 1

A

Find the Taylor series expansion of e^{-x} in powers of $(x + 4)$ up to and including the term in $(x + 4)^3$.

Let $f(x) = e^{-x}$ and $a = -4$.

$$f(x) = f(-4) + f'(-4)(x+4) + \frac{f''(-4)}{2!}(x+4)^2 + \frac{f'''(-4)}{3!}(x+4)^3 + \dots$$

Use the Taylor series expansion (B).

$$f(x) = e^{-x} \Rightarrow f(-4) = e^4$$
$$f'(x) = -e^{-x} \Rightarrow f'(-4) = -e^4$$
$$f''(x) = e^{-x} \Rightarrow f''(-4) = e^4$$
$$f'''(x) = -e^{-x} \Rightarrow f'''(-4) = -e^4$$

You need to find $f(-4)$, $f'(-4)$, $f''(-4)$ and $f'''(-4)$.

Substituting the values in the series expansion gives

$$e^{-x} = e^4 - e^4(x+4) + \frac{e^4}{2!}(x+4)^2 - \frac{e^4}{3!}(x+4)^3 + \dots$$
$$e^{-x} = e^4\left(1 - (x+4) + \tfrac{1}{2}(x+4)^2 - \tfrac{1}{6}(x+4)^3 + \dots\right)$$

Take a factor of e^4 out of each term on the right-hand side.

Example 2

Express $\tan\left(x + \frac{\pi}{4}\right)$ as a series in ascending powers of x up to and including the term x^3.

Let $f(x) = \tan x$, then $\tan\left(x + \frac{\pi}{4}\right) = f\left(x + \frac{\pi}{4}\right)$.

You need to use the Taylor series expansion (A) with $f(x) = \tan x$ and $a = \frac{\pi}{4}$

$$f(x) = \tan x \Rightarrow f\left(\frac{\pi}{4}\right) = 1$$
$$f'(x) = \sec^2 x \Rightarrow f'\left(\frac{\pi}{4}\right) = 2$$
$$f''(x) = 2 \times \sec x \times (\sec x \tan x)$$
$$= 2 \times \sec^2 x \times \tan x \Rightarrow f''\left(\frac{\pi}{4}\right) = 2 \times 2 \times 1 = 4$$
$$f'''(x) = 2 \times \sec^2 x \times \sec^2 x + 2 \times \tan x\,(2 \times \sec^2 x \tan x)$$
$$\Rightarrow f'''\left(\frac{\pi}{4}\right) = 2 \times 2 \times 2 + 2 \times 4 = 16$$

Using $f(x + a) = f(a) + f'(a)x + \frac{f''(a)}{2!}x^2 + \frac{f'''(a)}{3!}x^3 + \dots$

$$\tan\left(x + \frac{\pi}{4}\right) = 1 + 2x + \frac{4}{2!}x^2 + \frac{16}{3!}x^3 + \dots$$
$$= 1 + 2x + 2x^2 + \tfrac{8}{3}x^3 + \dots$$

Online Explore the Taylor series expansion of $f(x) = \tan x$ using GeoGebra.

Watch out Make sure you simplify your coefficients as much as possible.

Example 3

a Show that the Taylor series about $\frac{\pi}{6}$ of $\sin x$ in ascending powers of $\left(x - \frac{\pi}{6}\right)$ up to and including the term $\left(x - \frac{\pi}{6}\right)^2$ is $\sin x = \frac{1}{2} + \frac{\sqrt{3}}{2}\left(x - \frac{\pi}{6}\right) - \frac{1}{4}\left(x - \frac{\pi}{6}\right)^2$

b Using the series in part **a** find, in terms of π, an approximation for $\sin 40°$.

A

a $f(x) = \sin x$, $f'(x) = \cos x$, $f''(x) = -\sin x$,

so $f\left(\frac{\pi}{6}\right) = \frac{1}{2}$, $f'\left(\frac{\pi}{6}\right) = \frac{\sqrt{3}}{2}$, $f''\left(\frac{\pi}{6}\right) = -\frac{1}{2}$ — Find f(a), f′(a) and f″(a) where $a = \frac{\pi}{6}$

so $\sin x = \frac{1}{2} + \frac{\sqrt{3}}{2}\left(x - \frac{\pi}{6}\right) - \frac{1}{2 \times 2!}\left(x - \frac{\pi}{6}\right)^2 - \ldots$ — Substitute into Taylor series expansion (B) with $a = \frac{\pi}{6}$

$= \frac{1}{2} + \frac{\sqrt{3}}{2}\left(x - \frac{\pi}{6}\right) - \frac{1}{4}\left(x - \frac{\pi}{6}\right)^2 - \ldots$

b $\sin 40° = \sin\left(\frac{2\pi}{9}\right)$, so substituting $x = \frac{2\pi}{9}$

in to the series from part **a** gives

$\sin 40° \approx \frac{1}{2} + \frac{\sqrt{3}}{2}\left(\frac{\pi}{18}\right) - \frac{1}{4}\left(\frac{\pi}{18}\right)^2$

$\approx \frac{1}{2} + \frac{\pi\sqrt{3}}{36} - \frac{\pi^2}{1296}$ — The percentage error in this approximation is about 0.1%.

Exercise 6A

1 a Find the Taylor series expansion of $\sqrt{x}$ in ascending powers of $(x - 1)$ as far as the term in $(x - 1)^4$.

b Use your answer in a to obtain an estimate for $\sqrt{1.2}$, giving your answer to 3 decimal places.

2 Use Taylor series expansion to express each the following as a series in ascending powers of $(x - a)$ as far as the term in $(x - a)^k$, for the given values of a and k.

a $\ln x$ $(a = \mathrm{e}, k = 2)$ **b** $\tan x\left(a = \frac{\pi}{3}, k = 3\right)$ **c** $\cos x (a = 1, k = 4)$

3 a Use Taylor series expansion to express each of the following as a series in ascending powers of x as far as the term in x^4.

i $\cos\left(x + \frac{\pi}{4}\right)$ **ii** $\ln(x + 5)$ **iii** $\sin\left(x - \frac{\pi}{3}\right)$

b Use your result in **ii** to find an approximation for ln 5.2, giving your answer to 4 significant figures.

E **4** Given that $y = x\mathrm{e}^x$,

a show that $\frac{\mathrm{d}^n y}{\mathrm{d}x^n} = (n + x)\mathrm{e}^x$ **(3 marks)**

b find the Taylor series expansion of $x\mathrm{e}^x$ in ascending powers of $(x + 1)$ up to and including the term in $(x + 1)^4$. **(3 marks)**

E **5 a** Find the Taylor series for $x^3 \ln x$ in ascending powers of $(x - 1)$ up to and including the term in $(x - 1)^4$. **(4 marks)**

b Using your series from part **a**, find an approximation for ln 1.5, giving your answer to 4 decimal places. **(2 marks)**

A 6 Find the Taylor series expansion of $\tan(x - \alpha)$ about 0, where $\alpha = \arctan\left(\frac{3}{4}\right)$, in ascending powers of x up to and including the term in x^2. **(4 marks)**

7 Find the Taylor series expansion of $\sin 2x$ about $\frac{\pi}{6}$ in ascending powers of $\left(x - \frac{\pi}{6}\right)$ up to and including the term in $\left(x - \frac{\pi}{6}\right)^4$. **(4 marks)**

8 Given that $y = \frac{1}{\sqrt{(1 + x)}}$,

a find the values of $\frac{dy}{dx}$ and $\frac{d^2y}{dx^2}$ when $x = 3$ **(3 marks)**

b find the Taylor series of $\frac{1}{\sqrt{(1 + x)}}$, in ascending powers of $(x - 3)$ up to and including the term in $(x - 3)^2$. **(4 marks)**

9 Find the Taylor series expansion of $\cosh x$ about $x = \ln 5$ in ascending powers of $(x - \ln 5)$ up to and including the term in $(x - \ln 5)^4$. **(5 marks)**

P 10 a Given that the coefficient of $(x - \ln 2)$ in the Taylor series expansion of $\sinh ax$ about $\ln 2$ is $\frac{17}{4}$, find the value of a. **(3 marks)**

b Find the Taylor series of $\sinh ax$ about $\ln 2$ in terms up to the term $(x - \ln 2)^3$. **(3 marks)**

P 11 Show that the Taylor series of $\ln x$ in powers of $(x - 2)$ is

$$\ln 2 + \sum_{n=1}^{\infty}(-1)^{n-1}\frac{(x-2)^n}{n2^n}$$

(6 marks)

Challenge

a Find the Taylor series expansion of $\ln(\cos 2x)$ about π in ascending powers of $(x - \pi)$ up to and including the term in $(x - \pi)^4$.

b Hence obtain an estimate for $\ln\left(\frac{\sqrt{3}}{2}\right)$.

6.2 Finding limits

In your A level maths course, you considered **limits** of a function as x approaches 0, or infinity. By looking at how different parts of the function behave, you can evaluate the limit. Here is a very simple example:

$$\lim_{h\to 0}(2 + h) = 2$$

It is clear that as h gets closer to 0, $(2 + h)$ gets closer to 2.

Links Results like this are used when differentiating from first principles.
← Pure Year 1, Chapter 12

In general, we say that $f(x) \to L$ as $x \to a$ if we can make $f(x)$ arbitrarily close to L by choosing a value of x sufficiently close to a. If this is possible then we write

$$\lim_{x\to a} f(x) = L$$

Notation L is the limit of $f(x)$ as x approaches a. You sometimes say that 'f(x) tends to L as x tends to a'.

You can use some simple properties of limits to evaluate certain limits. These rules are sometimes referred to as the **algebra of limits**:

- **Given $\lim_{x \to a} f(x) = L$ and $\lim_{x \to a} g(x) = M$, then:**
  - **$\lim_{x \to a} (f(x) + g(x)) = L + M$**
  - **If c is a constant, then $\lim_{x \to a} cf(x) = cL$**
  - **$\lim_{x \to a} f(x)g(x) = LM$**
  - **If $M \neq 0$, then $\lim_{x \to a} \frac{f(x)}{g(x)} = \frac{L}{M}$**

Example 4

Find $\lim_{x \to 0} \frac{5 - x}{2 + x}$

$\lim_{x \to 0} (5 - x) = 5$ and $\lim_{x \to 0} (2 + x) = 2$

So $\lim_{x \to 0} \frac{5 - x}{2 + x} = \frac{5}{2}$

Use $\lim_{x \to a} \frac{f(x)}{g(x)} = \frac{L}{M}$

Example 5

Find $\lim_{x \to \infty} \frac{2 - 3x}{1 + x}$

$$\lim_{x \to \infty} \frac{2 - 3x}{1 + x} = \lim_{x \to \infty} \left(\frac{\frac{2}{x} - 3}{\frac{1}{x} + 1} \right)$$

$$= \frac{\lim_{x \to \infty} \left(\frac{2}{x} - 3 \right)}{\lim_{x \to \infty} \left(\frac{1}{x} + 1 \right)} = \frac{-3}{1} = -3$$

Problem-solving

$2 - 3x \to -\infty$ and $1 + x \to \infty$ as $x \to \infty$, so it is not possible to evaluate the limit directly. However, by dividing each term in the numerator and denominator by x, you can determine the limit.

In many cases, the above methods will not allow you to calculate a limit. Suppose you wanted to find $\lim_{x \to 0} \frac{\sin x}{x}$. Both the numerator and denominator tend to 0 as $x \to 0$, so you cannot use $\lim_{x \to 0} \frac{f(x)}{g(x)} = \frac{L}{M}$

Notation When a function tends to $\frac{0}{0}$, it is known as an **indeterminate form**. Other examples are functions which tend to $\frac{\infty}{\infty}$, $0 \times \infty$ or 0^0.

In order to evaluate this limit, we need a more precise way of comparing $\sin x$ and x for values of x close to 0. You can use a Maclaurin series to do this. The Maclaurin series expansion (or in other words, the Taylor series expansion at $x = 0$) of $\sin x$ is

$$\sin x = x - \frac{x^3}{3!} + \frac{x^5}{5!} - \frac{x^7}{7!} + \dots$$

A Therefore we have

$$\frac{\sin x}{x} = 1 - \frac{x^2}{3!} + \frac{x^4}{5!} - \ldots$$

You can cancel a factor of x from each term.

and so

$$\lim_{x \to 0} \frac{\sin x}{x} = \lim_{x \to 0}\left(1 - \frac{x^2}{3!} + \frac{x^4}{5!} - \ldots\right)$$

Each of the terms containing a positive power of x tends to 0 as $x \to 0$, so $\lim_{x \to 0} \frac{\sin x}{x} = 1$.

Example 6

Find $\lim_{x \to 0} \frac{\sin x - x}{x^3}$

$$\sin x = x - \frac{x^3}{3!} + \frac{x^5}{5!} - \ldots$$

This is the Taylor series expansion of x at $x = 0$, or the Maclaurin series expansion.

$$\text{So } \lim_{x \to 0} \frac{\sin x - x}{x^3} = \lim_{x \to 0}\left(\frac{\frac{-x^3}{3!} + \frac{x^5}{5!} - \ldots}{x^3}\right)$$

$$= \lim_{x \to 0}\left(-\frac{1}{3!} + \frac{x^2}{5!} - \ldots\right)$$

Subtracting x in the numerator leaves terms in x^3 and higher. You can now cancel a factor of x^3 from each term to leave you with one constant term, and terms in positive powers of x, which all tend to 0.

$$\text{Therefore } \lim_{x \to 0} \frac{\sin x - x}{x^3} = -\frac{1}{6}$$

Example 7

Find $\lim_{x \to \frac{\pi}{2}}\left(x - \frac{\pi}{2}\right)\tan x$

Problem-solving

You cannot expand $\tan x$ about $x = \frac{\pi}{2}$, so rewrite the function as a quotient using the fact that $\tan x = \frac{1}{\cot x}$

$$\left(x - \frac{\pi}{2}\right)\tan x = \frac{x - \frac{\pi}{2}}{\cot x}$$

$f(x) = \cot x$

Then $f'(x) = -\operatorname{cosec}^2 x$, $f''(x) = 2\cot x \operatorname{cosec}^2 x$,

$f'''(x) = -2(\cos 2x + 2)\operatorname{cosec}^4 x$

$$\cot x = f\left(\frac{\pi}{2}\right) + f'\left(\frac{\pi}{2}\right)\left(x - \frac{\pi}{2}\right) + \frac{f''\left(\frac{\pi}{2}\right)}{2!}\left(x - \frac{\pi}{2}\right)^2 + \frac{f'''\left(\frac{\pi}{2}\right)}{3!}\left(x - \frac{\pi}{2}\right)^3$$

$$= 0 - \left(x - \frac{\pi}{2}\right) + \frac{0}{2!}\left(x - \frac{\pi}{2}\right)^2 - \frac{2}{3!}\left(x - \frac{\pi}{2}\right)^3 + \ldots$$

$$= -\left(x - \frac{\pi}{2}\right) - \frac{1}{3}\left(x - \frac{\pi}{2}\right)^3 + \ldots$$

Find the Taylor series expansion of $\cot x$ about $x = \frac{\pi}{2}$

The subsequent terms will have higher powers of $\left(x - \frac{\pi}{2}\right)$

$$\lim_{x \to \frac{\pi}{2}}\left(x - \frac{\pi}{2}\right)\tan x = \lim_{x \to \frac{\pi}{2}}\left(\frac{x - \frac{\pi}{2}}{-\left(x - \frac{\pi}{2}\right) - \frac{1}{3}\left(x - \frac{\pi}{2}\right)^3 + \ldots}\right)$$

$$= \lim_{x \to \frac{\pi}{2}}\left(\frac{1}{-1 - \frac{1}{3}\left(x - \frac{\pi}{2}\right)^2 + \ldots}\right) = -1$$

Example 8

A

Find $\lim_{x \to 1} \dfrac{3 \ln x}{x^2 + 2x - 3}$

$f(x) = \ln x$ then $f'(x) = \frac{1}{x}$, $f''(x) = -\frac{1}{x^2}$, $f'''(x) = \frac{2}{x^3}$

$\ln x = f(1) + f'(1)(x - 1) + \frac{f''(1)}{2!}(x - 1)^2 + \ldots$

$= (x - 1) - \frac{1}{2}(x - 1)^2 + \ldots$

Calculate the Taylor series expansion of $\ln x$ around $x = 1$.

Also $x^2 + 2x - 3 = (x - 1)(x + 3)$

Factorise the denominator and cancel $(x - 1)$ in the numerator and denominator of each term.

Hence $\lim_{x \to 1} \dfrac{3 \ln x}{x^2 + 2x - 3} = \lim_{x \to 1} \left(\dfrac{3}{x + 3} - \dfrac{3(x - 1)}{2(x + 3)} + \ldots \right) = \dfrac{3}{4}$

The second and subsequent terms will all contain a factor of $(x - 1)$, which $\to 0$ as $x \to 1$.

Exercise 6B

1 Evaluate the following limits.

a $\lim_{x \to 0} \dfrac{7 + x}{5 - x}$ **b** $\lim_{x \to 0} \dfrac{3 - 2x}{x + 2}$ **c** $\lim_{x \to \infty} \dfrac{4 - 2x}{2 + x}$ **d** $\lim_{x \to \infty} \dfrac{4x + 1}{3 + 2x}$

Hint In parts **c** and **d**, divide the numerator and denominator by x.

2 Evaluate the following limits.

a $\lim_{x \to 0} \dfrac{\sin 4x}{x}$ **b** $\lim_{x \to 0} \dfrac{\cos x - 1}{x^2}$ **c** $\lim_{x \to 0} \dfrac{x}{e^{3x} - 1}$ **d** $\lim_{x \to 0} \dfrac{x}{\arctan 4x}$

3 Evaluate the following limits.

a $\lim_{x \to \pi} \dfrac{x - \pi}{\sin x}$ **b** $\lim_{x \to 2} \dfrac{\sin(x^2 - 4)}{x - 2}$

P **4** Evaluate the following limits.

a $\lim_{x \to 0} \dfrac{\ln(1 + x^2)}{x^2}$ **b** $\lim_{x \to 1} \dfrac{\ln x}{\sqrt{x} - 1}$

c $\lim_{x \to 0} \dfrac{e^x - e^{-x} - 2x}{x^2 - x \ln(1 + x)}$ **d** $\lim_{x \to 0} \dfrac{e^{x^2} \sin x - x}{x \ln(1 + x^2)}$

E/P **5** **a** Find the Taylor series expansions about $x = 0$ of $\sin x$ and e^{-x}. **(4 marks)**

b Hence evaluate $\lim_{x \to 0} \left(\dfrac{1}{\sin x} - \dfrac{1}{1 - e^{-x}} \right)$ **(4 marks)**

E/P **6** **a** Find the Taylor series expansions about $x = 1$ of $\ln x$ and $\sqrt{x}$. **(4 marks)**

b Hence evaluate $\lim_{x \to 1} \left(\dfrac{1}{\sqrt{x}} - \dfrac{1}{\ln x - 1} \right)$ **(4 marks)**

E/P **7** **a** Find the Taylor series expansion about $x = 0$ of $\sinh x$ up to the term in x^5. **(4 marks)**

b Hence find $\lim_{x \to 0} (2x \operatorname{cosech} 3x)$. **(4 marks)**

E/P **8** **a** Find the Taylor series expansion of $\sqrt{1 + 4x}$ at $x = 2$ up to the term in $(x - 2)^2$. **(4 marks)**

b Hence find $\lim_{x \to 2} \dfrac{\sqrt{1 + 4x} - 3}{x^4 - 2x^2 - 8}$ **(4 marks)**

Challenge

A **a** Find the Taylor series expansion of $\sqrt{1+5y}$ about $y = 0$ up to the term in y^3.

b Using part **a** and the substitution $y = \frac{1}{x}$, find $\lim_{x\to\infty} \sqrt{x^2+5x} - x$.

6.3 Series solutions of differential equations

You can use Taylor series to find **series solutions** of differential equations that can't be solved using other techniques. This can allow you to find useful approximate solutions, and to find solutions that cannot be expressed using elementary functions.

Links You can use integrating factors or auxiliary equations to solve some first and second-order differential equations directly.
← Core Pure Book 2, Chapter 7

Suppose you have a first-order differential equation of the form $\frac{dy}{dx} = f(x, y)$ and know the initial condition that at $x = x_0$, $y = y_0$, then you can calculate $\left.\frac{dy}{dx}\right|_{x_0}$ by substituting x_0 and y_0 into the original differential equation.

Watch out $f(x, y)$ denotes a function of both x and y, such as $x^2y + 1$, or e^{xy}. Such functions cannot always be written as a product of functions $g(x)h(y)$.

Notation $\left.\frac{dy}{dx}\right|_{x_0}$ is used to denote the value of $\frac{dy}{dx}$ when $x = x_0$.

By successive differentiation of the original differential equation, the values of $\left.\frac{d^2y}{dx^2}\right|_{x_0}$, $\left.\frac{d^3y}{dx^3}\right|_{x_0}$ and so on can be found by substituting previous results into the derived equations.

- **The series solution to the differential equation $\frac{dy}{dx} = f(x, y)$ is found using the Taylor series expansion in the form**

$$y = y_0 + (x - x_0)\left.\frac{dy}{dx}\right|_{x_0} + \frac{(x-x_0)^2}{2!}\left.\frac{d^2y}{dx^2}\right|_{x_0} + \frac{(x-x_0)^3}{3!}\left.\frac{d^3y}{dx^3}\right|_{x_0} + \dots \qquad \textbf{(C)}$$

- **In the situation where $x_0 = 0$, this reduces to the Maclaurin series**

$$y = y_0 + x\left.\frac{dy}{dx}\right|_0 + \frac{x^2}{2!}\left.\frac{d^2y}{dx^2}\right|_0 + \frac{x^3}{3!}\left.\frac{d^3y}{dx^3}\right|_0 + \dots \qquad \textbf{(D)}$$

Second order, and higher differential equations can be solved in the same manner.

Example 9

A Use the Taylor series method to find a series solution, in ascending powers of x up to and including the term in x^3, of

$$\frac{d^2y}{dx^2} = y - \sin x$$

given that when $x = 0$, $y = 1$ and $\frac{dy}{dx} = 2$.

The given conditions are $x_0 = 0$, $y_0 = 1$, $\left.\frac{dy}{dx}\right|_0 = 2$

Substituting $x_0 = 0$ and $y_0 = 1$, into $\frac{d^2y}{dx^2} = y - \sin x$

gives $\left.\frac{d^2y}{dx^2}\right|_0 = 1 - \sin 0 = 1$

First find $\left.\frac{d^2y}{dx^2}\right|_0$

$$\frac{d^3y}{dx^3} = \frac{dy}{dx} - \cos x \qquad (1)$$

Differentiate the given differential equation with respect to x.

Substituting $x_0 = 0$ and $\left.\frac{dy}{dx}\right|_0 = 2$ into (1)

gives $\left.\frac{d^3y}{dx^3}\right|_0 = 2 - \cos 0 = 1.$

Find $\left.\frac{d^3y}{dx^3}\right|_0$

Substituting the results into

$$y = y_0 + x\left.\frac{dy}{dx}\right|_0 + \frac{x^2}{2!}\left.\frac{d^2y}{dx^2}\right|_0 + \frac{x^3}{3!}\left.\frac{d^3y}{dx^3}\right|_0 + \ldots$$

gives $y = 1 + x \times 2 + \frac{x^2}{2!} \times 1 + \frac{x^3}{3!} \times 1 + \ldots$

$$= 1 + 2x + \frac{x^2}{2} + \frac{x^3}{6} + \ldots$$

Example 10

Given that $\frac{d^2y}{dx^2} + 2\frac{dy}{dx} = xy$ and that $y = 1$ and $\frac{dy}{dx} = 2$, at $x = 1$, express y as a series in ascending powers of $(x - 1)$ up to and including the term in $(x - 1)^4$.

The given conditions are $x_0 = 1$, $y_0 = 1$, $\left.\frac{dy}{dx}\right|_1 = 2$

You need to find $\left.\frac{d^2y}{dx^2}\right|_1$, $\left.\frac{d^3y}{dx^3}\right|_1$ and $\left.\frac{d^4y}{dx^4}\right|_1$.

Substituting $x_0 = 1$, $y_0 = 1$ and $\left.\frac{dy}{dx}\right|_1 = 2$ into

$\frac{d^2y}{dx^2} + 2\frac{dy}{dx} = xy$ gives $\left.\frac{d^2y}{dx^2}\right|_1 = -3$

A

$$\frac{d^3y}{dx^3} + 2\frac{d^2y}{dx^2} = y + x\frac{dy}{dx} \qquad (1)$$

Differentiate the given equation with respect to x.

Substituting $x_0 = 1$, $y_0 = 1$, $\left.\frac{dy}{dx}\right|_1 = 2$ and $\left.\frac{d^2y}{dx^2}\right|_1 = -3$ into (1)

gives $\left.\frac{d^3y}{dx^3}\right|_1 = 9$

$$\frac{d^4y}{dx^4} + 2\frac{d^3y}{dx^3} = \frac{dy}{dx} + x\frac{d^2y}{dx^2} + \frac{dy}{dx} \qquad (2)$$

Watch out The initial conditions are given at $x_0 = 1$ so you need to make sure you expand about this point in order to use the series solution.

Substituting $x_0 = 1$, $\left.\frac{dy}{dx}\right|_1 = 2$, $\left.\frac{d^2y}{dx^2}\right|_1 = -3$ and $\left.\frac{d^3y}{dx^3}\right|_1 = 9$

into (2) gives $\left.\frac{d^4y}{dx^4}\right|_1 = -17$

Substituting all the values into

$$y = y_0 + (x - x_0)\left.\frac{dy}{dx}\right|_1 + \frac{(x - x_0)^2}{2!}\left.\frac{d^2y}{dx^2}\right|_1 + \frac{(x - x_0)^3}{3!}\left.\frac{d^3y}{dx^3}\right|_1 + \ldots$$

Then use the Taylor series expansion (C).

gives $y = 1 + 2(x - 1) + \frac{-3}{2!}(x - 1)^2 + \frac{9}{3!}(x - 1)^3 + \frac{-17}{4!}(x - 1)^4 + \ldots$

$$y = 1 + 2(x - 1) - \tfrac{3}{2}(x - 1)^2 + \tfrac{3}{2}(x - 1)^3 - \tfrac{17}{24}(x - 1)^4 + \ldots$$

Example 11

Given that y satisfies the differential equation $\frac{dy}{dx} = y^2 - x$ and that $y = 1$ at $x = 0$, find a series solution for y in ascending powers of x up to and including the term in x^3.

The given conditions are $x_0 = 0$, $y_0 = 1$.

Substituting $x_0 = 0$ and $y_0 = 1$ into $\frac{dy}{dx} = y^2 - x$

gives $\left.\frac{dy}{dx}\right|_0 = 1^2 - 0 = 1$

You need to find $\left.\frac{d^2y}{dx^2}\right|_0$, $\left.\frac{d^3y}{dx^3}\right|_0$ and $\left.\frac{d^4y}{dx^4}\right|_0$.

$$\frac{d^2y}{dx^2} = 2y\frac{dy}{dx} - 1 \qquad (1)$$

Differentiate the given equation with respect to x.

Substituting $y_0 = 1$ and $\left.\frac{dy}{dx}\right|_0 = 1$ into (1)

gives $\left.\frac{d^2y}{dx^2}\right|_0 = 2y_0\left.\frac{dy}{dx}\right|_0 - 1 = 2 \times 1 \times 1 - 1 = 1$

$$\frac{d^3y}{dx^3} = 2y\frac{d^2y}{dx^2} + 2\left(\frac{dy}{dx}\right)^2 \qquad (2)$$

Differentiate (1).

Substituting $y_0 = 1$, $\left.\frac{dy}{dx}\right|_0 = 1$ and $\left.\frac{d^2y}{dx^2}\right|_0 = 1$ into (2)

gives $\left.\frac{d^3y}{dx^3}\right|_0 = 2y_0\left.\frac{d^2y}{dx^2}\right|_0 + 2\left.\frac{dy}{dx}\right|_0^2 = 2 \times 1 \times 1 + 2 \times 1^2 = 4$

A

Substituting all of the values into

$$y = y_0 + x\left.\frac{dy}{dx}\right|_0 + \frac{x^2}{2!}\left.\frac{d^2y}{dx^2}\right|_0 + \frac{x^3}{3!}\left.\frac{d^3y}{dx^3}\right|_0 + \dots$$

Use Taylor series expansion (D).

gives $y = 1 + x + \frac{1}{2}x^2 + \frac{2}{3}x^3 + \dots$

Exercise 6C

E **1** Find a series solution, in ascending powers of x up to and including the term in x^4, for the differential equation $\frac{d^2y}{dx^2} = x + 2y$, given that at $x = 0$, $y = 1$ and $\frac{dy}{dx} = \frac{1}{2}$. **(8 marks)**

E **2** The variable y satisfies $(1 + x^2)\frac{d^2y}{dx^2} + x\frac{dy}{dx} = 0$ and at $x = 0$, $y = 0$ and $\frac{dy}{dx} = 1$.

Use the Taylor series method to find a series expansion for y in powers of x up to and including the term in x^3. **(8 marks)**

E **3** Given that y satisfies the differential equation $\frac{dy}{dx} + y - e^x = 0$, and that $y = 2$ at $x = 0$, find a series solution for y in ascending powers of x up to and including the term in x^3. **(6 marks)**

E **4** Use the Taylor series method to find a series solution for $\frac{d^2y}{dx^2} + x\frac{dy}{dx} + y = 0$, given that $x = 0$, $y = 1$ and $\frac{dy}{dx} = 2$, giving your answer in ascending powers of x up to and including the term in x^4. **(8 marks)**

E **5** The variable y satisfies the differential equation $\frac{d^2y}{dx^2} + 2\frac{dy}{dx} = 3xy$, and $y = 1$ and $\frac{dy}{dx} = -1$ at $x = 1$.

Express y as a series in powers of $(x - 1)$ up to and including the term in $(x - 1)^3$. **(8 marks)**

E **6** Find a series solution, in ascending powers of x up to and including the term x^4, to the differential equation $\frac{d^2y}{dx^2} + 2y\frac{dy}{dx} + y^3 = 1 + x$, given that at $x = 0$, $y = 1$ and $\frac{dy}{dx} = 1$. **(8 marks)**

E/P **7** $(1 + 2x)\frac{dy}{dx} = x + 2y^2$

a Show that $(1 + 2x)\frac{d^3y}{dx^3} + 4(1 - y)\frac{d^2y}{dx^2} = 4\left(\frac{dy}{dx}\right)^2$ **(4 marks)**

b Given that $y = 1$ at $x = 0$, find a series solution of $(1 + 2x)\frac{dy}{dx} = x + 2y^2$, in ascending powers of x up to and including the term in x^3. **(4 marks)**

A P **8** Find the series solution in ascending powers of $\left(x - \frac{\pi}{4}\right)$ up to and including the term in $\left(x - \frac{\pi}{4}\right)^2$ for the differential equation $\sin x \frac{dy}{dx} + y \cos x = y^2$ given that $y = \sqrt{2}$ at $x = \frac{\pi}{4}$ **(6 marks)**

P **9** The variable y satisfies the differential equation $\frac{dy}{dx} - x^2 - y^2 = 0$.

a Show that:

i $\frac{d^2y}{dx^2} - 2y\frac{dy}{dx} - 2x = 0$ **(2 marks)**

ii $\frac{d^3y}{dx^3} - 2y\frac{d^2y}{dx^2} - 2\left(\frac{dy}{dx}\right)^2 = 2$ **(2 marks)**

b Derive a similar equation involving $\frac{d^4y}{dx^4}, \frac{d^3y}{dx^3}, \frac{d^2y}{dx^2}, \frac{dy}{dx}$ and y. **(3 marks)**

c Given also that $y = 1$ at $x = 0$, express y as a series in ascending powers of x in powers of x up to and including the term in x^4. **(4 marks)**

P **10** Given that $\cos x \frac{dy}{dx} + y \sin x + 2y^3 = 0$, and that $y = 1$ at $x = 0$, use the Taylor series method to show that, close to $x = 0$, $y \approx 1 - 2x + \frac{11}{2}x^2 - \frac{56}{3}x^3$. **(6 marks)**

P **11** $\frac{d^2y}{dx^2} = 4x\frac{dy}{dx} - 2y \qquad (1)$

a Show that

$$\frac{d^5y}{dx^5} = px\frac{d^4y}{dx^4} + q\frac{d^3y}{dx^3},$$

where p and q are integers to be determined. **(4 marks)**

b Hence find a series solution, in ascending powers of $(x - 1)$ up to the term in x^5 of differential equation (1), given that $y = \frac{dy}{dx} = 2$ when $x = 1$. **(5 marks)**

Mixed exercise 6

1 Using Taylor series, show that the first three terms in the series expansion of $\left(x - \frac{\pi}{4}\right)\cot x$, in powers of $\left(x - \frac{\pi}{4}\right)$, are $\left(x - \frac{\pi}{4}\right) - 2\left(x - \frac{\pi}{4}\right)^2 + 2\left(x - \frac{\pi}{4}\right)^3$.

2 **a** For the function $f(x) = \ln(1 + e^x)$, find the values of $f'(0)$ and $f''(0)$.

b Show that $f'''(0) = 0$.

c Find the Taylor series expansion of $\ln(1 + e^x)$, in ascending powers of x up to and including the term in x^2.

A **3** **a** Write down the Taylor series for $\cos 4x$ in ascending powers of x, up to and including the term in x^6.

b Hence, or otherwise, show that the first three non-zero terms in the series expansion of $\sin^2 2x$ are $4x^2 - \frac{16}{3}x^4 + \frac{128}{45}x^6$.

P **4** Given that terms in x^5 and higher powers may be neglected, use the Taylor series for e^x and $\cos x$, to show that $e^{\cos x} \approx e\left(1 - \frac{x^2}{2} + \frac{x^4}{6}\right)$.

E **5** $\frac{dy}{dx} = 2 + x + \sin y$

a Given that $y = 0$, when $x = 0$, use the Taylor series method to obtain y as a series in ascending powers of x up to and including the term in x^3. **(5 marks)**

b Hence obtain an approximate value for y at $x = 0.1$. **(1 mark)**

E **6** Given that $|2x| < 1$, find the first two non-zero terms in the Taylor series expansion of $\ln((1 + x)^2(1 - 2x))$ in ascending powers of x. **(5 marks)**

E **7** Find the series solution, in ascending powers of x up to and including the term in x^3, of the differential equation $\frac{d^2y}{dx^2} - (x + 2)\frac{dy}{dx} + 3y = 0$, given that at $x = 0$, $y = 2$ and $\frac{dy}{dx} = 4$. **(5 marks)**

E/P **8** **a** Use differentiation and Maclaurin series expansion, to express $\ln(\sec x + \tan x)$ as a series in ascending powers of x up to and including the term in x^3. **(4 marks)**

b Hence find $\lim_{x \to 0} \frac{\sin x - \ln(\sec x + \tan x)}{x(\cos x - 1)}$ **(4 marks)**

P **9** Find an expression in terms of n for the nth term in the Taylor series expansion of $\cosh x$ about $\ln 2$ in the case when:

a n is even **b** n is odd

P **10** Find $\lim_{x \to \pi} \frac{(x - \pi)^2}{1 + \cos x}$

P **11** Find $\lim_{x \to 0} \frac{\arctan x - x}{\sin x - x}$

12 Show that the results of differentiating the following series expansions

$$e^x = 1 + x + \frac{x^2}{2!} + \frac{x^3}{3!} + \ldots + \frac{x^r}{r!} + \ldots,$$

$$\sin x = x - \frac{x^3}{3!} + \frac{x^5}{5!} - \frac{x^7}{7!} + \ldots + \frac{(-1)^r x^{2r+1}}{(2r+1)!} + \ldots$$

$$\cos x = 1 - \frac{x^2}{2!} + \frac{x^4}{4!} - \frac{x^6}{6!} + \ldots + (-1)^r \frac{x^{2r}}{(2r)!} + \ldots$$

agree with the results

a $\frac{d}{dx}(e^x) = e^x$ **b** $\frac{d}{dx}(\sin x) = \cos x$ **c** $\frac{d}{dx}(\cos x) = -\sin x$

A E 13 $\frac{d^2y}{dx^2} + y\frac{dy}{dx} = x$, and at $x = 1$, $y = 0$ and $\frac{dy}{dx} = 2$.

Find a series solution of the differential equation, in ascending powers of $(x - 1)$ up to and including the term in $(x - 1)^3$. **(8 marks)**

E 14 **a** Given that $\cos x = 1 - \frac{x^2}{2!} + \frac{x^4}{4!} - \ldots$, show that $\sec x = 1 + \frac{x^2}{2} + \frac{5}{24}x^4 + \ldots$. **(4 marks)**

b Using the result found in part **a**, and given that $\sin x = x - \frac{x^3}{3!} + \frac{x^5}{5!} - \ldots$, find the first three non-zero terms in the series expansion, in ascending powers of x, for $\tan x$. **(4 marks)**

c Find $\lim_{x \to 0} \frac{\tan x}{e^{2x} - 1}$ **(4 marks)**

E/P 15 **a** By using the Taylor series expansions of e^x and $\cos x$, or otherwise, find the expansion of $e^x \cos 3x$ in ascending powers of x up to and including the term in x^3. **(4 marks)**

b Hence find $\lim_{x \to 0} \frac{e^x \cos 3x - \sin x - \cos x}{x^3 + 2x^2}$ **(4 marks)**

E 16 $\frac{d^2y}{dx^2} + x^2\frac{dy}{dx} + y = 0$ with $y = 2$ at $x = 0$ and $\frac{dy}{dx} = 1$ at $x = 0$.

a Use the Taylor series method to express y as a polynomial in x up to and including the term in x^3. **(4 marks)**

b Show that at $x = 0$, $\frac{d^4y}{dx^4} = 0$. **(3 marks)**

E 17 **a** Find the first three derivatives of $(1 + x)^2 \ln(1 + x)$. **(4 marks)**

b Hence, or otherwise, find the Taylor series expansion of $(1 + x)^2 \ln(1 + x)$ in ascending powers of x up to and including the term in x^3. **(4 marks)**

E 18 **a** Expand $\ln(1 + \sin x)$ in ascending powers of x up to and including the term in x^4. **(6 marks)**

b Hence find an approximation for $\int_0^{\frac{\pi}{6}} \ln(1 + \sin x)\,dx$ giving your answer to 3 decimal places. **(3 marks)**

E/P 19 **a** Using the first two terms, $x + \frac{x^3}{3}$, in the Taylor series of $\tan x$, show that

$e^{\tan x} = 1 + x + \frac{x^2}{2} + \frac{x^3}{2} + \ldots$ **(4 marks)**

b Deduce the first four terms in the Taylor series of $e^{-\tan x}$, in ascending powers of x. **(2 marks)**

c Hence find $\lim_{x \to 0} \frac{e^{\tan x} - e^x}{\sin x - x}$ **(4 marks)**

E/P 20 Find $\lim_{x \to 0} \frac{x^2(x - \sin x)^2}{2\cos x^2 - 2 + x^4}$

A E/P **21** $y\dfrac{d^2y}{dx^2} + \left(\dfrac{dy}{dx}\right)^2 + y = 0$

a Find an expression for $\dfrac{d^3y}{dx^3}$ **(5 marks)**

Given that $y = 1$ and $\dfrac{dy}{dx} = 1$ at $x = 0$,

b find the series solution for y, in ascending powers of x, up to an including the term in x^3. **(5 marks)**

c Comment on whether it would be sensible to use your series solution from part **b** to give estimates for y at $x = 0.2$ and at $x = 50$. **(2 marks)**

P **22** **a** Using Maclaurin series, and differentiation, show that $\ln \cos x = -\dfrac{x^2}{2} - \dfrac{x^4}{12} + \ldots$

b Using $\cos x = 2\cos^2\left(\dfrac{x}{2}\right) - 1$, and the result in part **a**, show that

$$\ln(1 + \cos x) = \ln 2 - \frac{x^2}{4} - \frac{x^4}{96} + \ldots$$

c Find $\lim\limits_{x\to 0} \dfrac{\ln(1 + \cos x) - \ln(2\cos x)}{1 - \cos x}$

P **23** **a** By writing $3^x = e^{x\ln 3}$, find the first four terms in the Taylor series of 3^x.

b Using your answer from part **a**, with a suitable value of x, find an approximation for $\sqrt{3}$, giving your answer to 3 significant figures.

24 Given that $f(x) = \operatorname{cosec} x$,

a show that:

i $f''(x) = \operatorname{cosec} x(2\operatorname{cosec}^2 x - 1)$

ii $f'''(x) = -\operatorname{cosec} x \cot x\,(6\operatorname{cosec}^2 x - 1)$

b Find the Taylor series expansion of $\operatorname{cosec} x$ in ascending powers of $\left(x - \dfrac{\pi}{4}\right)$ up to and including the term $\left(x - \dfrac{\pi}{4}\right)^3$.

E/P **25** **a** Given that $f(x) = \ln\left(1 + 2\cos\left(\dfrac{\pi x}{2}\right)\right)$, find f' and f''. **(4 marks)**

b Hence, using Taylor series, show that the first two non-zero terms, in ascending powers of $(x - 1)$, of $\ln\left(1 + 2\cos\left(\dfrac{\pi x}{2}\right)\right)$ are $-\pi(x - 1) - \dfrac{\pi^2}{2}(x - 1)^2$. **(2 marks)**

c Find $\lim\limits_{x\to 1} \dfrac{\ln\left(1 + 2\cos\left(\dfrac{\pi x}{2}\right)\right)}{3\ln(2 - x)}$ **(4 marks)**

Challenge

A

a Use induction to prove that the nth derivative of $\ln x$ is given by

$$\frac{d^n}{dx^n}\ln x = (-1)^{n+1}\frac{(n-1)!}{x^n}$$

b Hence write down the Taylor series expansion about $x = a$ of $\ln(x)$, where $a > 0$.

The **ratio test** is a sufficient condition for convergence of an infinite series. It says that a series $\sum_{n=1}^{\infty} a_n$ converges if $\lim_{n\to\infty}\left|\frac{a_{n+1}}{a_n}\right| < 1$ and diverges if $\lim_{n\to\infty}\left|\frac{a_{n+1}}{a_n}\right| > 1$.

(If the limit is 1 or doesn't exist then the test is inconclusive.)

c Using the ratio test, show that the Taylor series expansion of $\ln x$ about $x = a$ converges for x such that $0 < x < 2a$.

When the ratio test is inconclusive, one possible alternative is the **alternating series test**. It states that for a series of the form $\sum_{n=1}^{\infty}(-1)^{n+1}b_n$, if the coefficients b_n satisfy:

- $b_n \geqslant 0$ for each n
- $b_n \geqslant b_{n+1}$ for each n
- $\lim_{n\to\infty} b_n = 0$

then the series converges to a finite limit.

d Use the alternating series test and the result from part **c** to show that the Taylor series expansion of $\ln x$ about $x = a$ converges for each x such that $0 < x \leqslant 2a$.

Summary of key points

A

1 $f(x + a) = f(a) + f'(a)x + \frac{f''(a)}{2!}x^2 + \frac{f'''(a)}{3!}x^3 + \ldots + \frac{f^{(r)}(a)}{r!}x^r + \ldots$ (A)

$f(x) = f(a) + f'(a)(x - a) + \frac{f''(a)}{2!}(x - a)^2 + \frac{f'''(a)}{3!}(x - a)^3 + \ldots + \frac{f^{(r)}(a)}{r!}(x - a)^r + \ldots$ (B)

The expansions (A) and (B) given above are known as **Taylor series** expansions of $f(x)$ at (or about) the point $x = a$.

The Taylor series expansion is valid only if $f^{(n)}(a)$ exists and is finite for all $n \in \mathbb{N}$, and for values of x for which the infinite series converges.

2 Given $\lim_{x \to a} f(x) = L$ and $\lim_{x \to a} g(x) = M$, then:

- $\lim_{x \to a}(f(x) + g(x)) = L + M$
- If c is a constant, then $\lim_{x \to a} cf(x) = cL$
- $\lim_{x \to a} f(x)\,g(x) = LM$
- If $M \neq 0$, then $\lim_{x \to a} \frac{f(x)}{g(x)} = \frac{L}{M}$

3
- The series solution to the differential equation $\frac{dy}{dx} = f(x, y)$ is found using the Taylor series expansion in the form

  $y = y_0 + (x - x_0)\left.\frac{dy}{dx}\right|_{x_0} + \frac{(x - x_0)^2}{2!}\left.\frac{d^2y}{dx^2}\right|_{x_0} + \frac{(x - x_0)^3}{3!}\left.\frac{d^3y}{dx^3}\right|_{x_0} + \ldots$ (C)

- In the situation where $x_0 = 0$, this reduces to the Maclaurin series

  $y = y_0 + x\left.\frac{dy}{dx}\right|_0 + \frac{x^2}{2!}\left.\frac{d^2y}{dx^2}\right|_0 + \frac{x^3}{3!}\left.\frac{d^3y}{dx^3}\right|_0 + \ldots.$ (D)

Methods in calculus

7

Objectives

After completing this chapter you should be able to:

- Apply Leibnitz's theorem for differentiating products → pages 150–152
- Understand the use of derivatives to evaluate limits of indeterminate forms using L'Hospital's rule → pages 152–156
- Use tangent half-angle substitutions to find definite and indefinite integrals using Weierstrass substitution → pages 156–158

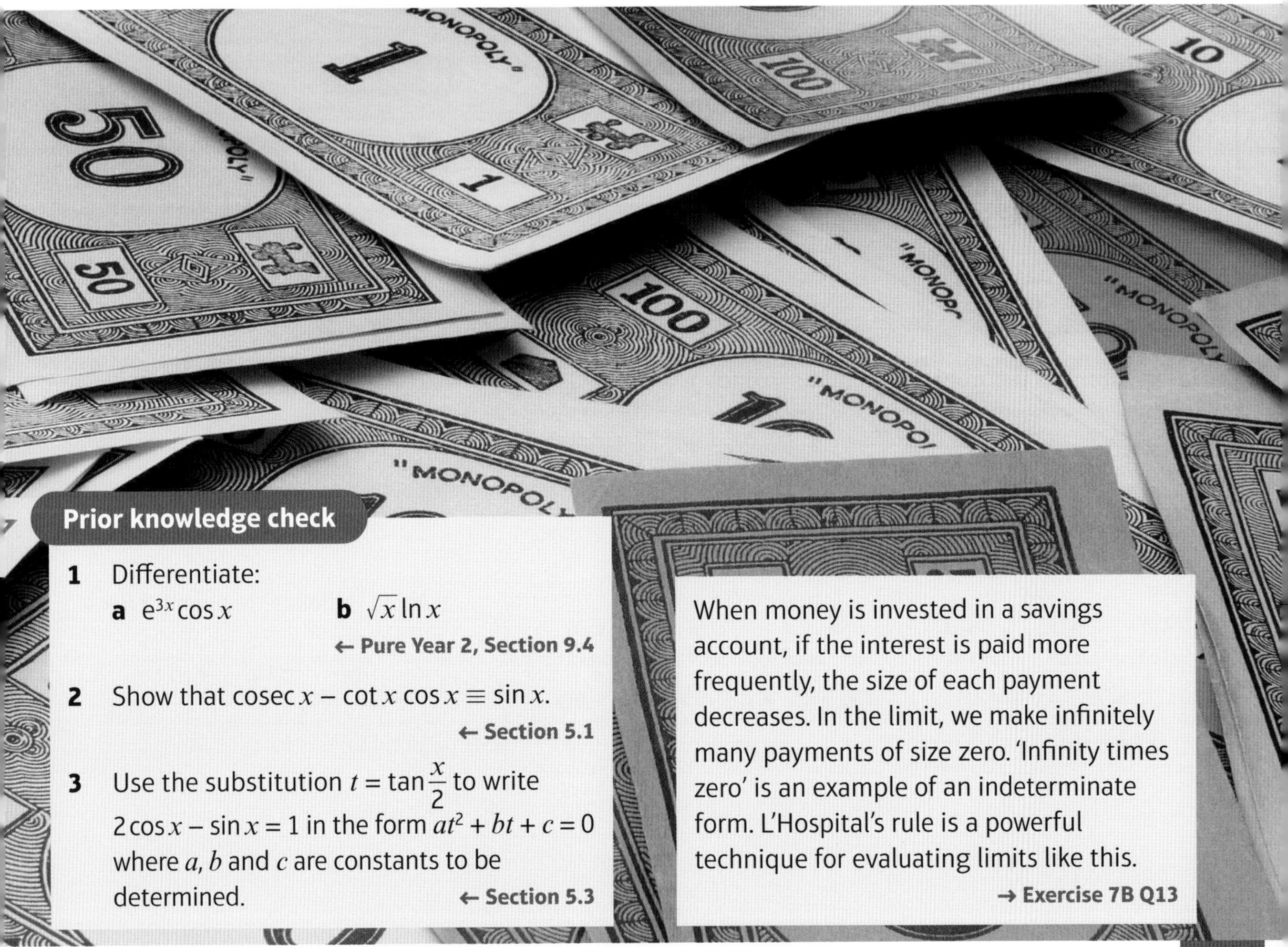

Prior knowledge check

1 Differentiate:

a $e^{3x}\cos x$ **b** $\sqrt{x}\ln x$

← Pure Year 2, Section 9.4

2 Show that $\operatorname{cosec} x - \cot x \cos x \equiv \sin x$.

← Section 5.1

3 Use the substitution $t = \tan\frac{x}{2}$ to write $2\cos x - \sin x = 1$ in the form $at^2 + bt + c = 0$ where a, b and c are constants to be determined.

← Section 5.3

When money is invested in a savings account, if the interest is paid more frequently, the size of each payment decreases. In the limit, we make infinitely many payments of size zero. 'Infinity times zero' is an example of an indeterminate form. L'Hospital's rule is a powerful technique for evaluating limits like this.

→ Exercise 7B Q13

7.1 Leibnitz's theorem and nth derivatives

A You can use the product rule to differentiate the product of two functions.

If $y = uv$, where u and v are functions of x, then

$$\frac{dy}{dx} = u\frac{dv}{dx} + v\frac{du}{dx}$$

The product rule generates two terms.

Applying the product rule again gives

$$\frac{d^2y}{dx^2} = u\frac{d^2v}{dx^2} + \frac{du}{dx}\frac{dv}{dx} + \frac{du}{dx}\frac{dv}{dx} + v\frac{d^2u}{dx^2}$$
$$= u\frac{d^2v}{dx^2} + 2\frac{du}{dx}\frac{dv}{dx} + v\frac{d^2u}{dx^2}$$

Repeatedly applying the product rule gives

$$\frac{d^3y}{dx^3} = u\frac{d^3v}{dx^3} + 3\frac{du}{dx}\frac{d^2v}{dx^2} + 3\frac{d^2u}{dx^2}\frac{dv}{dx} + v\frac{d^3u}{dx^3}$$
$$\frac{d^4y}{dx^4} = u\frac{d^4v}{dx^4} + 4\frac{du}{dx}\frac{d^3v}{dx^3} + 6\frac{d^2u}{dx^2}\frac{d^2v}{dx^2} + 4\frac{d^3u}{dx^3}\frac{dv}{dx} + v\frac{d^4u}{dx^4}$$ and so on.

The coefficients follow the same pattern as the binomial expansion.

The nth derivative of a product of two functions follows a pattern that is summarised by Leibnitz's theorem.

- **Leibnitz's theorem gives a general formula for the nth derivative of the product of two functions. If $y = uv$, where u and v are functions of x, then**

$$\frac{d^ny}{dx^n} = \sum_{k=0}^{n}\binom{n}{k}\frac{d^ku}{dx^k}\frac{d^{n-k}v}{dx^{n-k}}$$

Notation $\binom{n}{k} = \frac{n!}{k!(n-k)!}$ is the binomial coefficient.
$\frac{d^0u}{dx^0} = u$ and $\frac{d^0v}{dx^0} = v$.

Example 1

Use Leibnitz's theorem to calculate $\frac{d^4y}{dx^4}$ for $y = e^x \sin x$.

Let $u = e^x$ and $v = \sin x$

$\frac{d^nu}{dx^n} = e^x$ for every n — The derivative of e^x is e^x.

$\frac{dv}{dx} = \cos x$, $\frac{d^2v}{dx^2} = -\sin x$, $\frac{d^3v}{dx^3} = -\cos x$, $\frac{d^4v}{dx^4} = \sin x$

$\frac{d^4y}{dx^4} = u\frac{d^4v}{dx^4} + 4\frac{du}{dx}\frac{d^3v}{dx^3} + 6\frac{d^2u}{dx^2}\frac{d^2v}{dx^2} + 4\frac{d^3u}{dx^3}\frac{dv}{dx} + v\frac{d^4u}{dx^4}$ — Apply Leibnitz's theorem for $n = 4$: $\frac{d^4y}{dx^4} = \sum_{k=0}^{4}\binom{4}{k}\frac{d^ku}{dx^k}\frac{d^{4-k}v}{dx^{4-k}}$

$= e^x\sin x - 4e^x\cos x - 6e^x\sin x + 4e^x\cos x + e^x\sin x$

$= -4e^x\sin x$ — Simplify.

Example 2

Use Leibniz's theorem to calculate $\frac{d^3y}{dx^3}$ for $y = x^3\cosh 2x$.

Let $u = x^3$ and $v = \cosh 2x$

$\frac{du}{dx} = 3x^2, \frac{d^2u}{dx^2} = 6x, \frac{d^3u}{dx^3} = 6$

$\frac{dv}{dx} = 2\sinh 2x, \frac{d^2v}{dx^2} = 4\cosh 2x, \frac{d^3v}{dx^3} = 8\sinh 2x$

Use the chain rule to differentiate $\cosh 2x$ and $\sinh 2x$.

$\frac{d^3y}{dx^3} = u\frac{d^3v}{dx^3} + 3\frac{du}{dx}\frac{d^2v}{dx^2} + 3\frac{d^2u}{dx^2}\frac{dv}{dx} + v\frac{d^3u}{dx^3}$

$= 8x^3\sinh 2x + 36x^2\cosh 2x + 36x\sinh 2x + 6\cosh 2x$

Apply Leibniz's theorem for $n = 3$:

$\frac{d^3y}{dx^3} = \sum_{k=0}^{3}\binom{3}{k}\frac{d^ku}{dx^k}\frac{d^{3-k}v}{dx^{3-k}}$

You can use Leibniz's theorem to find an expression for the nth derivative of a product.

Example 3

Given that $y = x^2e^{-x}$, use Leibniz's theorem to show that for $n > 2$,

$$\frac{d^ny}{dx^n} = (-1)^ne^{-x}(x^2 - 2nx + n(n-1))$$

Let $u = x^2$ and $v = e^{-x}$

$\frac{du}{dx} = 2x, \frac{d^2u}{dx^2} = 2, \frac{d^ku}{dx^k} = 0$ for $k > 2$

$\frac{dv}{dx} = -e^{-x}, \frac{d^2v}{dx^2} = e^{-x}, \frac{d^3v}{dx^3} = -e^{-x}, \frac{d^kv}{dx^k} = (-1)^ke^{-x}$

Problem-solving

Write a general expression for $\frac{d^kv}{dx^k}$ using the fact that $(-1)^k$ generates the sequence 1, –1, 1, –1, …

$\frac{d^ny}{dx^n} = u\frac{d^nv}{dx^n} + \binom{n}{1}\frac{du}{dx}\frac{d^{n-1}v}{dx^{n-1}} + \binom{n}{2}\frac{d^2u}{dx^2}\frac{d^{n-2}v}{dx^{n-2}} + \dots$

$= x^2(-1)^ne^{-x} + n(2x)(-1)^{n-1}e^{-x} + \frac{n(n-1)}{2}(2)(-1)^{n-2}e^{-x}$

$= (-1)^ne^{-x}(x^2 - 2nx + n(n-1))$

Apply Leibniz's theorem. As $\frac{d^ku}{dx^k} = 0$ for $k > 2$, the remaining terms disappear.

Use the fact that $(-1)^{n-1} = (-1)(-1)^n$ and $(-1)^{n-2} = (-1)^n$ to simplify.

Exercise 7A

1 For each of the following functions:

i find $\frac{dy}{dx}, \frac{d^2y}{dx^2}$ and $\frac{d^3y}{dx^3}$

ii deduce an expression for $\frac{d^ny}{dx^n}$

a $y = e^{5x}$

b $y = e^{-x}$

c $y = x^m$

d $y = xe^{-x}$

2 Use Leibniz's theorem to calculate the following:

a $\frac{d^2y}{dx^2}$ for $y = (2x^2 + x - 2)(4x^2 - 3x + 8)$

b $\frac{d^2y}{dx^2}$ for $y = \ln x\sin x$

c $\frac{d^2y}{dx^2}$ for $y = e^{3x}\cos 2x$

d $\frac{d^2y}{dx^2}$ for $y = x^3\ln(2x + 1)$

e $\frac{d^3y}{dx^3}$ for $y = (x^2 - x + 2)(x^3 - 1)$

f $\frac{d^3y}{dx^3}$ for $y = \sqrt{2x}\sinh 3x$

g $\frac{d^4y}{dx^4}$ for $y = (x^2 - x)\cosh 2x$

h $\frac{d^4y}{dx^4}$ for $y = \cos x\sinh x$

A P **3** Use Leibniz's theorem to calculate the following:

a $\frac{d^2y}{dx^2}$ for $y = \frac{\sqrt{x}}{\ln x}$ **b** $\frac{d^3y}{dx^3}$ for $y = \frac{\ln x}{x+3}$

c $\frac{d^3y}{dx^3}$ for $y = \frac{e^x + 1}{e^x - 1}$ **d** $\frac{d^4y}{dx^4}$ for $y = \frac{\sin x}{4x^2}$

Hint Rewrite each quotient as a product.

E/P **4** Show that $y = e^x \cos x$ satisfies $\frac{d^6y}{dx^6} + 8\frac{dy}{dx} - 8y = 0$. **(4 marks)**

E/P **5** Given that $y = 2x^3e^{2x}$, use Leibniz's theorem to show that

$$\frac{d^ny}{dx^n} = 2^{n-2}e^{2x}(8x^3 + 12nx^2 + 6n(n-1)x + n(n-1)(n-2))$$ **(4 marks)**

E/P **6** **a** Using proof by induction, or otherwise, show that if $y = \frac{1}{x}$, then $\frac{d^ny}{dx^n} = (-1)^n\frac{n!}{x^{n+1}}$ **(4 marks)**

b Hence use Leibniz's theorem to show that if $y = x^3 \ln x$, then

$$\frac{d^ny}{dx^n} = \frac{6(-1)^n(n-4)!}{x^{n-3}}, \text{ for } n \geqslant 4$$ **(5 marks)**

E/P **7** Given that $y = x^2 \sinh kx$, where k is a constant, show that for any even integer n,

$$\frac{d^ny}{dx^n} = k^{n-2}\sinh kx(k^2x^2 + n(n-1)) + 2nk^{n-1}x\cosh kx$$

and for any odd integer n,

$$\frac{d^ny}{dx^n} = k^{n-2}\cosh kx(k^2x^2 + n(n-1)) + 2nk^{n-1}x\sinh kx$$ **(5 marks)**

Challenge

a Given that $F(x) = f(x)g(x)$, show that when $n = 1$, the formula

$$F^{(n)}(x) = \sum_{k=0}^{n}\binom{n}{k}f^{(k)}(x)g^{(n-k)}(x)$$

reduces to $F'(x) = f(x)g'(x) + g(x)f'(x)$.

b Hence use the product rule and proof by induction to prove Leibniz's theorem for all $n \in \mathbb{Z}^+$.

Problem-solving

You can change the summation limits to simplify an expression:

$$\sum_{k=0}^{n} f(k) = \sum_{k=1}^{n+1} f(k-1)$$

You can also use the following result when simplifying binomial coefficients:

$$\binom{k}{r-1} + \binom{k}{r} = \binom{k+1}{r}$$

7.2 L'Hospital's rule

You can use L'Hospital's rule to find limits of some indeterminate forms. It allows you to find a limit of a function that can be written in the form $\frac{f(x)}{g(x)}$, where f and g are differentiable functions, at points where $f(x)$ and $g(x)$ both tend to 0, or both tend to $\pm\infty$.

- **L'Hospital's rule states that for two functions $f(x)$ and $g(x)$, if either:**
  - $\lim_{x\to a} f(x) = \lim_{x\to a} g(x) = 0$**, or**
  - $\lim_{x\to a} f(x) = \pm\infty$ **and** $\lim_{x\to a} g(x) = \pm\infty$

  then provided that $\lim_{x\to a}\frac{f(x)}{g(x)}$ **exists,** $\lim_{x\to a}\frac{f(x)}{g(x)} = \lim_{x\to a}\frac{f'(x)}{g'(x)}$

Watch out $0 \times \infty$, $\infty - \infty$, 0^0, 1^∞ and ∞^0 are also indeterminate forms, but you can only apply L'Hospital's rule to the forms $\frac{0}{0}$ and $\frac{\pm\infty}{\pm\infty}$

Example 4

A

Calculate $\lim_{x \to 0} \frac{\sin x}{x}$

Let $f(x) = \sin x$ and $g(x) = x$

$f(0) = \sin 0 = 0$ and $g(0) = 0$, so we can apply L'Hospital's rule.

$f'(x) = \cos x$ and $g'(x) = 1$

By L'Hospital's rule,

$$\lim_{x \to 0} \frac{\sin x}{x} = \lim_{x \to 0} \frac{\cos x}{1} = \frac{1}{1} = 1$$

$\frac{\sin x}{x}$ is in the form $\frac{f(x)}{g(x)}$

Check that this expression is an indeterminate form by considering how the value of each function behaves at or near $x = 0$.

Find the derivatives of $f(x)$ and $g(x)$.

Links It is often easier to use L'Hospital's rule to evaluate a limit than to use Taylor series.

← **Section 6.2**

Watch out Do not differentiate the whole expression of $\frac{f(x)}{g(x)}$. You need to differentiate $f(x)$ and $g(x)$ separately.

If the limit of the derivatives is also indeterminate, then you can apply L'Hospital's rule a second time.

- **In general, you can apply L'Hospital's rule repeatedly, provided that the conditions are met at each step, and that the numerator and denominator can both be differentiated the required number of times.**

Example 5

Calculate $\lim_{x \to 0} \frac{1 - \cos x}{x^2}$

Let $f(x) = 1 - \cos x$ and $g(x) = x^2$

$f(0) = 1 - \cos 0 = 0$ and $g(0) = 0^2 = 0$, so we can apply L'Hospital's rule.

$f'(x) = \sin x$ and $g'(x) = 2x$

By L'Hospital's rule,

$$\lim_{x \to 0} \frac{1 - \cos x}{x^2} = \lim_{x \to 0} \frac{\sin x}{2x}$$

$f'(0) = \sin 0 = 0$ and $g'(0) = 2 \times 0 = 0$

$f''(x) = \cos x$ and $g''(x) = 2$

By L'Hospital's rule,

$$\lim_{x \to 0} \frac{1 - \cos x}{x^2} = \lim_{x \to 0} \frac{\sin x}{2x} = \lim_{x \to 0} \frac{\cos x}{2} = \frac{1}{2}$$

When $x = 0$, $\frac{f(x)}{g(x)}$ is the indeterminate form $\frac{0}{0}$

Find the derivatives of $f(x)$ and $g(x)$.

When $x = 0$, $\frac{\sin x}{2x}$ is an indeterminate form, so we can apply L'Hospital's rule a second time.

This limit is not indeterminate.

You may be able to rewrite functions as a quotient and apply L'Hospital's rule.

Example 6

Calculate $\lim_{x \to 0}(\operatorname{cosec} x - \cot x)$

$$\lim_{x \to 0}(\operatorname{cosec} x - \cot x) = \lim_{x \to 0}\left(\frac{1}{\sin x} - \frac{\cos x}{\sin x}\right)$$

$$= \lim_{x \to 0}\frac{1 - \cos x}{\sin x}$$

Let $f(x) = 1 - \cos x$ and $g(x) = \sin x$

$f(0) = 1 - \cos 0 = 0$ and $g(0) = \sin 0 = 0$, so we can apply L'Hospital's rule.

$f'(x) = \sin x$ and $g'(x) = \cos x$

By L'Hospital's rule,

$$\lim_{x \to 0}\frac{1 - \cos x}{\sin x} = \lim_{x \to 0}\frac{\sin x}{\cos x} = \frac{0}{1} = 0$$

Problem-solving

Use trigonometric relationships to rewrite the function as a quotient.

When $x = 0$, $\frac{f(x)}{g(x)}$ is the indeterminate form $\frac{0}{0}$

Find the derivatives of $f(x)$ and $g(x)$.

You can use the following rule, together with L'Hospital's rule, to evaluate the limits of some indeterminate forms.

- **If $\lim_{x \to a} f(x)$ exists, then $\lim_{x \to a} e^{f(x)} = e^{\lim_{x \to a} f(x)}$**

Example 7

Use the relationship $t = e^{\ln t}$ to calculate $\lim_{x \to \infty}\left(1 + \frac{a}{x}\right)^x$

$\lim_{x \to \infty}\left(1 + \frac{a}{x}\right)^x$ has the indeterminate form 1^∞.

$$\lim_{x \to \infty}\left(1 + \frac{a}{x}\right)^x = \lim_{x \to \infty} e^{\ln\left(1 + \frac{a}{x}\right)^x}$$

$$= e^{\lim_{x \to \infty}\ln\left(1 + \frac{a}{x}\right)^x}$$

Use the relationship given in the question to rewrite the function as an exponential.

Apply the rule $\lim_{x \to a} e^{f(x)} = e^{\lim_{x \to a} f(x)}$

$$\lim_{x \to \infty}\ln\left(1 + \frac{a}{x}\right)^x = \lim_{x \to \infty} x\ln\left(1 + \frac{a}{x}\right)$$

$$= \lim_{x \to \infty}\frac{\ln\left(1 + \frac{a}{x}\right)}{\frac{1}{x}}$$

Consider $\lim_{x \to \infty}\ln\left(1 + \frac{a}{x}\right)^x$ and use the rule $\ln a^n = n\ln a$.

Let $f(x) = \ln\left(1 + \frac{a}{x}\right)$ and $g(x) = \frac{1}{x}$

$\lim_{x \to \infty} f(x) = \ln 1 = 0$ and $\lim_{x \to \infty} g(x) = 0$

So apply L'Hospital's rule:

Problem-solving

$\lim_{x \to \infty} x\ln\left(1 + \frac{a}{x}\right)$ has the indeterminate form $\infty \times 0$. Writing the function as quotient will allow you to apply L'Hospital's rule.

$$f'(x) = \frac{-\frac{a}{x^2}}{1 + \frac{a}{x}} \text{ and } g'(x) = -\frac{1}{x^2}$$

Find the derivatives of $f(x)$ and $g(x)$. Use the chain rule to differentiate $f(x)$.

By L'Hospital's rule,

$$\lim_{x \to \infty}\frac{\ln\left(1 + \frac{a}{x}\right)}{\frac{1}{x}} = \lim_{x \to \infty}\frac{\frac{-\frac{a}{x^2}}{1 + \frac{a}{x}}}{-\frac{1}{x^2}} = \lim_{x \to \infty}\frac{a}{1 + \frac{a}{x}}$$

$$= \frac{a}{1 + 0} = a$$

Simplify and evaluate the limit.

Online Explore the graph of this function using GeoGebra.

So $\lim_{x\to\infty} \ln\left(1+\frac{a}{x}\right)^x = a$

Hence $\lim_{x\to\infty} \left(1+\frac{a}{x}\right)^x = e^a$

Use the fact that if $\lim_{x\to\infty} \ln\left(1+\frac{a}{x}\right)^x = a$ then $e^{\lim_{x\to\infty}\ln\left(1+\frac{a}{x}\right)^x} = e^a$.

Exercise 7B

1 Use L'Hospital's rule to calculate the following limits.

a $\lim_{x\to1}\frac{x^2-1}{x^2+3x-4}$ **b** $\lim_{x\to4}\frac{x-4}{\sqrt{x}-2}$ **c** $\lim_{x\to\infty}\frac{\ln x}{x^2}$

d $\lim_{x\to0}\frac{x^2-x}{x^2-\sin\pi x}$ **e** $\lim_{x\to0}\frac{e^{4x}-4x-1}{e^x-\cos x}$ **f** $\lim_{x\to0}\frac{\arctan 4x}{\arctan 5x}$

2 Use L'Hospital's rule to calculate the following limits.

a $\lim_{x\to0}\frac{x\sin x}{e^x-1}$ **b** $\lim_{x\to0}\frac{\sqrt{x}}{\tan x}$ **c** $\lim_{x\to\infty}\frac{x^2+x+1}{e^x}$

3 Evaluate the following limits, giving your answers as exact values where appropriate.

a $\lim_{x\to0} x^x$ **b** $\lim_{x\to\infty} x^{\frac{1}{x}}$ **c** $\lim_{x\to0} 1-x^{\frac{1}{x}}$

Hint Use $\lim_{x\to a} e^{f(x)} = e^{\lim_{x\to a} f(x)}$

4 a Show that $\frac{2x^2+x-1}{3x^2-2x-1} \equiv A + \frac{B}{3x+1} + \frac{C}{x-1}$, where A, B and C are rational constants to be found.

b Hence write down the value of $\lim_{x\to\infty}\frac{2x^2+x-1}{3x^2-2x-1}$

c Use two applications of L'Hospital's rule to evaluate $\lim_{x\to\infty}\frac{2x^2+x-1}{3x^2-2x-1}$

5 Anton uses L'Hospital's rule to find the value of $\lim_{x\to3}\frac{x^2-5x+6}{4x}$. He writes down the following working:

$$\lim_{x\to3}\frac{x^2-5x+6}{4x} = \lim_{x\to3}\frac{2x-5}{4} = \frac{6-5}{4} = \frac{1}{4}$$

a Explain the mistake Anton has made.

b Write down the correct value of this limit.

6 a Explain why you cannot use L'Hospital's rule to evaluate $\lim_{x\to0}\frac{\cosh x}{2x^2}$ **(1 mark)**

b Find $\lim_{x\to0}\frac{\cosh x-1}{2x^2}$ **(5 marks)**

7 Use L'Hospital's rule to find the value of $\lim_{x\to0}\frac{\sin^2 x}{x\tan x}$ **(5 marks)**

8 Use L'Hospital's rule to find the value of $\lim_{x\to0}\frac{x^2e^x}{\tan^2 x}$ **(5 marks)**

9 Use L'Hospital's rule to find the value of $\lim_{x\to0}(\ln x\sin x)$. **(5 marks)**

A E/P **10** Use L'Hospital's rule to show that $\lim_{x\to k} \frac{\sqrt{x} - \sqrt{k}}{\sqrt[3]{x} - \sqrt[3]{k}} = \frac{3\sqrt[6]{k}}{2}$ **(5 marks)**

E/P **11** Use L'Hospital's rule to find the value of $\lim_{x\to 0} (\cos x)^{\frac{1}{x}}$ **(6 marks)**

P **12** Use the definition of the derivative, $f'(x) = \lim_{h\to 0} \frac{f(x+h) - f(x)}{h}$ and L'Hospital's rule, to show that if $f(x) = \sin x$ then $f'(x) = \cos x$.

P **13** When savings accounts pay interest, they often compound. That is, they pay interest on previous interest payments.

a Show that a £1000 savings account paying 5% interest each year will contain £1276.28 after 5 years.

Usually interest rates are quoted annually. If the interest is paid more frequently than annually, then the effects of compounding mean that the interest rate can be measured in two different ways. The **nominal interest rate** is the interest paid as a percentage of the initial sum *ignoring* the effects of compounding. The **effective interest rate** is the interest paid as a percentage of the initial sum, *including* the effects of compounding. For example, if the nominal interest rate is 5%, and the interest payments are made monthly, then the savings account will pay $\frac{5}{12}$% interest each month.

b Show that if the nominal interest rate is 10%, and payments are made monthly, then the effective interest rate is approximately 10.47%.

c Suppose that the initial amount is A, the nominal rate of interest is $100r$%, and it is paid in n equal payments throughout the year. Write down a formula for $A_n(r)$, the amount after 1 year, in terms of A, r and n.

d Hence show that $A_\infty(r) = \lim_{n\to\infty} A_n(r) = Ae^r$.

Hint This is known as 'continuous compounding'. It is the result of letting the time between interest payments go to zero while maintaining a fixed nominal rate.

7.3 The Weierstrass substitution

You can use the substitution $t = \tan\frac{x}{2}$ to simplify integrals.

Example 8

Find $\int \operatorname{cosec} x \, dx$.

Let $t = \tan\frac{x}{2}$, then $\operatorname{cosec} x = \frac{1+t^2}{2t}$

Use the t-formula $\sin x = \frac{2t}{1+t^2}$ and $\operatorname{cosec} x \equiv \frac{1}{\sin x}$ ← Chapter 5

Also $\frac{dt}{dx} = \frac{d}{dx}\left(\tan\frac{x}{2}\right) = \frac{1}{2}\sec^2\frac{x}{2}$

$= \frac{1}{2}\left(1 + \tan^2\frac{x}{2}\right) = \frac{1+t^2}{2}$

Use the identity $\sec^2\theta \equiv 1 + \tan^2\theta$.

So $dx = \frac{2}{1+t^2}dt$.

Hence

$\int \operatorname{cosec} x \, dx = \int \frac{1+t^2}{2t}\,\frac{2}{1+t^2}\,dt$

$= \int \frac{1}{t}\,dt$

$= \ln|t| + c$

$= \ln\left|\tan\frac{x}{2}\right| + c$

A

- **The Weierstrass substitution is $t = \tan\frac{x}{2}$, and the corresponding substitution for dx is d$x = \frac{2}{1+t^2}$dt**

When using the Weierstrass substitution, you replace each trigonometric function by the corresponding function of t using the ***t*-formulae**:

Function	Substitution
$\sin x$	$\frac{2t}{1+t^2}$
$\cos x$	$\frac{1-t^2}{1+t^2}$
$\tan x$	$\frac{2t}{1-t^2}$
$\sec x$	$\frac{1+t^2}{1-t^2}$
$\text{cosec}\, x$	$\frac{1+t^2}{2t}$
$\cot x$	$\frac{1-t^2}{2t}$

Links The t-formulae can be used to prove trigonometric identities and solve equations.
← Chapter 5

After using the Weierstrass substitution, you are usually left with a **rational function**, i.e. a quotient of polynomials. You can integrate this using **partial fractions** or any other appropriate technique, then reverse the substitution to get the answer.

Example 9

Find $\int_{\frac{\pi}{3}}^{\frac{\pi}{2}} \frac{1}{1+\sin x - \cos x}\,dx$

Using the Weierstrass substitution,

$$\int_{\frac{\pi}{3}}^{\frac{\pi}{2}} \frac{1}{1+\sin x - \cos x}\,dx$$

$$= \int_{\frac{1}{\sqrt{3}}}^{1} \left(\frac{1}{1 + \left(\frac{2t}{1+t^2}\right) - \left(\frac{1-t^2}{1+t^2}\right)}\right)\left(\frac{2}{1+t^2}\right) dt$$

$$= \int_{\frac{1}{\sqrt{3}}}^{1} \frac{1+t^2}{1+t^2+2t-1+t^2}\,\frac{2}{1+t^2}\,dt$$

$$\int_{\frac{1}{\sqrt{3}}}^{1} \frac{1}{t(t+1)}\,dt$$

Rewriting the integrand, $\frac{1}{t(t+1)} = \frac{1}{t} - \frac{1}{t+1}$

So

$$\int_{\frac{\pi}{3}}^{\frac{\pi}{2}} \frac{1}{1+\sin x - \cos x}\,dx = \int_{\frac{1}{\sqrt{3}}}^{1} \left(\frac{1}{t} - \frac{1}{t+1}\right) dt$$

$$= \left[\ln\left|\frac{t}{t+1}\right|\right]_{\frac{1}{\sqrt{3}}}^{1}$$

$$= \ln\frac{1}{2} - \ln\frac{1}{1+\sqrt{3}}$$

$$= \ln(1+\sqrt{3}) - \ln 2$$

$$= \ln\left(\frac{1+\sqrt{3}}{2}\right)$$

The limits have been transformed by the substitution. When $x = \frac{\pi}{2}$, $t = \tan\frac{\pi}{4} = 1$, and when $x = \frac{\pi}{3}$, $t = \tan\frac{\pi}{6} = \frac{1}{\sqrt{3}}$

Write $\frac{1}{t(t+1)} = \frac{A}{t} + \frac{B}{t+1}$, so $1 = A(t+1) + Bt$.
Then $t = 0$ implies $A = 1$ and $t = -1$ implies $B = -1$.

$\ln|t| - \ln|t+1| = \ln\left|\frac{t}{t+1}\right|$.

Problem-solving

You don't need to make the substitution back into x for a definite integral as long as you transform the limits.

Exercise 7C

A

1 Use the substitution $t = \tan\frac{x}{2}$ to integrate the following:

a $\displaystyle\int \frac{1}{1 + 3\cos x}\,dx$ **b** $\displaystyle\int \sec x\,dx$ **c** $\displaystyle\int \frac{1}{\sin x + \tan x}\,dx$ **d** $\displaystyle\int \frac{2}{1 - \sin x}\,dx$

2 Use the substitution $t = \tan\frac{x}{2}$ to evaluate the following:

a $\displaystyle\int_0^{\frac{\pi}{2}} \frac{\sec x}{1 + \tan x}\,dx$ **b** $\displaystyle\int_0^{\frac{\pi}{2}} \frac{1 - \cos x}{1 + \sin x + 2\cos x}\,dx$ **c** $\displaystyle\int_{\frac{\pi}{3}}^{\frac{\pi}{2}} \frac{\sin x}{1 + \cos^2 x}\,dx$ **d** $\displaystyle\int_{\frac{\pi}{2}}^{\frac{2\pi}{3}} \frac{\cot x}{1+\operatorname{cosec} x}\,dx$

E **3 a** Using the substitution, $t = \tan\frac{x}{2}$, show that the integral

$$\int \frac{1}{12 - 13\sin x}\,dx$$

can be written as

$$\int \frac{1}{6t^2 - 13t + 6}\,dt$$

(3 marks)

b Hence evaluate $\displaystyle\int_0^{\frac{\pi}{3}} \frac{1}{12 - 13\sin x}\,dx$ **(5 marks)**

P **4** $\displaystyle\int_0^{\frac{\pi}{2}} \frac{1}{a + \cos x}\,dx = \frac{\pi}{3\sqrt{3}}$ where a is a positive integer.

Find the value of a.

P **5** Show that $\displaystyle\int_0^1 \frac{\arccos\left(\frac{1 - x^2}{1 + x^2}\right)}{1 + x^2}\,dx = \frac{\pi^2}{16}$

Challenge

Evaluate $\displaystyle\int_{-1}^{1} \frac{\sqrt{1 - x^2}}{1 + x^2}\,dx$

Hint Consider the substitution $x = \sin\theta$.

Mixed exercise 7

1 Use Leibnitz's theorem to calculate the following:

a $\dfrac{d^2y}{dx^2}$ for $y = (3x^2 - 2x)(x^3 + 2x - 6)$ **b** $\dfrac{d^2y}{dx^2}$ for $y = e^{4x}\tan 2x$ **c** $\dfrac{d^3y}{dx^3}$ for $y = x^{\frac{3}{2}}\arctan 2x$

2 By writing $\tan x = \dfrac{\sin x}{\cos x}$, use Leibnitz's theorem to compute $\dfrac{d^2}{dx^2}\tan x$.

A **3** $y = \mathrm{fgh}$ where f, g and h are functions of x.

a Use Leibnitz's theorem to show that

$$\frac{\mathrm{d}^2y}{\mathrm{d}x^2} = \frac{\mathrm{d}^2\mathrm{f}}{\mathrm{d}x^2}\mathrm{gh} + \mathrm{f}\frac{\mathrm{d}^2\mathrm{g}}{\mathrm{d}x^2}\mathrm{h} + \mathrm{fg}\frac{\mathrm{d}^2\mathrm{h}}{\mathrm{d}x^2} + 2\left(\frac{\mathrm{df}}{\mathrm{d}x}\frac{\mathrm{dg}}{\mathrm{d}x}\mathrm{h} + \frac{\mathrm{df}}{\mathrm{d}x}\mathrm{g}\frac{\mathrm{dh}}{\mathrm{d}x} + \mathrm{f}\frac{\mathrm{dg}}{\mathrm{d}x}\frac{\mathrm{dh}}{\mathrm{d}x}\right)$$ **(4 marks)**

b Hence compute $\frac{\mathrm{d}^2y}{\mathrm{d}x^2}$ for $y = \mathrm{e}^x \sin 2x \cos 3x$. **(2 marks)**

P **4** Use Leibnitz's theorem to calculate $\frac{\mathrm{d}^3y}{\mathrm{d}x^3}$ for $\frac{\sqrt{3x+2}}{\cos x}$ **(5 marks)**

P **5** **a** Using proof by induction, or otherwise, show that if $y = \sin x$, then

$$\frac{\mathrm{d}^ny}{\mathrm{d}x^n} = \sin\left(\frac{n\pi}{2} + x\right)$$ **(4 marks)**

b Hence use Leibnitz's theorem to show that if $y = x^2 \sin x$, then

$$\frac{\mathrm{d}^ny}{\mathrm{d}x^n} = \sin\left(\frac{n\pi}{2} + x\right)(x^2 + n - n^2) - 2nx\cos\left(\frac{n\pi}{2} + x\right)$$ **(4 marks)**

6 Use L'Hospital's rule to find:

a $\lim_{x\to 0}\frac{\mathrm{e}^{2x} - 1}{3x}$ **b** $\lim_{x\to 0}\frac{\tan x}{2x + \sin x}$ **c** $\lim_{x\to 1}\frac{x^2 + x - 2}{x\ln x}$ **d** $\lim_{x\to 1}\frac{\sin \pi x}{x^2 + 7x - 8}$

P **7** Use L'Hospital's rule to find the value of $\lim_{x\to\pi}\frac{\cos\left(\frac{1}{2}x\right)}{x - \pi}$ **(3 marks)**

P **8** Given that n is a positive integer, find $\lim_{x\to 2}\frac{x - 2}{x^n - 2^n}$, giving your answer in terms of n. **(4 marks)**

P **9** Use L'Hospital's rule to find the value of $\lim_{x\to\infty}(1 + ax)^{\frac{1}{x}}$ **(6 marks)**

10 Use the substitution $t = \tan\frac{x}{2}$ to integrate the following:

a $\int \frac{3}{2 + 4\cos x}\,\mathrm{d}x$ **b** $\int \frac{\sec x}{\sin x + 2\cos x}\,\mathrm{d}x$ **c** $\int \frac{2}{\sin x + \cos x}\,\mathrm{d}x$

11 Use the substitution $t = \tan\frac{x}{2}$ to evaluate the following:

a $\int_0^{\frac{\pi}{2}} \frac{2}{7 + 2\sin x + 8\cos x}\,\mathrm{d}x$ **b** $\int_{\frac{2\pi}{3}}^{\pi} \frac{1}{2 - 2\cos x}\,\mathrm{d}x$

E **12** **a** Using the substitution, $t = \tan\frac{x}{2}$, show that the integral

$$\int \frac{1}{4\cos x - 3\sin x}\,\mathrm{d}x$$

can be written as

$$\int \frac{-1}{2t^2 + 3t - 2}\,\mathrm{d}t$$ **(3 marks)**

b Hence evaluate $\int_{\frac{\pi}{3}}^{\frac{\pi}{2}} \frac{1}{4\cos x - 3\sin x}\,\mathrm{d}x$. **(5 marks)**

P **13** Show that $\int_{\frac{\pi}{3}}^{\frac{\pi}{2}} \frac{1 - \operatorname{cosec} x}{\sin x}\,\mathrm{d}x = \ln\sqrt{k} - \frac{1}{\sqrt{k}}$ where k is a positive integer to be found. **(8 marks)**

Challenge

A

Use proof by induction to prove that $\lim_{x\to\infty} \frac{x^n}{e^x} = 0$ for all $n \in \mathbb{N}$.

Summary of key points

1 **Leibnitz's theorem** gives a general formula for the nth derivative of the product of two functions. It states that if $y = uv$, where u and v are functions of x, then

$$\frac{d^n y}{dx^n} = \sum_{k=0}^{n} \binom{n}{k} \frac{d^k u}{dx^k} \frac{d^{n-k} v}{dx^{n-k}}$$

2 **L'Hospital's rule** states that for two functions f(x) and g(x) if either:

- $\lim_{x\to a} f(x) = \lim_{x\to a} g(x) = 0$, or
- $\lim_{x\to a} f(x) = \pm\infty$ and $\lim_{x\to a} f(x) = \pm\infty$

then provided that $\lim_{x\to a} \frac{f(x)}{g(x)}$ exists, $\lim_{x\to a} \frac{f(x)}{g(x)} = \lim_{x\to a} \frac{f'(x)}{g'(x)}$

3 In general, you can apply L'Hospital's rule repeatedly, provided that the conditions are met at each step, and that the numerator and denominator can both be differentiated the required number of times.

4 If $\lim_{x\to a} f(x)$ exists, then $\lim_{x\to a} e^{f(x)} = e^{\lim_{x\to a} f(x)}$

5 The **Weierstrass substitution** is $t = \tan\frac{x}{2}$, and the corresponding substitution for dx is $dx = \frac{2}{1+t^2}dt$

Numerical methods 8

Objectives

After completing this chapter you should be able to:

- Find numerical solutions to first-order differential equations using Euler's method and the midpoint method → pages 162–168
- Extend Euler's method to find numerical solutions to second-order differential equations → pages 169–172
- Use Simpson's rule to find an approximation for a given definite integral → pages 173–175

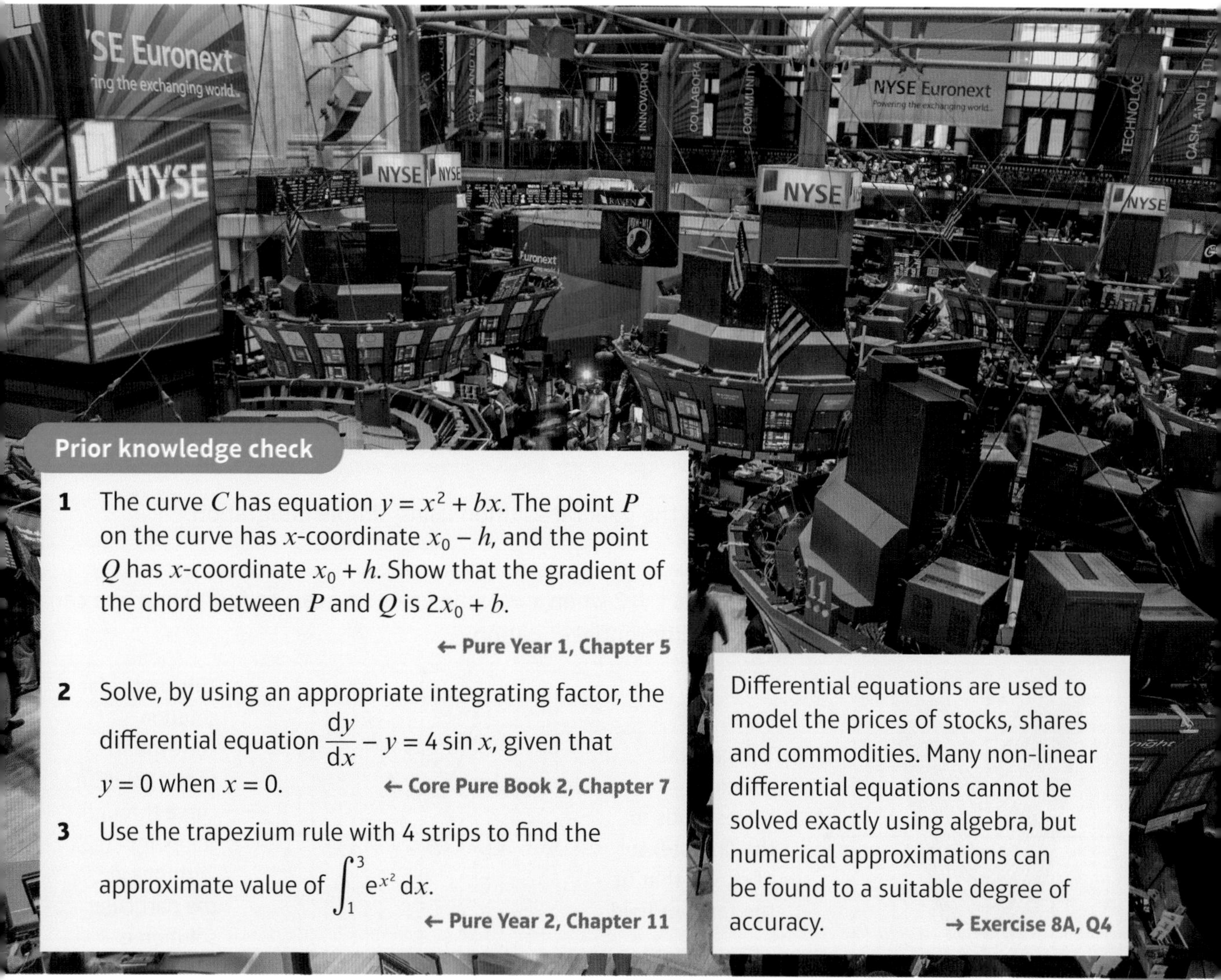

Prior knowledge check

1 The curve C has equation $y = x^2 + bx$. The point P on the curve has x-coordinate $x_0 - h$, and the point Q has x-coordinate $x_0 + h$. Show that the gradient of the chord between P and Q is $2x_0 + b$.

← Pure Year 1, Chapter 5

2 Solve, by using an appropriate integrating factor, the differential equation $\frac{dy}{dx} - y = 4\sin x$, given that $y = 0$ when $x = 0$. ← Core Pure Book 2, Chapter 7

3 Use the trapezium rule with 4 strips to find the approximate value of $\int_1^3 e^{x^2}\,dx$.

← Pure Year 2, Chapter 11

Differential equations are used to model the prices of stocks, shares and commodities. Many non-linear differential equations cannot be solved exactly using algebra, but numerical approximations can be found to a suitable degree of accuracy. → Exercise 8A, Q4

8.1 Solving first-order differential equations

Some first-order differential equations of the form

$$\frac{dy}{dx} = f(x, y)$$

can be solved analytically.

Notation Solving analytically means to use an algebraic method. Some differential equations of the form $\frac{dy}{dx} = f(x, y)$ can be solved analytically by separating the variables or using an integrating factor.
← Pure Year 2, Chapter 11; Core Pure Book 2, Chapter 7;

However, in some cases this might be difficult or impossible. You can use numerical methods to find approximate solutions to differential equations in this situation.

Consider the first-order differential equation $\frac{dy}{dx} = 2x$. This is an equation which describes the relationship between a given x-value and the gradient of the curve at that point. For example, at the point where $x = 3$, the gradient of the curve is 6.

You can use the differential equation to sketch the gradient at any given point in the xy-plane.

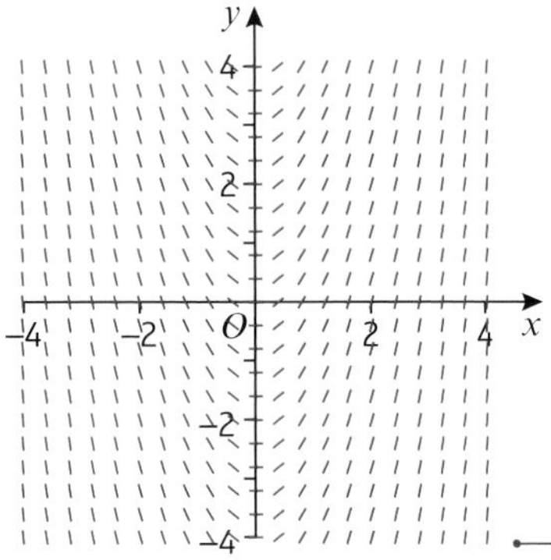

Each short line is part of a tangent to one member of the family of solution curves.

Notation Diagrams like this are sometimes called **tangent fields** or **compass point diagrams**.

Online Explore tangent fields using GeoGebra.

Links When you solve a differential equation analytically, you find the **general solution** first and then use given **initial** (or **boundary) conditions** to find the **particular solution**.
← Pure Year 2, Chapter 11

For the example given above, you can find the general solution using simple integration:

$$y = \int 2x \, dx = x^2 + c$$

If you are given an initial condition, such as $y = 2$ when $x = 1$, the value of c can be found and you can write down the particular solution to the differential equation.

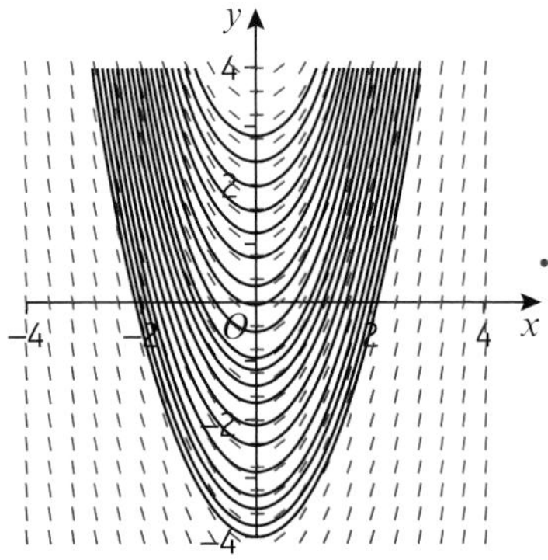

The general solution corresponds to the (infinite) set of curves that fit the tangent field.

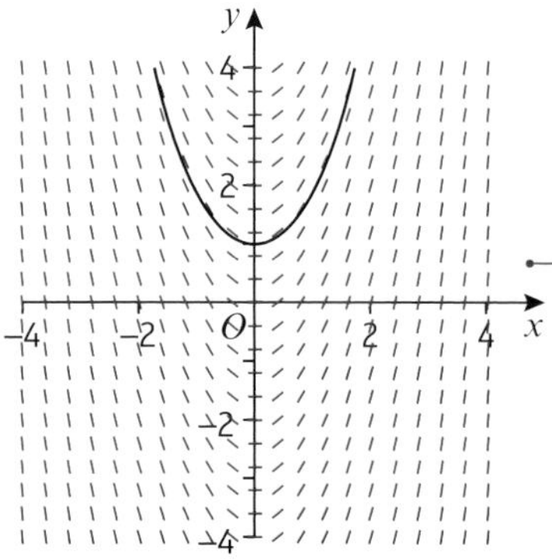

This particular solution corresponds to the solution curve through the point (1, 2). In this case, the particular solution is $y = x^2 + 1$.

If the given differential equation is not solvable using an analytical technique, some of these ideas can be adapted to find approximate solutions using numerical methods.

Consider the differential equation $\frac{dy}{dx} = x^2 + y^3$ with the initial condition that $y = 1$ when $x = 1$.

This equation cannot be solved analytically.

You can, however, work out the gradient of the curve at the given point by substituting into the expression for $\frac{dy}{dx}$:

$$\frac{dy}{dx} = 1^2 + 1^3 = 2$$

Using this information, it is possible to plot one line of the tangent field at the point $P_0(1, 1)$ for this differential equation.

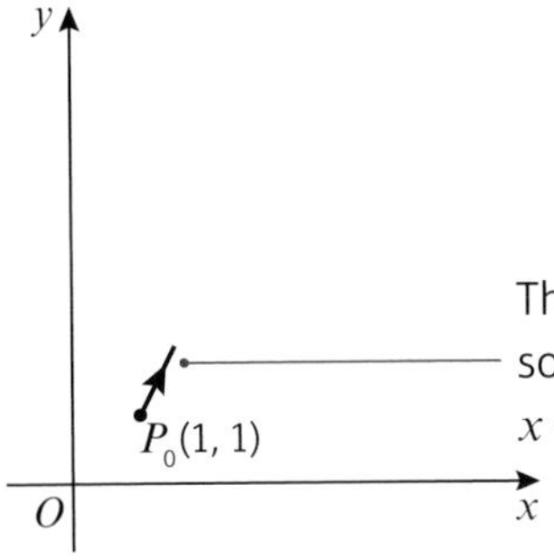

The gradient of the solution curve when $x = 1$ and $y = 1$ is 2.

A second point on the solution curve can be approximated by considering a small move along the tangent line.

Consider a small step of length h in the x-direction from the initial point.

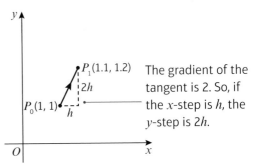

For example, if $h = 0.1$, the coordinates of the point P_1 at the end of the tangent line will be (1.1, 1.2). This represents the coordinates of the next point on the approximate solution curve.

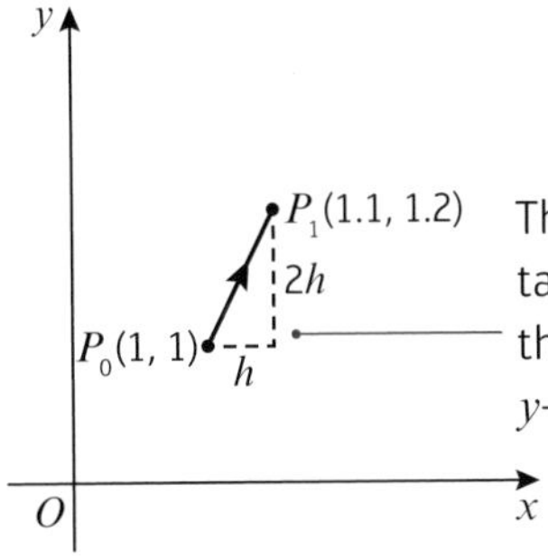

The gradient of the tangent is 2. So, if the x-step is h, the y-step is $2h$.

This process can be repeated with the new initial coordinate of (1.1, 1.2):

$$\frac{dy}{dx} = 1.1^2 + 1.2^3 = 2.938$$

For a step of 0.1, the coordinates of the next point on the approximate solution curve will be:

$x = 1.1 + 0.1 = 1.2$

$y = 1.2 + 0.1 \times 2.938 = 1.4938$

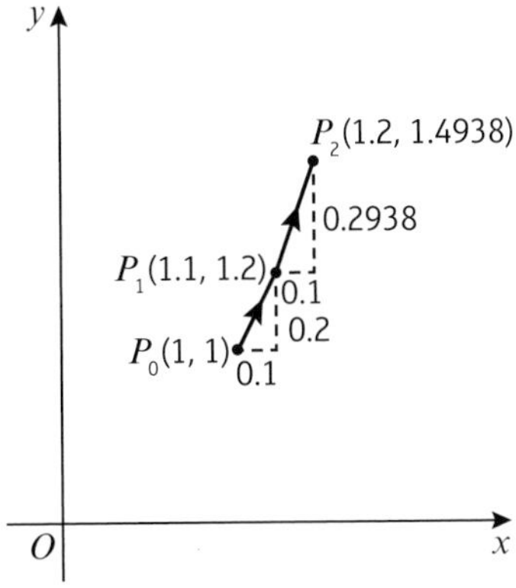

Links This method of approximation is **iterative**, because the first approximation is used as the starting point for the second approximation.

← Pure Year 2, Chapter 10

In general, if the coordinates of the initial given point are (x_0, y_0) and the next point is (x_1, y_1), where $x_1 = x_0 + h$, you can write $\left(\frac{dy}{dx}\right)_0 \approx \frac{y_1 - y_0}{x_1 - x_0} = \frac{y_1 - y_0}{h}$

Notation $\left(\frac{dy}{dx}\right)_0$ is used to denote the value of the gradient function $\frac{dy}{dx}$ when $x = x_0$.

- **Euler's method for finding approximate solutions to first-order differential equations uses the approximation**

$$\left(\frac{dy}{dx}\right)_0 \approx \frac{y_1 - y_0}{h}$$

It is often more useful to write this as an iterative formula:

$$y_{r+1} \approx y_r + h\left(\frac{dy}{dx}\right)_r, r = 0, 1, 2, \ldots$$

Example 1

$y = f(x)$ satisfies the differential equation $\frac{dy}{dx} = \frac{x^2 + y}{y^2 - x}$ and the initial condition, $f(3) = -1$.

Use two iterations of Euler's method to estimate the value of f(4), giving your answer correct to 2 decimal places.

$h = 0.5$

You need to use two iterations to get from $x_0 = 3$ to $x_2 = 4$, so your step length will be 0.5.

$(x_0, y_0) = (3, -1)$

$\left(\frac{dy}{dx}\right)_0 = \frac{3^2 - 1}{(-1)^2 - 3} = -4$

Substitute the values of x_0 and y_0 into the differential equation to find the value of $\left(\frac{dy}{dx}\right)_0$.

$y_1 \approx y_0 + h\left(\frac{dy}{dx}\right)_0$

$= -1 + 0.5 \times (-4)$

$= -3$

Your values of x_1 and y_1 determine the starting point for the next iteration. Use the differential equation to find the gradient at (x_1, y_1).

$(x_1, y_1) = (3.5, -3)$

$\left(\frac{dy}{dx}\right)_1 = \frac{3.5^2 - 3}{(-3)^2 - 3.5} = 1.6818\ldots$

$y_2 \approx y_1 + h\left(\frac{dy}{dx}\right)_1$

$= -3 + 0.5 \times 1.6818\ldots$

$= -2.15909\ldots$

So $f(4) \approx -2.16$ (2 d.p.)

Do not round any values until your final answer.

Exercise 8A

1 Use Euler's method to estimate the value at $x = 2$ of the particular solution to the differential equation

$$\frac{dy}{dx} = x^2 + y^2$$

which passes through the point (1, 2). Use a step length of 0.25.

Hint You will need to carry out 4 iterations.

2 $y = f(x)$ satisfies the differential equation

$$\frac{dy}{dx} = \sqrt{e^x + 2e^y}$$

Given that $f(2.5) = 0$, use five iterations of Euler's method to estimate the value of $f(3)$.

E/P **3** The differential equation

$$\frac{dy}{dx} = \ln(x + y^2)$$

has a particular solution that passes through the point (e, 2).

Use of the approximation formula, $\left(\frac{dy}{dx}\right)_0 \approx \frac{y_1 - y_0}{h}$, gives $y_1 = 2.4$.

a Determine the value that was used for the step length, h. **(2 marks)**

b Using this step length, calculate, correct to three decimal places, the values of y_2 and y_3. **(5 marks)**

E/P **4** The value, v thousand pounds, of a particular asset in a stock portfolio t days after it is purchased is modelled by the differential equation

$$\frac{dv}{dt} = \frac{v - t}{vt - t^3}$$

Given that the asset is worth £10 000 two days after it is purchased, use two iterations of the approximation formula $\left(\frac{dy}{dx}\right)_0 \approx \frac{y_1 - y_0}{h}$ to estimate, correct to the nearest hundred pounds, the value of the asset five days after it is purchased. **(6 marks)**

E/P **5** A pendulum consists of a light, inextensible string of length 1 m with a metal ball attached to one end. The other end is fixed to a point about which the pendulum is free to swing.
The pendulum swings in a vertical plane and the equation of motion is modelled using the differential equation

$$\frac{d\theta}{dt} = \sqrt{9.8(2\cos\theta - 1)}$$

where θ is the angle the string makes with the downward vertical.
Given that $\theta = 0$ when $t = 0$, use two iterations of the approximation formula $\left(\frac{dy}{dx}\right)_0 \approx \frac{y_1 - y_0}{h}$ to find the value of θ when $t = 0.3$. **(6 marks)**

You can visualise Euler's method by constructing a right-angled triangle with one vertex at (x_0, y_0) and with its hypotenuse as a tangent to the curve at (x_0, y_0).

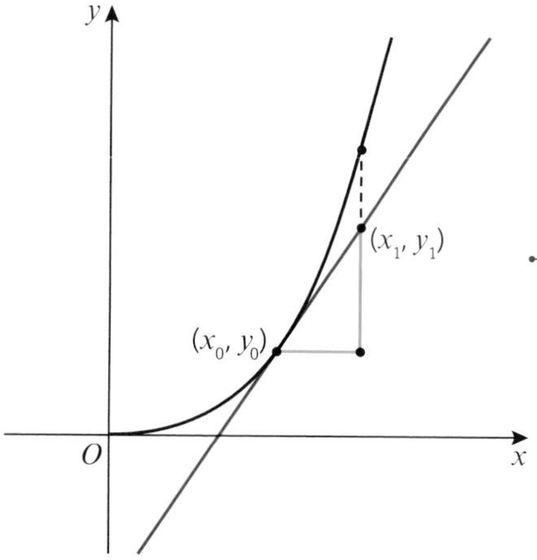

If the curve representing the particular solution is quite convex (or concave) near (x_0, y_0), the approximation is quite a long way from the true value.

The accuracy of Euler's method can be improved by reducing the step length. However, another way of addressing this issue is to use a different method. Consider the diagram to the right:

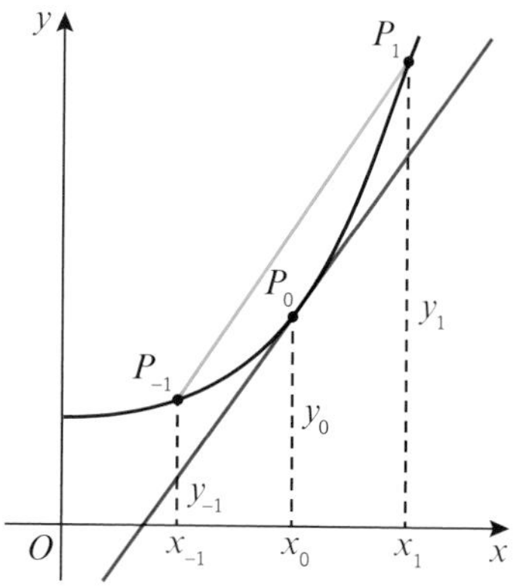

The gradient of the chord joining points P_{-1} and P_1 is approximately the same as the gradient of the tangent to the curve at P_0. Hence you can write

$$\left(\frac{dy}{dx}\right)_0 \approx \frac{y_1 - y_{-1}}{x_1 - x_{-1}} = \frac{y_1 - y_{-1}}{2h}$$, where h is the step length $x_1 - x_0$.

Generalising further, you can write:

- **The midpoint method for finding approximate solutions to first-order differential equations uses the formula**

$$\left(\frac{dy}{dx}\right)_0 \approx \frac{y_1 - y_{-1}}{2h}$$

It is often more useful to write this as an iterative formula:

$$y_{r+1} \approx y_{r-1} + 2h\left(\frac{dy}{dx}\right)_r, \ r = 0, 1, 2, \ldots$$

Example 2

Use the midpoint formula with a step length of 0.25 to estimate the value at $x = 0.5$ of the particular solution to the differential equation

$$\frac{dy}{dx} = \frac{xy + y}{y^2 + x^2}$$

which passes through the point (0, 2). Give your answer correct to 4 decimal places.

$y_2 \approx y_0 + 2h\left(\frac{dy}{dx}\right)_1$

Your initial condition will be (x_0, y_0), so rewrite the midpoint formula using y_2 and y_0.

$x_0 = 0$, $y_0 = 2$, $h = 0.25$

$x_1 = 0.25$

$x_2 = 0.5$

$\left(\frac{dy}{dx}\right)_0 = \frac{0 \times 2 + 2}{2^2 + 0^2} = \frac{1}{2}$

$y_1 \approx y_0 + \left(\frac{dy}{dx}\right)_0 h$

$= 2 + \frac{1}{2} \times 0.25$

$= 2.125$

Watch out Write down the information you know. You can't calculate $\left(\frac{dy}{dx}\right)_1$ without a value for y_1, so the first step in your method is to use Euler's method to find y_1.

$\left(\frac{dy}{dx}\right)_1 = \frac{0.25 \times 2.125 + 2.125}{2.125^2 + 0.25^2} = 0.58020...$

Calculate $\left(\frac{dy}{dx}\right)_1$ using your value of y_1.

$y_2 \approx y_0 + 2h\left(\frac{dy}{dx}\right)_1$

$= 2 + 2 \times 0.25 \times 0.58020...$

$= 2.2901$ (4 d.p.)

Use the midpoint formula to calculate y_2.

Exercise 8B

1 A particular solution to the differential equation $\frac{dy}{dx} = x^3 - y^2$ passes through the point (2, 2).

a Taking $(x_0, y_0) = (2, 2)$ and $x_1 = 2.25$, apply Euler's method once to obtain a value for y_1.

b Apply the midpoint method once to obtain an approximate value for the solution to the differential equation at $x = 2.5$.

2 Use the midpoint formula to estimate the value at $x = 1.5$ of the particular solution to the differential equation

$$\frac{dy}{dx} = \ln x + 3y$$

which passes through the point (1, 1). Use a step length of 0.1.

(E) **3** A particular solution to the differential equation

$$\frac{dy}{dx} = \sin xy$$

passes through the point (1, 2).

a Verify that the approximation formula $\left(\frac{dy}{dx}\right)_0 \approx \frac{y_1 - y_0}{h}$ with a step length of 0.2 gives $y_1 = 2.1819$ correct to five significant figures. **(3 marks)**

b Use the midpoint formula with a step length of 0.2 to obtain an estimate of the value of y when $x = 2$. Give your answer to four significant figures. **(3 marks)**

E/P **4** The population of a given species of rabbit, P, at time t months is modelled by the differential equation

$$\frac{dP}{dt} = 3P - 0.002P^2 - 100\cos(0.6t)$$

Given that the initial starting population of this species of rabbit is 700, use the approximation formula

$$\left(\frac{dy}{dx}\right)_0 \approx \frac{y_1 - y_{-1}}{2h}$$

with a step length of 0.5 to estimate, correct to the nearest 10 rabbits, the population after two months. **(6 marks)**

E/P **5** The velocity, v, of a bungee jumper, at the point where the bungee cord becomes taut, is modelled using the differential equation

$$\frac{dv}{dx} = \frac{1.5x - 24.8}{v} - 0.003v$$

where x is the displacement from the top of the crane from which the jump was made.

Given that $v = 12$ when $x = 10$, use the approximation formula $\left(\frac{dy}{dx}\right)_0 \approx \frac{y_1 - y_{-1}}{2h}$ with a step length of 0.5 to find the value of v when $x = 11.5$. **(6 marks)**

A E **6** A particular solution to the differential equation

$$\frac{dy}{dx} = x^2 - x + 1 - 2y \qquad (1)$$

passes through the point (1, 1).

a Verify that the approximation formula $\left(\frac{dy}{dx}\right)_0 \approx \frac{y_1 - y_0}{h}$ with a step length of 0.1 gives $y_1 = 0.9$. **(3 marks)**

b Use the formula $\left(\frac{dy}{dx}\right)_0 \approx \frac{y_1 - y_{-1}}{2h}$ with a step length of 0.1 to obtain an estimate of the value of y when $x = 1.2$. **(3 marks)**

c Using a suitable integrating factor, find the particular solution to differential equation (1) at the point where $x = 1$. **(4 marks)**

d Find the exact y-value on the solution curve in part **c** when $x = 1.2$ and hence find the percentage error in using the approximation in part **b**. **(3 marks)**

Challenge

A particular solution to the differential equation

$$\frac{dy}{dx} = \frac{2}{x - 1} + y$$

passes through the origin.

Use Euler's method once followed by the midpoint formula to obtain an estimate for the y-value of the particular solution when $x = 1$. Use a step length of 0.5.

Explain, with reference to the differential equation and the general solution curve, why this estimate is invalid.

8.2 Solving second-order differential equations

You can extend Euler's method to find approximate solutions to second-order differential equations of the form $\frac{\mathrm{d}^2y}{\mathrm{d}x^2} = \mathrm{f}\left(x, y, \frac{\mathrm{d}y}{\mathrm{d}x}\right)$.

Consider successive iterations of Euler's method:

Links In order to find a particular solution to a second-order differential equation, you need two initial conditions. These are often given as a value for y and a value for $\frac{\mathrm{d}y}{\mathrm{d}x}$ at some given value for x.

← **Core Pure Book 2, Chapter 7**

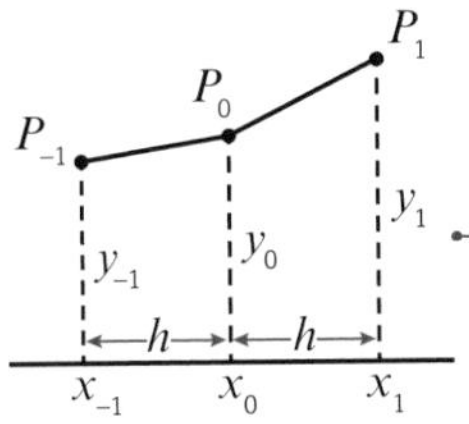

In this example you are interested in the behaviour of the solution near the point P_0. For convenience, this has been set as the 'middle' point, and the points on either side have been labelled as P_{-1} and P_1.

The gradient of the line segment $P_{-1}P_0$ is given by $\frac{y_0 - y_{-1}}{h} \approx \left(\frac{\mathrm{d}y}{\mathrm{d}x}\right)_{-1}$, and the gradient of the line segment P_0P_1 is given by $\frac{y_1 - y_0}{h} \approx \left(\frac{\mathrm{d}y}{\mathrm{d}x}\right)_0$.

The second derivative is a measure of the rate of change of the derivative. As such, you can estimate $\left(\frac{\mathrm{d}^2y}{\mathrm{d}x^2}\right)_0$, the value of $\frac{\mathrm{d}^2y}{\mathrm{d}x^2}$ at (x_0, y_0), by considering the change in $\frac{\mathrm{d}y}{\mathrm{d}x}$ across an interval of width h.

$$\left(\frac{\mathrm{d}^2y}{\mathrm{d}x^2}\right)_0 \approx \frac{\left(\frac{\mathrm{d}y}{\mathrm{d}x}\right)_0 - \left(\frac{\mathrm{d}y}{\mathrm{d}x}\right)_{-1}}{h}$$

$$= \frac{\frac{y_1 - y_0}{h} - \frac{y_0 - y_{-1}}{h}}{h}$$

$$= \frac{y_1 - 2y_0 + y_{-1}}{h^2}$$

Problem-solving

$$\left(\frac{\mathrm{d}^2y}{\mathrm{d}x^2}\right)_0 \approx \frac{(\text{Gradient of } P_0P_1) - (\text{Gradient of } P_{-1}P_0)}{h}$$

- **Euler's method can be extended to find approximate solutions to second-order differential equations using the formula**

$$\left(\frac{\mathrm{d}^2y}{\mathrm{d}x^2}\right)_0 \approx \frac{y_1 - 2y_0 + y_{-1}}{h^2}$$

It is often more useful to write this as an iterative formula:

$$y_{r+1} \approx 2y_r - y_{r-1} + h^2\left(\frac{\mathrm{d}^2y}{\mathrm{d}x^2}\right)_r, \; r = 0, 1, 2, \ldots$$

If a second-order differential equation is of the form $\frac{\mathrm{d}^2y}{\mathrm{d}x^2} = \mathrm{f}(x, y)$, you can use a single application of Euler's method to find y_1 before applying the above iterative formula.

Example 3

$\frac{d^2x}{dt^2} - \sin(x + t) = 0$. When $t = 0$, $x = -1$ and $\frac{dx}{dt} = 3$.

Use the approximations $\left(\frac{dx}{dt}\right)_0 \approx \frac{x_1 - x_0}{h}$ and $\left(\frac{d^2x}{dt^2}\right)_0 \approx \frac{x_1 - 2x_0 + x_{-1}}{h^2}$ to obtain estimates for x at $t = 0.1$ and $t = 0.2$, giving your answers correct to 4 decimal places.

$x_0 = -1$, $\left(\frac{dx}{dt}\right)_0 = 3$, $h = 0.1$

$x_1 \approx x_0 + h\left(\frac{dx}{dt}\right)_0$
$= -1 + 0.1 \times 3$
$= -0.7$

You need two values of x to substitute into the approximation for $\frac{d^2x}{dt^2}$. You are given x_0 and you can use Euler's formula to find x_1.

$\left(\frac{d^2x}{dt^2}\right)_1 = \sin(x_1 + t_1)$
$= \sin(-0.7 + 0.1)$
$= -0.5646...$

Rearrange the original equation to evaluate $\left(\frac{d^2x}{dt^2}\right)_1$, using the value of x_1 you have just found.

$x_2 \approx 2x_1 - x_0 + h^2\left(\frac{d^2x}{dt^2}\right)_1$
$= 2(-0.7) - (-1) + 0.1^2(-0.5646...)$
$= -0.4056$ (4 d.p.)

Watch out Be careful with the index numbers when using the approximation formula for $\frac{d^2x}{dt^2}$. The index number of $\frac{d^2x}{dt^2}$ should be **one less** than the index number of the value you are approximating.

If a second-order differential equation includes a term in $\frac{dy}{dx}$, you will also need to make use of the approximation $\left(\frac{dy}{dx}\right)_0 \approx \frac{y_1 - y_{-1}}{2h}$

Example 4

The curve $y = f(x)$ satisfies the differential equation $\frac{d^2y}{dx^2} = x^2 + y^2 + \frac{dy}{dx}$. When $x = 1$, $y = 4$ and $\frac{dy}{dx} = 3$.

Use the approximations $\left(\frac{dy}{dx}\right)_0 \approx \frac{y_1 - y_{-1}}{2h}$ and $\left(\frac{d^2y}{dx^2}\right)_0 \approx \frac{y_1 - 2y_0 + y_{-1}}{h^2}$ with $h = 0.2$ to estimate the value of y when $x = 1.2$.

$y_0 = 4$, $\left(\frac{dy}{dx}\right)_0 = 3$, $h = 0.2$

Write down the initial conditions and step length.

$\left(\frac{d^2y}{dx^2}\right)_0 = 1^2 + 4^2 + 3 = 20$

Evaluate $\left(\frac{d^2y}{dx^2}\right)_0$ using the original equation and the initial conditions.

$\left(\frac{dy}{dx}\right)_0 \approx \frac{y_1 - y_{-1}}{2h}$

$3 = \frac{y_1 - y_{-1}}{0.4} \Rightarrow y_1 - y_{-1} = 1.2$ (1)

Problem-solving
Use the approximations for $\frac{dy}{dx}$ and $\frac{d^2y}{dx^2}$ to form two simultaneous equations in y_1 and y_{-1}.

$$\left(\frac{d^2y}{dx^2}\right)_0 \approx \frac{y_1 - 2y_0 + y_{-1}}{h^2}$$

$$20 = \frac{y_1 - 8 + y_{-1}}{0.04} \Rightarrow y_1 + y_{-1} = 8.8 \qquad (2)$$

Adding (1) and (2),

$2y_1 = 10 \Rightarrow y_1 = 5$

Exercise 8C

1 Use the approximations $\left(\frac{dy}{dx}\right)_0 \approx \frac{y_1 - y_0}{h}$ and $\left(\frac{d^2y}{dx^2}\right)_0 \approx \frac{y_1 - 2y_0 + y_{-1}}{h^2}$ to obtain estimates for y_1, y_2 and y_3 for the following differential equations. In each case the initial conditions and step length are given.

a $\frac{d^2y}{dx^2} = x + y - 1$, given that when $x = 2$, $y = 4$ and $\frac{dy}{dx} = 1$, $h = 0.1$

b $\frac{d^2y}{dx^2} = x^2 + y^2$, given that when $x = 1$, $y = 1$ and $\frac{dy}{dx} = 2$, $h = 0.2$

c $\frac{d^2y}{dx^2} - 2xy + y^2 = 1$, given that when $x = 2$, $y = 1$ and $\frac{dy}{dx} = 1$, $h = 0.1$

d $\frac{d^2y}{dx^2} - \sin(xy) + 2 = 0$, given that when $x = 3$, $y = 2$ and $\frac{dy}{dx} = 2$, $h = 0.05$

2 Use the approximations $\left(\frac{dy}{dx}\right)_0 \approx \frac{y_1 - y_{-1}}{2h}$ and $\left(\frac{d^2y}{dx^2}\right)_0 \approx \frac{y_1 - 2y_0 + y_{-1}}{h^2}$ to estimate the value of y_1 for the following differential equations. In each case the initial conditions and step length are given.

a $\frac{d^2y}{dx^2} = x + y - \frac{dy}{dx}$, given that when $x = 1$, $y = 2$ and $\frac{dy}{dx} = 0.5$, $h = 0.1$

b $\frac{d^2y}{dx^2} = 3x^2 - \frac{dy}{dx}\sin y$, given that when $x = 2$, $y = 3$ and $\frac{dy}{dx} = 2$, $h = 0.05$

c $\frac{d^2y}{dx^2} - 3x\frac{dy}{dx} + y = 0$, given that when $x = 3$, $y = 1$ and $\frac{dy}{dx} = 1$, $h = 0.1$

d $\frac{d^2y}{dx^2} + 2xy\frac{dy}{dx} = \sin x$, given that when $x = 0$, $y = 1.5$ and $\frac{dy}{dx} = 0.8$, $h = 0.2$

(E) **3** A curve C satisfies the differential equation $\frac{d^2y}{dx^2} = x^3 - y^2$ and passes through the point (1, 1).

Given that the gradient of the curve at the point (1, 1) is –1,

a use an approximation of the form $\left(\frac{dy}{dx}\right)_0 \approx \frac{y_1 - y_0}{h}$ with $h = 0.1$ to find an estimate for the coordinates of the point on the curve where $x = 1.1$ **(2 marks)**

b use an approximation of the form $\left(\frac{d^2y}{dx^2}\right)_0 \approx \frac{y_1 - 2y_0 + y_{-1}}{h^2}$ with $h = 0.1$ to find further estimates, correct to 4 decimal places, for the coordinates of the points on the curve where $x = 1.2$ and $x = 1.3$. **(3 marks)**

(E) **4** $\frac{d^2y}{dx^2} - 3\sin x + y^2 = 1$

Given that at $x = 2$, $y = 1$ and $\frac{dy}{dx} = 0.6$, use the approximations $\left(\frac{dy}{dx}\right)_0 \approx \frac{y_1 - y_0}{h}$ and $\left(\frac{d^2y}{dx^2}\right)_0 \approx \frac{y_1 - 2y_0 + y_{-1}}{h^2}$ with a step length of 0.2 to obtain estimates for y at $x = 2.2$, $x = 2.4$ and $x = 2.6$. **(5 marks)**

(E) **5** The variable y satisfies the differential equation

$$\frac{d^2y}{dx^2} = \frac{x^2 - y}{3x}\frac{dy}{dx}$$

When $x = 2$, $y = 0$ and $\frac{dy}{dx} = 3$.

a Find the value of $\frac{d^2y}{dx^2}$ at $x = 2$ **(1 mark)**

b Use the approximations $\left(\frac{d^2y}{dx^2}\right)_0 \approx \frac{(y_1 - 2y_0 + y_{-1})}{h^2}$ and $\left(\frac{dy}{dx}\right)_0 \approx \frac{y_1 - y_{-1}}{2h}$, with $h = 0.1$, to find an estimate of y at $x = 2.1$. **(6 marks)**

(E) **6** The variable y satisfies the differential equation

$$\frac{d^2y}{dx^2} + \frac{dy}{dx} - \sin(xy) = 0$$

Given that $y = -3$ and $\frac{dy}{dx} = -0.5$ when $x = 1$, use the approximations

$\left(\frac{d^2y}{dx^2}\right)_0 \approx \frac{y_1 - 2y_0 + y_{-1}}{h^2}$ and $\left(\frac{dy}{dx}\right)_0 \approx \frac{y_1 - y_{-1}}{2h}$, with $h = 0.05$, to find an estimate of y at $x = 1.05$. **(7 marks)**

8.3 Simpson's rule

A

Simpson's rule is a numerical method for finding an approximate value of a definite integral of the form $I = \int_a^b \text{f}(x)\,\text{d}x$.

If you consider the definite integral to be the area beneath the curve $y = \text{f}(x)$ between limits a and b, then Simpson's rule works by splitting the area up into an even number of strips of equal width and then approximating each section of the curve by a quadratic function. The area of each strip can then be found.

Links Simpson's rule approximates each section of a curve as a curve, rather than a straight line. Because of this, it usually gives a more accurate estimate than the trapezium rule.
← Pure Year 2, Chapter 11

In the diagram below, four strips of width 0.5 are being used to estimate $\int_0^2 (\text{e}^{-x^2} + \cos x + 1)\,\text{d}x$.

The strips are paired off and a quadratic curve is used to approximate the curve for each pair.

Watch out Because the strips are paired off, Simpson's rule only works with an even number of strips.

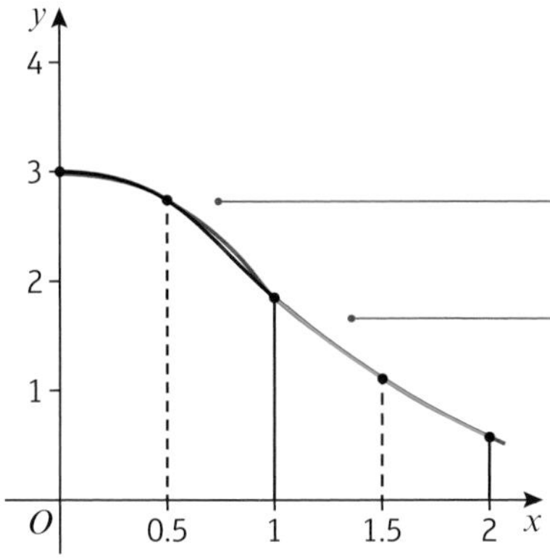

The section of the curve between $x = 0$ and $x = 1$ is approximated by a quadratic which passes through $(0, y_0)$, $(0.5, y_1)$ and $(1, y_2)$. There is only one quadratic curve which passes through three given distinct points, so the curve is unique.

A second quadratic is used to approximate the curve between $x = 1$ and $x = 2$. This curve passes through $(1, y_2)$, $(1.5, y_3)$ and $(2, y_4)$.

You find the corresponding y-coordinates by substituting these x-coordinates into the given function. You can then use a formula to find the approximation.

- **Simpson's rule for $2n$ strips of width h is given by**

$$\int_a^b \text{f}(x)\,\text{d}x \approx \tfrac{1}{3}h\big(y_0 + 4(y_1 + y_3 + \ldots + y_{2n-1}) + 2(y_2 + y_4 + \ldots + y_{2n-2}) + y_{2n}\big)$$

Note You can derive this formula by using the fact that the area under a quadratic curve which passes through the points (x_0, y_0), $(x_0 + h, y_1)$ and $(x_0 + 2h, y_2)$ is given by $\frac{1}{3}h(y_0 + 4y_1 + y_2)$.
→ Mixed exercise, Challenge

Informally, Simpson's rule is

$$\int_a^b \text{f}(x)\,\text{d}x \approx \tfrac{1}{3}h((\text{endpoints}) + 4(\text{odd values}) + 2(\text{even values}))$$

Watch out You need to learn Simpson's rule. It's not given in the formulae booklet.

Example 5

Use Simpson's rule with 4 intervals to estimate

$$\int_0^2 (e^{-x^2} + \cos x + 1)\,dx$$

Online Explore the use of Simpson's rule to estimate the integral using GeoGebra.

$h = 2 \div 4 = 0.5$

x_i	0	0.5	1	1.5	2
y_i	3	2.65638…	1.90818…	1.17613…	0.60216…

Make a table of values to show the x-coordinates for the endpoints of each strip, and the corresponding y-coordinates.

$$\int_0^2 (e^{-x^2} + \cos x + 1)\,dx$$
$$\approx \tfrac{1}{3} \times 0.5(3 + 4(2.65638\ldots + 1.17613\ldots) + 2(1.90818\ldots) + 0.60216\ldots)$$
$$= 3.791 \text{ (4 s.f.)}$$

Substitute the y-values into the formula for Simpson's rule and round your final answer to a sensible level of accuracy.

Exercise 8D

E **1** Use Simpson's rule with 4 intervals to estimate

$$\int_2^4 \frac{\ln x}{x}\,dx$$

(5 marks)

E **2** Use Simpson's rule with 6 intervals to estimate

$$\int_0^3 \sqrt{1 + x^5}\,dx$$

(5 marks)

E **3** The diagram shows the graph of $y = f(x)$ where $f(x) = \sqrt{\cos x + \tan x}$.

The shaded area is bounded by the curve, the x-axis and the lines $x = 0$ and $x = 1$.

a Use Simpson's rule with 4 intervals to estimate the shaded area. **(5 marks)**

b Suggest how you could improve your approximation using Simpson's rule. **(1 mark)**

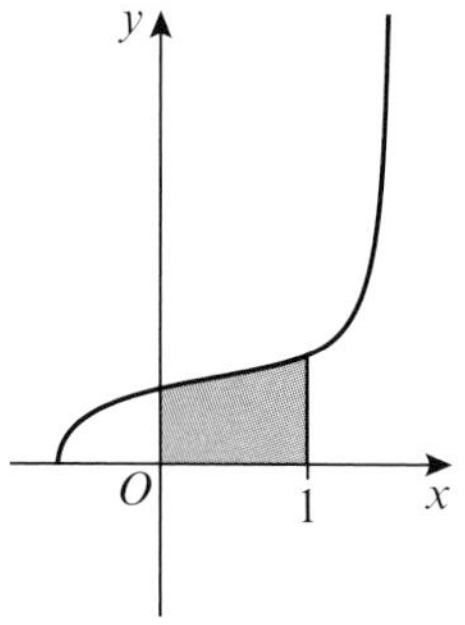

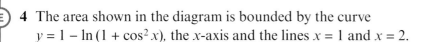

E **4** The area shown in the diagram is bounded by the curve $y = 1 - \ln(1 + \cos^2 x)$, the x-axis and the lines $x = 1$ and $x = 2$.

a Explain why you cannot use Simpson's rule with 7 intervals. **(1 mark)**

b Use Simpson's rule with 8 intervals to find an estimate for the shaded area. **(6 marks)**

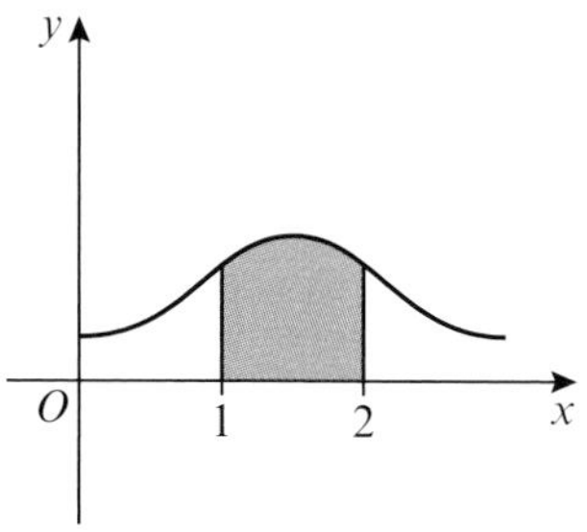

A **5** $f(x) = \dfrac{1}{\sin x + \tan x}$

P

a Use Simpson's rule with $h = 0.25$ to find an approximation for

$$\int_{0.5}^{1.5} f(x)\,dx$$ **(5 marks)**

b Use the Weierstrass substitution to find the value of the integral

$$\int_{0.5}^{1.5} f(x)\,dx$$

correct to 5 decimal places. **(6 marks)**

c Hence find, correct to 2 decimal places, the percentage error in using the method in part **a**. **(2 marks)**

P **6** $f(x) = x \sinh x$

a Use Simpson's rule with 4 intervals to find an approximation, to 2 decimal places, for

$$\int_{1}^{3} f(x)\,dx$$ **(5 marks)**

b Use integration by parts to show that

$$\int_{1}^{3} f(x)\,dx = e^3 + 2e^{-3} - e^{-1}$$ **(6 marks)**

c Hence find, correct to 2 significant figures, the percentage error in using the method in part **a** to approximate $\int_{1}^{3} f(x)\,dx$. **(2 marks)**

P **7** The diagram shows a curve defined parametrically as

$$x = t + t^2,\ y = t - t^2$$

The region enclosed by the curve and the x-axis is rotated 360° about the x-axis.

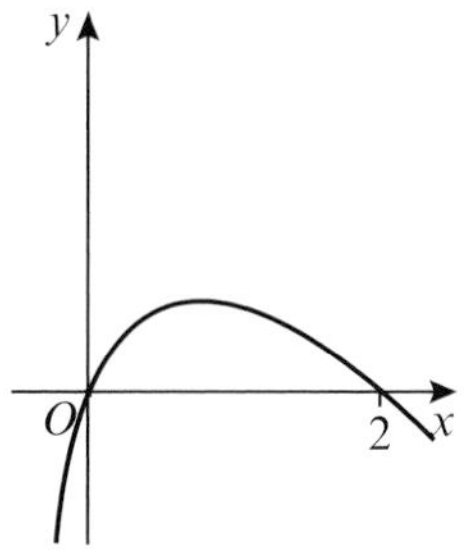

a Show that the volume of the solid generated is given by

$$\pi\int_{0}^{1} (t^2 - 3t^4 + 2t^5)\,dt$$ **(6 marks)**

b Use Simpson's rule with 4 intervals to estimate the volume of the solid. **(4 marks)**

c By calculating the exact volume of revolution, show that the percentage error in using Simpson's rule is less than 1.6%. **(4 marks)**

d Explain how your approximation in part **b** could be improved. **(1 mark)**

Mixed exercise 8

1 $y = f(x)$ satisfies the differential equation

$$\frac{dy}{dx} = ye^{2x} - x\ln y$$

Given that $f(2) = 1$, use two iterations of the approximation formula $\left(\dfrac{dy}{dx}\right)_0 \approx \dfrac{y_1 - y_0}{h}$ to estimate the value of $f(3)$, correct to three decimal places. **(5 marks)**

E/P **2** The variable y satisfies the differential equation

$$\frac{dy}{dx} = (x + y^2)^2 - (x^2 - y)^2$$

Given that a particular solution passes through the point (0, 2), use of the approximation formula $\left(\frac{dy}{dx}\right)_0 \approx \frac{y_1 - y_0}{h}$ gives $y_1 = 2.6$.

a Determine the value that was used for the step length. **(2 marks)**

b Using this step length, calculate, correct to three decimal places, the values of y_2 and y_3. **(5 marks)**

E/P **3** The value, x thousand pounds, of a particular tradeable commodity t days after it is purchased is modelled by the differential equation

$$\frac{dx}{dt} = \frac{x^2 - t}{xt - t^2}$$

If the commodity is worth £5000 two days after it is purchased, use two iterations of the approximation formula $\left(\frac{dy}{dx}\right)_0 \approx \frac{y_1 - y_0}{h}$ to estimate, correct to the nearest hundred pounds, the value of the commodity three days after it is purchased. **(6 marks)**

E/P **4** The velocity, $v\,\text{ms}^{-1}$, of a particle moving in a straight line, is modelled using the differential equation

$$\frac{dv}{dx} = \frac{2x - 25.6}{3v} - 0.001v$$

where x cm is the displacement of the particle from its starting position.

Given that $v = 8$ when $x = 5$, use the approximation formula $\left(\frac{dy}{dx}\right)_0 \approx \frac{y_1 - y_{-1}}{2h}$ with a step length of 0.25 to estimate the velocity of the particle when it is 5.75 cm from its starting position. **(6 marks)**

A E/P **5** A particular solution to the differential equation

$$\frac{dv}{dt} = 10t - 2v \qquad (1)$$

has $v = 2$ when $t = 0$.

a Verify that the approximation formula $\left(\frac{dy}{dx}\right)_0 \approx \frac{y_1 - y_0}{h}$ with a step length of 0.1 gives $v_1 = 1.6$. **(3 marks)**

b Use the formula $\left(\frac{dy}{dx}\right)_0 \approx \frac{y_1 - y_{-1}}{2h}$ with a step length of 0.1 to obtain an estimate of the value of v when $t = 0.2$. **(3 marks)**

c By using a suitable integrating factor, find the particular solution to differential equation (1) at the point where $t = 0$. **(4 marks)**

d Find the exact v-value for the particular solution found in part **c** when $t = 0.2$ and hence find the percentage error in using the approximation in part **b**. **(3 marks)**

P **6** $\frac{d^2y}{dx^2} - 2\cos x + y^3 = 3$

Given that at $x = 1$, $y = 2$ and $\frac{dy}{dx} = 0.5$, use the approximations $\left(\frac{dy}{dx}\right)_0 \approx \frac{y_1 - y_0}{h}$ and $\left(\frac{d^2y}{dx^2}\right)_0 \approx \frac{y_1 - 2y_0 + y_{-1}}{h^2}$ with a step length of 0.2 to obtain estimates for y at $x = 1.2$, $x = 1.4$ and $x = 1.6$. **(5 marks)**

P **7** The variable y satisfies the differential equation

$$\frac{d^2y}{dx^2} = \frac{x^3 - y^2}{3xy}\frac{dy}{dx}$$

Given that when $x = 1$, $y = 3$ and $\frac{dy}{dx} = 2$,

a find the value of $\frac{d^2y}{dx^2}$ at $x = 1$ **(1 mark)**

b use the approximations $\left(\frac{d^2y}{dx^2}\right)_0 \approx \frac{(y_1 - 2y_0 + y_{-1})}{h^2}$ and $\left(\frac{dy}{dx}\right)_0 \approx \frac{y_1 - y_{-1}}{2h}$, with $h = 0.1$, to find an estimate of y at $x = 1.1$. **(6 marks)**

A P **8** The diagram shows the graph of $y = f(x)$ where $f(x) = \sin(x^2) + x^2$. The shaded area is bounded by the curve, the x-axis and the lines $x = -2$ and $x = -1$.

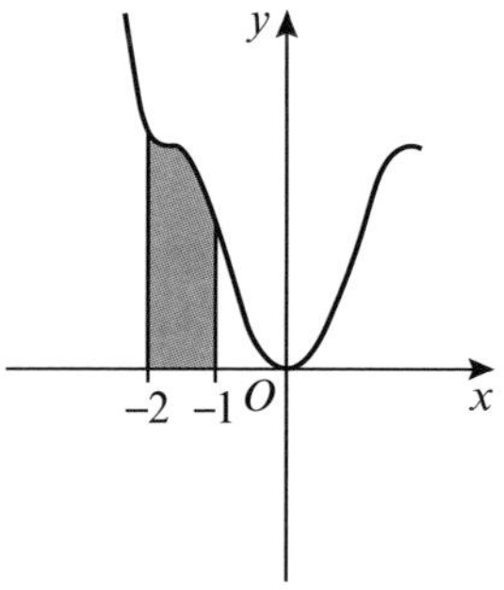

a Use Simpson's rule with 4 intervals to estimate the shaded area. **(5 marks)**

b Suggest how you could improve your approximation using Simpson's rule. **(1 mark)**

P **9** $f(x) = \frac{1}{1 + \sin x}$

a Use Simpson's rule with 4 intervals to find an approximation, to 3 significant figures, for

$$\int_0^1 f(x)\,dx$$ **(5 marks)**

b Use the Weierstrass substitution to find the value of the integral

$$\int_0^1 f(x)\,dx$$

correct to 5 decimal places. **(6 marks)**

c Hence find, correct to 2 significant figures, the percentage error in using the method in part **a**. **(2 marks)**

Challenge

A

The diagram shows three points, P_0, P_1 and P_2. The horizontal distance between each point is h. The points are joined with a parabola.

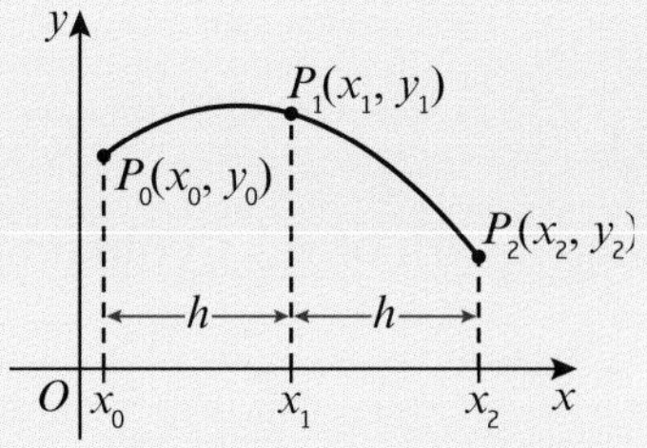

a Show that the area bounded by the parabola , the x-axis and the lines $x = x_0$ and $x = x_2$ is $\frac{1}{3}h(y_0 + 4y_1 + y_2)$.

b By considering further subdivisions of the interval $[x_0, x_2]$, derive the formula for Simpson's rule.

Summary of key points

1 **Euler's method** for finding approximate solutions to first-order differential equations uses the approximation

$$\left(\frac{dy}{dx}\right)_0 \approx \frac{y_1 - y_0}{h}$$

It is often more useful to write this as an iterative formula:

$$y_{r+1} \approx y_r + h\left(\frac{dy}{dx}\right)_r, \; r = 0, 1, 2, \ldots$$

2 The **midpoint method** for finding approximate solutions to first-order differential equations uses the formula

$$\left(\frac{dy}{dx}\right)_0 \approx \frac{y_1 - y_{-1}}{2h}$$

It is often more useful to write this as an iterative formula:

$$y_{r+1} \approx y_{r-1} + 2h\left(\frac{dy}{dx}\right)_r, \; r = 0, 1, 2, \ldots$$

3 Euler's method can be extended to find approximate solutions to second-order differential equations using the formula

$$\left(\frac{d^2y}{dx^2}\right)_0 \approx \frac{y_1 - 2y_0 + y_{-1}}{h^2}$$

It is often more useful to write this as an iterative formula:

$$y_{r+1} \approx 2y_r - y_{r-1} + h^2\left(\frac{d^2y}{dx^2}\right)_r, \; r = 0, 1, 2, \ldots$$

A

4 **Simpson's rule** for $2n$ strips of width h is given by

$$\int_a^b f(x)\,dx \approx \frac{1}{3}h(y_0 + 4(y_1 + y_3 + \ldots + y_{2n-1}) + 2(y_2 + y_4 + \ldots + y_{2n-2}) + y_{2n})$$

Reducible differential equations

9

Objectives

After completing this chapter you should be able to:

- Use a given substitution to transform a first-order differential equation into one that can be solved → pages 180–183
- Use a given substitution to transform a second-order differential equation into one that can be solved → pages 183–185

  Solve modelling problems involving reducible differential equations → pages 185–187

Prior knowledge check

1 Find the general solutions of these differential equations:

a $x\dfrac{dy}{dx} = 2(y - 1)$

← Pure Year 2, Chapter 11

b $\dfrac{dy}{dx} + \dfrac{y}{x} = 2x$

c $\dfrac{d^2y}{dx^2} - 4\dfrac{dy}{dx} + 3y = 0$

← Core Pure Book 2, Chapter 7

2 Given that $\dfrac{d^2y}{dx^2} - 4\dfrac{dy}{dx} + 4y = 2e^{-x}$,

a verify that $\frac{2}{9}e^{-x}$ is a particular integral of this differential equation

b find the general solution of this differential equation.

← Core Pure Book 2, Chapter 7

Many real-life situations can be modelled using differential equations: for example, the displacement of a point on a vibrating spring from a fixed point, or the distance fallen by a parachutist. → Mixed exercise, Q13

9.1 First-order differential equations

You can use a substitution to reduce a first-order differential equation into a form that you know how to solve, either by separating the variables, or by using an integrating factor.

Example 1

a Show that the substitution $y = xz$ transforms the differential equation

$$\frac{dy}{dx} = \frac{x^2 + 3y^2}{2xy}$$

into

$$x\frac{dz}{dx} = \frac{1 + z^2}{2z}$$

b Hence find the general solution to the original equation, giving y^2 in terms of x.

a $y = xz \quad (1)$

$\frac{dy}{dx} = x\frac{dz}{dx} + z \quad (2)$

Substituting into $\frac{dy}{dx} = \frac{x^2 + 3y^2}{2xy}$ gives

$x\frac{dz}{dx} + z = \frac{x^2 + 3x^2z^2}{2x^2z}$

$x\frac{dz}{dx} + z = \frac{x^2(1 + 3z^2)}{2x^2z}$

$x\frac{dz}{dx} = \frac{1 + 3z^2}{2z} - z$

$= \frac{1 + z^2}{2z}$ as required.

Substitute into the differential equation using equations (1) and (2).

Rearrange and simplify your equation.

b $\int \frac{2z}{1 + z^2} dz = \int \frac{1}{x} dx$

Separate the variables, then integrate including a constant of integration. ← Pure Year 2, Chapter 11

$\ln(1 + z^2) = \ln x + c$

$1 + z^2 = Ax$, where A is a positive constant

Take exponentials and let $A = e^c$.

$\left(1 + \left(\frac{y^2}{x^2}\right)\right) = Ax$

$y^2 = x^2(Ax - 1)$

Use the original substitution to transform the general solution in z back into a general solution in x and y.

$y = xz$, so $z = \frac{y}{x}$ and $z^2 = \left(\frac{y}{x}\right)^2$.

Watch out Using the substitution, differentiate to get $\frac{dy}{dx}$ in terms of $\frac{dz}{dx}$. Note that z is a function of x and y, not a constant, so you must use the product rule.

Example 2

a Use the substitution $z = y^{-1}$ to transform the differential equation $\frac{dy}{dx} + xy = xy^2$, into a differential equation in z and x.

b Solve the new equation, using an integrating factor.

c Find the general solution to the original equation, giving y in terms of x.

A

a As $z = y^{-1}$, $y = z^{-1}$

$$\frac{dy}{dx} = -\frac{1}{z^2}\frac{dz}{dx}$$

Substituting into $\frac{dy}{dx} + xy = xy^2$ gives

$$-\frac{1}{z^2}\frac{dz}{dx} + xz^{-1} = xz^{-2}$$

$$\Rightarrow \frac{dz}{dx} - xz = -x$$

b The integrating factor is $e^{\int -x\,dx} = e^{-\frac{x^2}{2}}$

$$e^{-\frac{x^2}{2}}\frac{dz}{dx} - xe^{-\frac{x^2}{2}}z = -xe^{-\frac{x^2}{2}}$$

$$\frac{d}{dx}(e^{-\frac{x^2}{2}}z) = -xe^{-\frac{x^2}{2}}$$

$$e^{-\frac{x^2}{2}}z = -\int xe^{-\frac{x^2}{2}}\,dx$$

$$e^{-\frac{x^2}{2}}z = e^{-\frac{x^2}{2}} + c$$

$$z = 1 + ce^{\frac{x^2}{2}}$$

c As $y = z^{-1}$,

$$y = \frac{1}{1 + ce^{\frac{x^2}{2}}}$$

Rearrange the substitution to make y the subject.

Differentiate to give $\frac{dy}{dx}$ in terms of $\frac{dz}{dx}$

Rearrange and simplify your equation.

To solve a differential equation in the form $\frac{dy}{dx} + P(x)y = Q(x)$, multiply every term in the equation by the integrating factor $e^{\int P(x)dx}$.

← **Core Pure Book 2, Chapter 7**

Integrate to give result then divide each term by the integrating factor.

Use the original substitution to write y in terms of x.

Example 3

a Use the substitution $u = y - x$ to transform the differential equation $\frac{dy}{dx} = \frac{y - x + 2}{y - x + 3}$ into a differential equation in u and x.

b By first solving this new equation, show that the general solution to the original equation may be written in the form $(y - x)^2 + 6y - 4x - 2c = 0$, where c is an arbitrary constant.

a Let $u = y - x$

Then $\frac{du}{dx} = \frac{dy}{dx} - 1$

Substituting into $\frac{dy}{dx} = \frac{y - x + 2}{y - x + 3}$ gives

$$\frac{du}{dx} + 1 = \frac{u + 2}{u + 3}$$

$$\Rightarrow \frac{du}{dx} = \frac{u + 2}{u + 3} - 1$$

$$\frac{du}{dx} = \frac{-1}{u + 3}$$

b $\int(u + 3)du = -\int dx$

$$\frac{1}{2}u^2 + 3u = -x + c$$

$$\frac{1}{2}(y - x)^2 + 3(y - x) = -x + c$$

$$\therefore\ (y - x)^2 + 6y - 4x - 2c = 0$$

Differentiate to give $\frac{du}{dx}$ in terms of $\frac{dy}{dx}$

Make $\frac{dy}{dx}$ the subject and substitute.

Rearrange and simplify your equation.

Separate the variables and integrate.

Substitute back to give your result in terms of x and y.

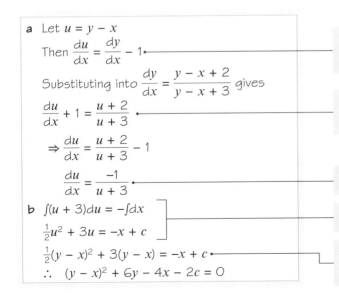

Exercise 9A

A

1 Use the substitution $z = \frac{y}{x}$ to transform each differential equation into a differential equation in z and x. By first solving the transformed equation, find the general solution to the original equation, giving y in terms of x.

a $\frac{dy}{dx} = \frac{y}{x} + \frac{x}{y}$, $x > 0, y > 0$

b $\frac{dy}{dx} = \frac{y}{x} + \frac{x^2}{y^2}$, $x > 0$

c $\frac{dy}{dx} = \frac{y}{x} + \frac{y^2}{x^2}$, $x > 0$

d $\frac{dy}{dx} = \frac{x^3 + 4y^3}{3xy^2}$, $x > 0$

E 2 a Use the substitution $z = y^{-2}$ to transform the differential equation

$$\frac{dy}{dx} + (\tfrac{1}{2}\tan x)\,y = -(2\sec x)y^3, \quad -\frac{\pi}{2} < x < \frac{\pi}{2}$$

into the differential equation $\frac{dz}{dx} - z\tan x = 4\sec x$. **(5 marks)**

b By first solving the transformed equation, find the general solution to the original equation, giving y in terms of x. **(6 marks)**

E 3 a Use the substitution $z = x^{\frac{1}{2}}$ to transform the differential equation

$$\frac{dx}{dt} + t^2x = t^2x^{\frac{1}{2}}$$

into the differential equation $\frac{dz}{dt} + \frac{1}{2}t^2z = \frac{1}{2}t^2$. **(4 marks)**

b By first solving the transformed equation, find the general solution to the original equation, giving x in terms of t. **(6 marks)**

E 4 a Use the substitution $z = y^{-1}$ to transform the differential equation

$$\frac{dy}{dx} - \frac{1}{x}y = \frac{(x+1)^3}{x}y^2$$

into the differential equation $\frac{dz}{dx} + \frac{1}{x}z = -\frac{(x+1)^3}{x}$ **(4 marks)**

b By first solving the transformed equation, find the general solution to the original equation, giving y in terms of x. **(6 marks)**

P 5 a Use the substitution $z = y^2$ to transform the differential equation

$$2(1 + x^2)\frac{dy}{dx} + 2xy = \frac{1}{y}$$

into a differential equation in z and x.

By first solving the transformed equation,

b find the general solution to the original equation, giving y in terms of x

c find the particular solution for which $y = 2$ when $x = 0$.

E/P 6 Show that the substitution $z = y^{-(n-1)}$ transforms the general equation

$$\frac{dy}{dx} + \mathrm{P}(x)y = \mathrm{Q}(x)y^n,$$

into the linear equation $\frac{dz}{dx} - \mathrm{P}(x)(n-1)z = -\,\mathrm{Q}(x)(n-1)$. **(5 marks)**

A P **7 a** Use the substitution $u = y + 2x$ to transform the differential equation

$$\frac{dy}{dx} = \frac{-(1 + 2y + 4x)}{1 + y + 2x}$$

into a differential equation in u and x. **(3 marks)**

b By first solving this new equation, show that the general solution to the original equation may be written as $4x^2 + 4xy + y^2 + 2y + 2x = k$, where k is a constant. **(6 marks)**

Challenge

$$x^2\frac{dy}{dx} - xy = y^2$$

By means of a suitable substitution, show that the general solution to the differential equation is given by

$$y = -\frac{x}{\ln x + C}$$

where C is a constant of integration.

9.2 Second-order differential equations

You can use a given substitution to reduce second-order differential equations into differential equations of the form $a\frac{d^2y}{dx^2} + b\frac{dy}{dx} + cy = f(x)$.

Example 4

Given that $x = e^u$, show that:

a $x\frac{dy}{dx} = \frac{dy}{du}$

b $x^2\frac{d^2y}{dx^2} = \frac{d^2y}{du^2} - \frac{dy}{du}$

c Hence find the general solution to the differential equation

$$x^2\frac{d^2y}{dx^2} + x\frac{dy}{dx} + y = 0$$

a As $x = e^u$, $\frac{dx}{du} = e^u = x$

From the chain rule,

$\frac{dy}{du} = \frac{dy}{dx} \times \frac{dx}{du} = e^u\frac{dy}{dx} = x\frac{dy}{dx}$, as required

Use the chain rule to express $\frac{dy}{dx}$ in terms of $\frac{dy}{du}$

b $\frac{d^2y}{du^2} = \frac{d}{du}\left(\frac{dy}{du}\right)$

$= \frac{d}{du}\left(e^u\frac{dy}{dx}\right)$

Differentiate this product using the product rule.

$= e^u\frac{dy}{dx} + e^u\frac{d^2y}{dx^2}\frac{dx}{du}$

$= \frac{dy}{du} + x^2\frac{d^2y}{dx^2}$, as $\frac{dx}{du} = e^u = x$

Use the chain rule to differentiate $\frac{dy}{dx}$ with respect to u, by differentiating with respect to x, giving $\frac{d^2y}{dx^2}$, and then multiplying by $\frac{dx}{du}$

So $x^2\frac{d^2y}{dx^2} = \frac{d^2y}{du^2} - \frac{dy}{du}$ as required.

A

c Substitute the results from parts **a** and **b** into the differential equation

$x^2\frac{d^2y}{dx^2} + x\frac{dy}{dx} + y = 0$

to obtain $\frac{d^2y}{du^2} - \frac{dy}{du} + \frac{dy}{du} + y = 0$

$\frac{d^2y}{du^2} + y = 0$

This is in the form $a\frac{d^2y}{du^2} + b\frac{dy}{du} + cy = 0$ with $a = 1$, $b = 0$ and $c = 1$. Find the general solution by considering the roots of the auxiliary equation. **← Core Pure Book 2, Section 7.2**

$m^2 + 1 = 0$

$m = \text{i}$ or $m = -\text{i}$

The roots are complex, so the general solution will be in the form $y = e^{pu}(A\cos qu + B\sin qu)$, with $p = 0$ and $q = 1$. **← Core Pure Book 2, Section 7.2**

So the general solution in terms of u is

$y = A\cos u + B\sin u$

where A and B are arbitrary constants.

$x = e^u \Rightarrow u = \ln x$ and the general solution to the differential equation $x^2\frac{d^2y}{dx^2} + x\frac{dy}{dx} + y = 0$ is $y = A\cos(\ln x) + B\sin(\ln x)$

Use $u = \ln x$ to give y in terms of x.

Exercise 9B

1 Find the general solution to each differential equation using the substitution $x = e^u$, where u is a function of x.

a $x^2\frac{d^2y}{dx^2} + 6x\frac{dy}{dx} + 4y = 0$ **b** $x^2\frac{d^2y}{dx^2} + 5x\frac{dy}{dx} + 4y = 0$ **c** $x^2\frac{d^2y}{dx^2} + 6x\frac{dy}{dx} + 6y = 0$

d $x^2\frac{d^2y}{dx^2} + 4x\frac{dy}{dx} - 28y = 0$ **e** $x^2\frac{d^2y}{dx^2} - 4x\frac{dy}{dx} - 14y = 0$ **f** $x^2\frac{d^2y}{dx^2} + 3x\frac{dy}{dx} + 2y = 0$

E **2** **a** Show that the transformation $y = \frac{z}{x}$ transforms the differential equation

$$x\frac{d^2y}{dx^2} + (2 - 4x)\frac{dy}{dx} - 4y = 0 \quad (1)$$

into the differential equation

$$\frac{d^2z}{dx^2} - 4\frac{dz}{dx} = 0 \quad (2)$$

(6 marks)

b Find the general solution to differential equation (2), giving z as a function of x. **(4 marks)**

c Hence obtain the general solution to differential equation (1) **(1 mark)**

E **3** **a** Show that the substitution $y = \frac{z}{x^2}$ transforms the differential equation

$$x^2\frac{d^2y}{dx^2} + 2x(x + 2)\frac{dy}{dx} + 2(x + 1)^2y = e^{-x} \quad (1)$$

into the differential equation

$$\frac{d^2z}{dx^2} + 2\frac{dz}{dx} + 2z = e^{-x} \quad (2)$$

(6 marks)

Hint Use a particular integral of the form λe^{-x}. **← Core Pure Book 2, Section 7.3**

b Find the general solution to differential equation (2), giving z as a function of x. **(7 marks)**

c Hence obtain the general solution to differential equation (1) **(1 mark)**

A **4** **a** Use the substitution $z = \sin x$ to transform the differential equation

$$\cos x \frac{d^2y}{dx^2} + \sin x \frac{dy}{dx} - 2y\cos^3 x = 2\cos^5 x$$

into the equation

$$\frac{d^2y}{dz^2} - 2y = 2(1 - z^2)$$ **(6 marks)**

b Hence solve the equation $\cos x \frac{d^2y}{dx^2} + \sin x \frac{dy}{dx} - 2y\cos^3 x = 2\cos^5 x$, giving y in terms of x. **(8 marks)**

5 **a** Show that the transformation $x = ut$ transforms the differential equation

$$t^2\frac{d^2x}{dt^2} - 2t\frac{dx}{dt} = -2(1 - 2t^2)x \quad (1)$$

into the differential equation

$$\frac{d^2u}{dt^2} - 4u = 0 \quad (2)$$ **(6 marks)**

b By solving differential equation (2), find the general solution to differential equation (1) in the form $x = f(t)$. **(8 marks)**

Given that $x = 2$ and $\frac{dx}{dt} = 1$ at $t = 1$,

c find the particular solution to differential equation (1). **(5 marks)**

Challenge

Use the substitution $u = \frac{dy}{dx}$ to find the general solution to the differential equation

$$x\frac{d^2y}{dx^2} + \frac{dy}{dx} = 12x$$

9.3 Modelling with differential equations

Differential equations can be used to model many real-life situations.

Example 5

A particle is moving along the x-axis and its displacement, x metres, is modelled using the differential equation

$$t\frac{dx}{dt} + x = 2t^3x^2, \quad 0 < t < 1.5$$

where t is the time in seconds.

a Use the substitution $u = xt$ to show that the differential equation can be expressed as

$$\frac{du}{dt} = 2u^2t$$

b Hence show that the general solution to the differential equation is

$$x = \frac{1}{t(A - t^2)}$$

where A is an arbitrary constant.

c Given that $x = 1$ when $t = 0.5$, find the displacement after 1.2 seconds.

A

a $u = xt \Rightarrow \frac{du}{dt} = t\frac{dx}{dt} + x$ and $x = \frac{u}{t}$ — Find expressions for $\frac{du}{dt}$ and for x.

$t\frac{dx}{dt} + x = 2t^3x^2$ — The left-hand side of the differential equation is the same as the expression for $\frac{du}{dt}$

$\frac{du}{dt} = 2t^3\left(\frac{u}{t}\right)^2$

$\frac{du}{dt} = 2u^2t$ as required.

b $\int \frac{1}{u^2}du = \int 2t\,dt$ — Separate the variables.

$-\frac{1}{u} = t^2 + c$

$u = -\frac{1}{t^2 + c}$

$xt = -\frac{1}{t^2 + c}$

$xt = \frac{1}{A - t^2}$

$x = \frac{1}{t(A - t^2)}$ as required.

Problem-solving

In order to obtain the equation in the form given, you need to change the constant from c to $-A$.

c $x = 1$ when $t = 0.5 \Rightarrow A = 2.25$

Hence when $t = 1.2$,

$x = \frac{1}{1.2(2.25 - 1.2^2)} = 1.0288...$

The displacement after 1.2 seconds is 1.03 m (3 s.f.).

Exercise 9C

E/P **1** A particle is moving along the x-axis and its displacement, x, at time t seconds, is modelled using the differential equation

$$tx\frac{dx}{dt} - x^2 = 3t^4$$

a Use the substitution $x = ut$ to show that the differential equation can be expressed as

$$u\frac{du}{dt} = 3t$$ **(4 marks)**

b Given that $x = 3$ when $t = 1$, show that the particular solution to the differential equation can be written as $x = t\sqrt{3t^2 + 6}$ **(5 marks)**

c Explain the behaviour of the particle as t becomes very large. **(2 marks)**

E/P **2** The velocity of a particle, v, at time t seconds, is modelled using the differential equation

$$3v^2t\frac{dv}{dt} = v^3 + t^3$$

a Use the substitution $v = zt$ to show that the differential equation can be expressed as

$$3z^2t\frac{dz}{dt} = 1 - 2z^3$$ **(3 marks)**

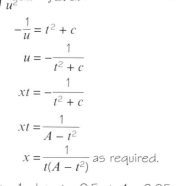

A **b** Given that $v = 2$ when $t = 1$, show that the particular solution to the differential equation can be written as

$$v = \sqrt[3]{\frac{t^3 + 15t}{2}}$$ **(8 marks)**

c Hence find, correct to 3 decimal places, the velocity and acceleration of the particle when $t = 2$. **(4 marks)**

P **3** The displacement of a particle, s, at time t seconds, is modelled using the differential equation

$$t\frac{d^2s}{dt^2} + (2 - t)\frac{ds}{dt} - (1 + 2t)s = e^{2t} \quad (1)$$

a Show that the substitution $v = st$ transforms the differential equation into

$$\frac{d^2v}{dt^2} - \frac{dv}{dt} - 2v = e^{2t} \quad (2)$$ **(8 marks)**

b Show that the general solution to differential equation (2) can be written as

$$v = Ae^{2t} + Be^{-t} + f(t)$$

where $f(t)$ is a particular integral to be found. **(8 marks)**

c Find the general solution to differential equation (1) in the form $s = g(t)$ and state one condition on t for the model to be valid. **(3 marks)**

P **4** A spring, fixed at one end, has an external force acting on it such that the other end moves in a straight line. At time t seconds, the displacement of the end of the spring from a fixed point O is x millimetres.

The displacement from O is modelled by the differential equation

$$t\frac{d^2x}{dt^2} - 2\frac{dx}{dt} + \left(\frac{2 + t^2}{t}\right)x = t^4 \quad (1)$$

a Show that the transformation $x = ut$ transforms equation (1) into the equation

$$\frac{d^2u}{dt^2} + u = t^2 \quad (2)$$ **(5 marks)**

b Hence find the general equation for the displacement of the end of the spring from O at time t seconds. **(8 marks)**

c State what happens to the displacement as t becomes large and comment on the model with reference to this behaviour. **(2 marks)**

Mixed exercise 9

E **1 a** Show that the transformation $z = y^{-1}$ transforms the differential equation

$$x\frac{dy}{dx} + y = y^2 \ln x \quad (1)$$

into the differential equation

$$\frac{dz}{dx} - \frac{z}{x} = -\frac{\ln x}{x} \quad (2)$$ **(4 marks)**

b By solving differential equation (1), find the general solution to differential equation (2). **(6 marks)**

A E **2 a** Show that the substitution $z = y^2$ transforms the differential equation

$$2\cos x\frac{\mathrm{d}y}{\mathrm{d}x} - y\sin x + y^{-1} = 0 \qquad (1)$$

into the differential equation

$$\cos x\frac{\mathrm{d}z}{\mathrm{d}x} - z\sin x = -1 \qquad (2)$$

(4 marks)

b Solve differential equation (2) to find z as a function of x. **(6 marks)**

c Hence write down the general solution to differential equation (1) in the form $y^2 = \mathrm{f}(x)$. **(1 mark)**

E **3 a** Show that the substitution $z = \frac{y}{x}$ transforms the differential equation

$$(x^2 - y^2)\frac{\mathrm{d}y}{\mathrm{d}x} - xy = 0 \qquad (1)$$

into the differential equation

$$x\frac{\mathrm{d}z}{\mathrm{d}x} = \frac{z^3}{1 - z^2} \qquad (2)$$

(4 marks)

b Solve equation (2) and hence obtain the general solution to equation (1). **(6 marks)**

E **4 a** Show that the transformation $z = \frac{y}{x}$ transforms the differential equation

$$\frac{\mathrm{d}y}{\mathrm{d}x} = \frac{y(x + y)}{x(y - x)} \qquad (1)$$

into the differential equation

$$x\frac{\mathrm{d}z}{\mathrm{d}x} = \frac{2z}{z - 1} \qquad (2)$$

(4 marks)

b Solve equation (2) and hence obtain the general solution to equation (1). **(6 marks)**

E **5 a** Show that the substitution $z = \frac{y}{x}$ transforms the differential equation

$$\frac{\mathrm{d}y}{\mathrm{d}x} = \frac{-3xy}{y^2 - 3x^2} \qquad (1)$$

into the equation

$$x\frac{\mathrm{d}z}{\mathrm{d}x} = -\frac{z^3}{z^2 - 3} \qquad (2)$$

(4 marks)

b By solving equation (2), find the general solution to equation (1). **(6 marks)**

E **6 a** Use the substitution $u = x + y$ to show that the differential equation

$$\frac{\mathrm{d}y}{\mathrm{d}x} = (x + y + 1)(x + y - 1)$$

can be written as

$$\frac{\mathrm{d}u}{\mathrm{d}x} = u^2$$

(3 marks)

b Hence find the general solution to the original differential equation. **(4 marks)**

A E 7 a Show that the transformation $u = y - x - 2$ can be used to transform the differential equation

$$\frac{dy}{dx} = (y - x - 2)^2 \qquad (1)$$

into the differential equation

$$\frac{du}{dx} = u^2 - 1 \qquad (2)$$ **(3 marks)**

b Solve equation (2) and hence find the general solution to equation (1). **(4 marks)**

P 8 A particle is moving with velocity v at time t such that

$$t\frac{dv}{dt} + v = 2t^3v^3, \quad 0 < t < \sqrt{3} \qquad (1)$$

a Use the substitution $u = v^{-2}$ to show that the differential equation can be transformed into

$$\frac{du}{dt} - \frac{2u}{t} = -4t^2$$ **(5 marks)**

b Given that $v = \frac{1}{2}$ when $t = 1$, show that the solution to differential equation (1) can be written as

$$v = \sqrt{\frac{1}{t^2(c - 4t)}}$$

where c is a constant to be found. **(8 marks)**

P 9 a Find the general solution to the differential equation

$$x^2\frac{d^2y}{dx^2} + 4x\frac{dy}{dx} + 2y = \ln x, \qquad x > 0$$

using the substitution $x = e^u$. **(10 marks)**

b Find the equation of the solution curve passing through the point (1, 1) with gradient 1. **(3 marks)**

P 10 Solve the equation $\frac{d^2y}{dx^2} + \tan x\frac{dy}{dx} + y\cos^2 x = \cos^2 x\, e^{\sin x}$, using the substitution $z = \sin x$.

Find the solution for which $y = 1$ and $\frac{dy}{dx} = 3$ at $x = 0$. **(13 marks)**

P 11 The displacement of a particle, x, at time t seconds is modelled by the differential equation

$$t^2\frac{d^2x}{dt^2} - 2t\frac{dx}{dt} + 2x = 4\ln t \qquad (1)$$

a Show that the substitution $t = e^u$ transforms equation (1) into

$$\frac{d^2x}{du^2} - 3\frac{dx}{du} + 2x = 4u \qquad (2)$$ **(6 marks)**

b By first solving equation (2), obtain the general solution to differential equation (1) giving your answer in the form $x = f(t)$. **(7 marks)**

c Describe the behaviour of the particle as t gets very large. **(1 mark)**

A E/P **12** A particle is subject to an external variable force such that the particle moves in the direction of the x-axis. The displacement, in cm, of the particle from a fixed point O at time t seconds is modelled by the differential equation

$$2t^2\frac{d^2x}{dt^2} - 4t\frac{dx}{dt} + (4 - 2t^2)x = t^4 \qquad (1)$$

a Show that the transformation $x = tv$ transforms equation (1) into the differential equation

$$2\frac{d^2v}{dt^2} - 2v = t \qquad (2)$$ **(6 marks)**

b Hence find the general equation of the displacement of the particle from O after t seconds. **(7 marks)**

E/P **13** The velocity of a skydiver, $v\,\text{m s}^{-1}$, at a time t seconds after jumping out of a stationary helicopter is modelled using the differential equation

$$1000\frac{dv}{dt} - 500v + tv^2 = 0, \quad 0 \leqslant t \leqslant 10 \qquad (1)$$

a By means of the substitution $u = v^{-1}$, show that differential equation (1) can be transformed into the differential equation

$$\frac{du}{dt} + 0.5u = 0.001t \qquad (2)$$ **(5 marks)**

b Find the general solution to differential equation (2), and hence find the general solution to differential equation (1) in the form $v = f(t)$. **(6 marks)**

c Given that the initial velocity of the skydiver is $2\,\text{m s}^{-1}$, find a particular solution to differential equation (1). **(3 marks)**

d By considering $\frac{1}{v}$, or otherwise, describe the behavior of v for large values of t, and comment on the validity of the model in these situations. **(2 marks)**

Challenge

By means of a suitable substitution, show that the general solution to the differential equation

$$\frac{d^2y}{dx^2} = \left(\frac{dy}{dx}\right)^2$$

is given by $y = A - \ln(x + B)$, where A and B are arbitrary constants.

Summary of key points

1 You can use a substitution to reduce a first-order differential equation into a form that you know how to solve, either by separating the variables, or by using an integrating factor.

2 You can use a given substitution to reduce second-order differential equations into differential equations of the form

$$a\frac{d^2y}{dx^2} + b\frac{dy}{dx} + cy = f(x)$$

Review exercise 2

E 1 $\sin\left(\frac{x}{2}\right) = \frac{12}{13}, 0 < \frac{x}{2} < \frac{\pi}{2}$

Show, without use of a calculator, that

$\cot x = -\frac{119}{120}$ **(3)**

← Section 5.1

E 2 $\sin\theta = \frac{\sqrt{6}+\sqrt{2}}{4}, \frac{\pi}{2} < \theta < \pi$

a Show, without use of a calculator, that $\tan\theta = -2 - \sqrt{3}$. **(3)**

b Using the t-formulae, find $\sin 2\theta$ and $\cos 2\theta$. **(3)**

c Hence deduce the value of θ. **(1)**

← Section 5.1

E 3 **a** Use the substitution $t = \tan\frac{x}{2}$ to show that $\sec x + \tan x = \frac{1+t}{1-t}$,

$x \neq (2n+1)\frac{\pi}{2}, n \in \mathbb{Z}$ **(3)**

b Hence show that

$\sec x + \tan x \equiv \tan\left(\frac{\pi}{4} + \frac{x}{2}\right)$ **(3)**

← Section 5.2

E 4 Use the t-formulae to show that

$2\cos^2\left(\frac{\theta}{2}\right) - 1 \equiv \cos\theta$. **(3)**

← Section 5.2

E 5 **a** Use the substitution $t = \tan\frac{x}{2}$ to show that the equation

$3\cos x - 4\sin x = 4$ (1)

can be written as $7t^2 + 8t + 1 = 0$. **(3)**

b Hence find all the solutions to equation (1) in the interval $0 < x < 2\pi$. **(3)**

← Section 5.3

E 6 **a** Use the substitution $t = \tan\frac{x}{2}$ to show that the equation

$2\sin x + \cos x = 1$ (1)

can be written as $t^2 - 2t = 0$. **(3)**

b Hence find all the solutions to equation (1) in the interval $0 \leqslant x \leqslant 2\pi$. **(3)**

← Section 5.3

A E/P 7 The displacement, s m, of a particle at time x seconds is given by

$s = 2\sin 4x + 4\sin 2x + 1, 0 \leqslant x \leqslant 2\pi$

a Show that the velocity of the particle, v m s^{-1} at time x seconds is given by

$v = \frac{16}{(1+t^2)^2}(1 - 3t^2)$ where $t = \tan x$. **(6)**

b Hence find the least value of s in the given interval, justifying that it is a minimum. **(4)**

← Section 5.4

E/P 8

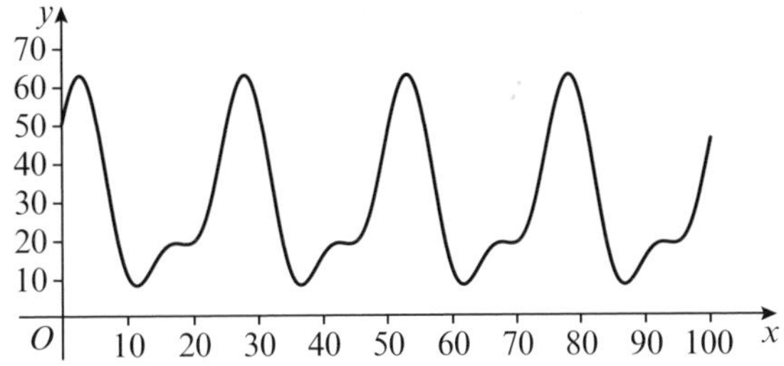

The diagram above shows the graph of $y = f(x)$ for the function

$f(x) = 30 + 10\sin\frac{x}{2} + 11\sin\frac{x}{4} + 20\cos\frac{x}{4}$,

$x \in [0, 100]$

a Show that

$f'(x) = \frac{(t+1)(9t^3 - 49t^2 - 71t + 31)}{4(1+t^2)^2}$

where $t = \tan\frac{x}{8}$ **(6)**

b Hence find the smallest exact multiple of π for which the graph has a stationary point. **(2)**

A

The function $kf(x)$ is used to model an electric pump which extracts L litres of water at time x seconds from a flooded mine shaft.

The maximum amount of water pumped is 300 litres.

c Suggest a suitable value of k. **(1)**

d Describe the point in the pumping cycle when x is equal to the value found in part **b** and estimate the amount of water being pumped at this point. **(2)**

← Section 5.4

E **9** **a** Find the Taylor series of $\cos 2x$ in ascending powers of $\left(x - \frac{\pi}{4}\right)$ up to and including the term in $\left(x - \frac{\pi}{4}\right)^5$. **(4)**

b Use your answer to part **a** to obtain an estimate of $\cos 2$, giving your answer to 6 decimal places. **(2)**

← Section 6.1

E **10** **a** Find the Taylor series of $\ln(\sin x)$ in ascending powers of $\left(x - \frac{\pi}{6}\right)$ up to and including the term in $\left(x - \frac{\pi}{6}\right)^3$. **(4)**

b Use your answer to part **a** to obtain an estimate of $\ln(\sin 0.5)$, giving your answer to 6 decimal places. **(2)**

← Section 6.1

E **11** Given that $y = \tan x$,

a find $\frac{dy}{dx}$, $\frac{d^2y}{dx^2}$ and $\frac{d^3y}{dx^3}$ **(3)**

b Find the Taylor series of $\tan x$ in ascending powers of $\left(x - \frac{\pi}{4}\right)$ up to and including the term in $\left(x - \frac{\pi}{4}\right)^3$ **(4)**

c Hence show that

$$\tan\frac{3\pi}{10} \approx 1 + \frac{\pi}{10} + \frac{\pi^2}{200} + \frac{\pi^3}{3000}$$ **(3)**

← Section 6.1

A E **12** **a** Find the Taylor series of $\ln x$ about $x = 1$. **(3)**

b Hence find the value of

$$\lim_{x\to1}\left(\frac{2\ln x}{x^2 - 3x + 2}\right)$$ **(4)**

← Section 6.2

E **13** **a** Find the Taylor series expansion about $x = 0$ for $\sinh x$. **(3)**

b Hence find the value of $\lim_{x\to0}(x\,\text{cosech}(2x))$ **(4)**

← Section 6.2

E **14** $(1 - x^2)\frac{d^2y}{dx^2} - x\frac{dy}{dx} + 2y = 0$

At $x = 0$, $y = 2$ and $\frac{dy}{dx} = -1$.

a Find the value of $\frac{d^3y}{dx^3}$ at $x = 0$. **(4)**

b Express y as a series in ascending powers of x, up to and including the term in x^3. **(4)**

← Section 6.3

E **15** $(1 + 2x)\frac{dy}{dx} = x + 4y^2$

a Show that

$$(1 + 2x)\frac{d^2y}{dx^2} = 1 + 2(4y - 1)\frac{dy}{dx} \quad (1)$$ **(4)**

b Differentiate equation (1) with respect to x to obtain an equation involving $\frac{d^3y}{dx^3}$, $\frac{d^2y}{dx^2}$, $\frac{dy}{dx}$, x and y. **(4)**

Given that $y = \frac{1}{2}$ at $x = 0$,

c find a series solution for y, in ascending powers of x, up to and including the term in x^3. **(4)**

← Section 6.3

E/P **16** $\frac{dy}{dx} = y^2 + xy + x$, $y = 1$ at $x = 0$

a Use the Taylor series method to find y as a series in ascending powers of x, up to and including the term in x^3. **(6)**

b Use your series to find y at $x = 0.1$, giving your answer to 2 decimal places. **(4)**

← Section 6.3

A P **17** $y\frac{dy}{dx} = \frac{x+3}{y+1}$

Given that $y = 1.5$ at $x = 0$,

a use the Taylor series method to find the series solution for y, in ascending powers of x, up to and including the term in x^3. **(6)**

b Use your result to part **a** to estimate, to 3 decimal places, the value of y at $x = 0.1$. **(4)**

← Section 6.3

E/P **18** $y\frac{d^2y}{dx^2} + \left(\frac{dy}{dx}\right)^2 + y = 0$

a Find an expression for $\frac{d^3y}{dx^3}$ **(4)**

Given that $y = 1$ and $\frac{dy}{dx} = 1$ at $x = 0$,

b find the series solution for y, in ascending powers of x, up to and including the term in x^3. **(4)**

c Comment on whether it would be sensible to use your series solution to give estimates for y at $x = 0.2$ and at $x = 50$. **(2)**

← Section 6.3

E/P **19** $\frac{d^2y}{dx^2} - 4\frac{dy}{dx} + 3y^2 = 6,$

with $y = 1$ and $\frac{dy}{dx} = 0$ at $x = 0$.

a Use the Taylor series method to obtain y as a series of ascending powers of x, up to and including the term in x^4. **(6)**

b Hence find the approximate value of y when $x = 0.2$. **(3)**

← Section 6.3

E/P **20** Given that $y = x^3e^{3x}$, use Leibnitz's theorem to show that

$$\frac{d^ny}{dx^n} = 3^{n-3}e^{3x}(27x^3 + 27nx^2 + 9n(n-1)x + n(n-1)(n-2))$$ **(4)**

← Section 7.1

E/P **21** Use Leibnitz's theorem to show that $y = e^x \sin x$ satisfies $\frac{d^6y}{dx^6} + 8\frac{dy}{dx} - 8y = 0.$ **(4)**

← Section 7.1

A E **22** Use L'Hospital's rule to evaluate

$$\lim_{x\to1}\left(\frac{\ln x}{x^2-1}\right)$$ **(4)**

← Section 7.2

E/P **23** Show, using L'Hospital's rule, that $\lim_{x\to0}(x\ln x) = 0.$ (4)

← Section 7.2

E/P **24** Use L'Hospital's rule to evaluate

$$\lim_{x\to0}\left(\frac{xe^x}{2\sin x}\right)$$ **(4)**

← Section 7.2

E/P **25** Show, using L'Hospital's rule, that

$$\lim_{x\to0}\left(\frac{e^x - \cos x}{x}\right) = 1$$ **(4)**

← Section 7.2

E **26** **a** Use the substitution $t = \tan\frac{x}{2}$ to show that the integral

$$\int\frac{1}{1-\sin x+\cos x}dx$$

can be written as

$$\int\frac{1}{1-t}dt$$ **(4)**

b Hence evaluate $\int_0^{\frac{\pi}{4}}\frac{1}{1-\sin x+\cos x}dx.$ **(4)**

← Section 7.3

E/P **27** Use the substitution $t = \tan\frac{x}{2}$ to show that

$$\int_{\frac{\pi}{2}}^{\frac{7\pi}{6}}\frac{1}{3\sin x - 4\cos x}dx = \tfrac{1}{5}\ln(a + b\sqrt{3})$$

where a and b are rational constants to be found. **(7)**

← Section 7.3

E **28** $y = f(x)$ satisfies the differential equation

$$\frac{dy}{dx} = x^3 - y^3$$

Given that $f(1) = 2$, use two iterations of the approximation formula $\left(\frac{dy}{dx}\right)_0 \approx \frac{y_1 - y_0}{h}$ to estimate the value of $f(1.5)$. **(5)**

← Section 8.1

(E/P) **29** The differential equation

$$\frac{dy}{dx} = 2e^x - y^2$$

has a particular solution that passes through the point (ln2, 1).
Use of the approximation formula,

$$\left(\frac{dy}{dx}\right)_0 \approx \frac{y_1 - y_0}{h}, \text{ gives } y_1 = 1.6.$$

a Determine the value that was used for the step length, h. **(2)**

b Using this step length, calculate, correct to three decimal places, the values of y_2 and y_3. **(5)**

← Section 8.1

(E/P) **30** The value, v thousand pounds, of a financial derivative t days after it is purchased is modelled by the differential equation

$$\frac{dv}{dt} = \frac{2v - 3t}{v^2t - t^3}$$

If the derivative is worth £8000 three days after it is purchased, use two iterations of the approximation formula $\left(\frac{dy}{dx}\right)_0 \approx \frac{y_1 - y_0}{h}$ to estimate, correct to the nearest pound, the value of the derivative five days after it is purchased. **(6)**

← Section 8.1

(E) **31** Use the approximation formula $\left(\frac{dy}{dx}\right)_0 \approx \frac{y_1 - y_{-1}}{2h}$ to estimate the value at $x = 1.3$ of the particular solution to the differential equation

$$\frac{dy}{dx} = 2\ln x - y$$

which passes through the point (1, 2). Use a step length of 0.1. **(6)**

← Section 8.1

(E) **32** A particular solution to the differential equation $\frac{dy}{dx} = \cos(x^2y)$ passes through the point (1, 1).

a Verify that the approximation formula $\left(\frac{dy}{dx}\right)_0 \approx \frac{y_1 - y_0}{h}$ with a step length of 0.2 gives $y_1 = 1.108$ correct to three decimal places. **(3)**

b Use the approximation formula $\left(\frac{dy}{dx}\right)_0 \approx \frac{y_1 - y_{-1}}{2h}$ with a step length of 0.2 to obtain an estimate of the value of y when $x = 1.6$. Give your answer to three decimal places. **(5)**

← Section 8.1

(E/P) **33** The population of a bacteria, P, at time t days is modelled by the differential equation

$$\frac{dP}{dt} = P - 0.00002P^2 - 0.5\cos(0.8t)$$

Given that the initial starting population of this bacteria is 1000, use the approximation formula $\left(\frac{dy}{dx}\right)_0 \approx \frac{y_1 - y_{-1}}{2h}$ with a step length of 1 to estimate, correct to the nearest bacteria, the population after three days. **(6)**

← Section 8.1

(E) **34** $\frac{d^2y}{dx^2} = x^4 + \frac{1}{y}$

Given that the gradient of the curve at the point (0, 1) is −2,

a use the approximation formula $\left(\frac{dy}{dx}\right)_0 \approx \frac{y_1 - y_0}{h}$ with $h = 0.1$ to find an estimate for the y-value of the particular solution when $x = 0.1$. **(3)**

b use the approximation formula $\left(\frac{d^2y}{dx^2}\right)_0 \approx \frac{y_1 - 2y_0 + y_{-1}}{h^2}$ with $h = 0.1$ to find a further estimate, correct to 4 decimal places, for the y-value when $x = 0.3$. **(5)**

← Section 8.2

35 $\frac{d^2y}{dx^2} + \sin x - 2\cos y = 2$

Given that at $x = 1$, $y = 2$ and $\frac{dy}{dx} = 0.5$,

use the approximations $\left(\frac{dy}{dx}\right)_0 \approx \frac{y_1 - y_0}{h}$

and $\left(\frac{d^2y}{dx^2}\right)_0 \approx \frac{y_1 - 2y_0 + y_{-1}}{h^2}$ with a step length of 0.2 to obtain, estimates for y at $x = 1.2$, $x = 1.4$ and $x = 1.6$. **(6)**

← Section 8.2

(P) **36** $\frac{d^2y}{dx^2} = 2xy + \left(\frac{dy}{dx}\right)^2$

When $x = 2$, $y = 2$ and $\frac{dy}{dx} = 3$

Use the approximations

$\left(\frac{d^2y}{dx^2}\right)_0 \approx \frac{y_1 - 2y_0 + y_{-1}}{h^2}$ and

$\left(\frac{dy}{dx}\right)_0 \approx \frac{y_1 - y_{-1}}{2h}$, with $h = 0.2$ to estimate the value of y when $x = 2.2$. **(6)**

← Section 8.2

(A/P) **37**

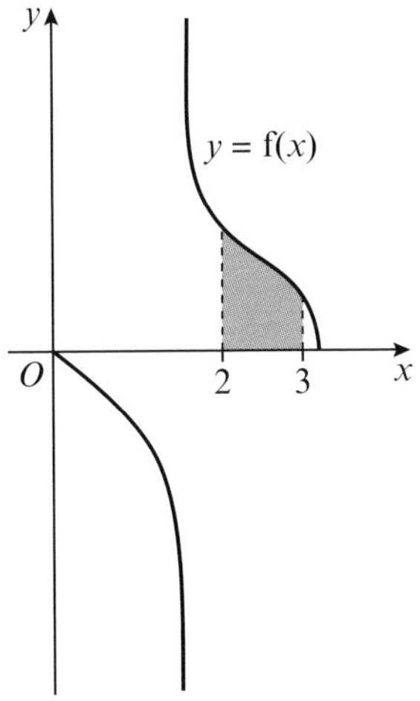

The diagram shows the graph of $y = f(x)$ where $f(x) = \sqrt[3]{\sin x - \tan x}$,

$0 \leqslant x \leqslant \pi$, $x \neq \frac{\pi}{2}$

The shaded area is bounded by the curve, the x-axis and the lines $x = 2$ and $x = 3$.

a Use Simpson's rule with 4 intervals to estimate the shaded area. **(5)**

b Suggest how you could improve your approximation using Simpson's rule. **(1)**

← Section 8.3

(A, E/P) **38** $f(x) = x\cosh x$

a Use Simpson's rule with 2 intervals to find an approximation for

$\int_1^2 f(x)\,dx$ **(4)**

b Use integration by parts to show that

$\int_1^2 f(x)\,dx = -\frac{3}{2}e^{-2} + e^{-1} + \frac{1}{2}e^2$ **(6)**

c Hence find, correct to 2 significant figures, the percentage error in using Simpson's rule to approximate

$\int_1^2 f(x)\,dx.$ **(2)**

← Section 8.3

(E) **39** **a** By using the substitution $y = \frac{1}{2}(u - x)$, or otherwise, find the general solution of the differential equation

$$\frac{dy}{dx} = x + 2y$$ **(4)**

Given that $y = 2$ at $x = 0$,

b express y in terms of x. **(3)**

← Section 9.1

(E) **40** **a** Use the substitution $y = vx$ to transform the equation

$$\frac{dy}{dx} = \frac{(4x + y)(x + y)}{x^2}, \ x > 0 \quad (1)$$

into the equation

$$x\frac{dv}{dx} = (2 + v)^2 \quad (2)$$ **(4)**

b Solve differential equation (2) to find v in terms of x. **(4)**

c Hence show that

$y = -2x - \frac{x}{\ln x + c}$, where c is an arbitrary constant, is the general solution to differential equation (1). **(3)**

← Section 9.1

(E) **41** **a** Show that the substitution $y = vx$ transforms the differential equation

$$\frac{dy}{dx} = \frac{3x - 4y}{4x + 3y} \quad (1)$$

into the differential equation

$$x\frac{dv}{dx} = -\frac{3v^2 + 8v - 3}{3v + 4} \quad (2)$$ **(4)**

A

b Find the general solution of differential equation (2). **(4)**

c Given that $y = 7$ at $x = 1$, show that the particular solution to differential equation (1) can be written as

$$(3y - x)(y + 3x) = 200 \qquad \textbf{(3)}$$

← Section 9.1

E **42 a** Use the substitution $\mu = y^{-2}$ to transform the differential equation

$$\frac{dy}{dx} + 2xy = xe^{-x^2}y^3 \qquad (1)$$

into the differential equation

$$\frac{d\mu}{dx} - 4x\mu = -2xe^{-x^2} \qquad (2) \quad \textbf{(4)}$$

b Find the general solution to differential equation (2). **(4)**

c Hence obtain the solution to differential equation (1) for which $y = 1$ at $x = 0$. **(3)**

← Section 9.1

E **43 a** Show that the transformation $y = xv$ transforms the equation

$$x^2\frac{d^2y}{dx^2} - 2x\frac{dy}{dx} + (2 + 9x^2)y = x^5 \qquad (1)$$

into the equation

$$\frac{d^2v}{dx^2} + 9v = x^2 \qquad (2)$$

(6)

b Solve differential equation (2) to find v as a function of x. **(4)**

c Hence state the general solution to differential equation (1). **(2)**

← Section 9.2

A E **44** Given that $x = t^{\frac{1}{2}}$, $x > 0$, $t > 0$, and that y is a function of x,

a find $\frac{dy}{dx}$ in terms of $\frac{dy}{dt}$ and t. **(2)**

Assuming that $\frac{d^2y}{dx^2} = 4t\frac{d^2y}{dt^2} + 2\frac{dy}{dt}$,

b show that the substitution $x = t^{\frac{1}{2}}$ transforms the differential equation

$$\frac{d^2y}{dx^2} + \left(6x - \frac{1}{x}\right)\frac{dy}{dx} - 16x^2y = 4x^2e^{2x^2} \qquad (1)$$

into the differential equation

$$\frac{d^2y}{dt^2} + 3\frac{dy}{dt} - 4y = e^{2t} \qquad (2) \quad \textbf{(6)}$$

c Hence find the general solution to (1) giving y in terms of x. **(6)**

← Section 9.2

E **45** Given that $x = \ln t$, $t > 0$, and that y is a function of x,

a find $\frac{dy}{dx}$ in terms of $\frac{dy}{dt}$ and t **(2)**

b show that $\frac{d^2y}{dx^2} = t^2\frac{d^2y}{dt^2} + t\frac{dy}{dt}$ **(4)**

c Show that the substitution $x = \ln t$ transforms the differential equation

$$\frac{d^2y}{dx^2} - (1 - 6e^x)\frac{dy}{dx} + 10ye^{2x}$$
$$= 5e^{2x}\sin 2e^x \qquad (1)$$

into the differential equation

$$\frac{d^2y}{dt^2} + 6\frac{dy}{dt} + 10y = 5\sin 2t \qquad (2) \quad \textbf{(6)}$$

d Hence find the general solution to (1), giving your answer in the form $y = f(x)$. **(6)**

← Section 9.2

A P **46** A scientist is modelling the amount of a chemical in the human bloodstream. The amount x of the chemical, measured in mg l^{-1}, at time t hours satisfies the differential equation

$$2x\frac{d^2x}{dt^2} - 6\left(\frac{dx}{dt}\right)^2 = x^2 - 3x^4, \quad x > 0$$

a Show that the substitution $y = \frac{1}{x^2}$ transforms this differential equation into

$$\frac{d^2y}{dt^2} + y = 3 \qquad (1)$$ **(5)**

A **b** Find the general solution to differential equation (1). **(7)**

Given that at time $t = 0$, $x = \frac{1}{2}$ and $\frac{dx}{dt} = 0$,

c find an expression for x in terms of t **(3)**

d write down the maximum value of x as t varies. **(2)**

← Section 9.3

Challenge

1 Use the substitutions $t = \tan\frac{x}{2}$ and $s = \tan\frac{y}{2}$ to prove that

$$\frac{\tan x + \tan y}{\cot x + \cot y} \equiv \tan x \tan y$$

← Section 5.2

2 $y = x^3 e^x \cosh x$

Use Leibnitz's theorem to show that

$$\frac{d^n y}{dx^n} = e^{2x}2^{n-4}(8x^3 + 12nx^2 + 6n(n-1)x + n(n-1)(n-2))$$

← Section 7.1

3 The function f(x) satisfies the differential equation $f''(x) = (f'(x))^3$.

a Use the substitution $u = f'(x)$ to show that $f(x) = A - \sqrt{B - 2x}$, where A and B are arbitrary constants.

b Given that f(0) = 0 and f(1) = 1, find the exact values of A and B.

← Section 9.2

Exam-style practice

Further Mathematics AS Level Further Pure 1

Time: 50 minutes

You must have: Mathematical Formulae and Statistical Tables, Calculator

1 Use algebra to find the set of values of x for which

$$\frac{1}{x+1} < \frac{x}{x+3}$$ **(6)**

2 a Use the substitution $t = \tan\frac{x}{2}$ to show that the equation

$$2\sin x - 5\cos x = 2 \qquad (1)$$

can be written as $3t^2 + 4t - 7 = 0$. **(3)**

b Hence find all the solutions to equation (1) in the interval $0 < x < 2\pi$.
Give your answers correct to 2 decimal places where appropriate. **(3)**

3 The variable y satisfies the differential equation

$$\frac{d^2y}{dx^2} = e^{xy} - \frac{dy}{dx}$$

When $x = 0$, $y = 1$ and $\frac{dy}{dx} = 2$

Use the approximations

$$\left(\frac{d^2y}{dx^2}\right)_0 \approx \frac{y_1 - 2y_0 + y_{-1}}{h^2}$$

and $\left(\frac{dy}{dx}\right)_0 \approx \frac{y_1 - y_{-1}}{2h}$

with $h = 0.1$ to estimate the value of y when $x = 0.1$. **(6)**

4

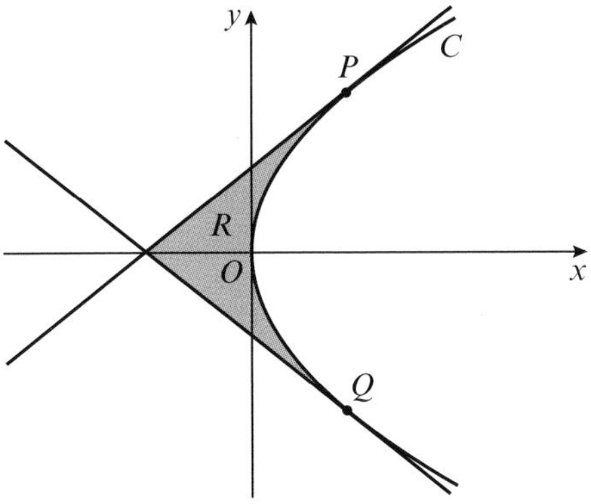

[You may quote without proof that for the general parabola $y^2 = 4ax$, $\frac{dy}{dx} = \frac{2a}{y}$]

The diagram shows the graph of the parabola C with equation $y^2 = 40x$.
The line $x = k$ intersects the parabola at the points P and Q.
The tangent to the curve at P intersects the y-axis at $(0, 2\sqrt{10})$.

a Find the value of k. **(4)**

b Write down the x-coordinate of the point of intersection of the two tangents. **(1)**

The finite region R, shown shaded in the diagram, is bounded by the tangents to the curve at P and Q and by the parabola C.

c Find, correct to three significant figures, the area of R. **(7)**

5 The diagram shows a model for a new kind of solid tetrahedral dice.

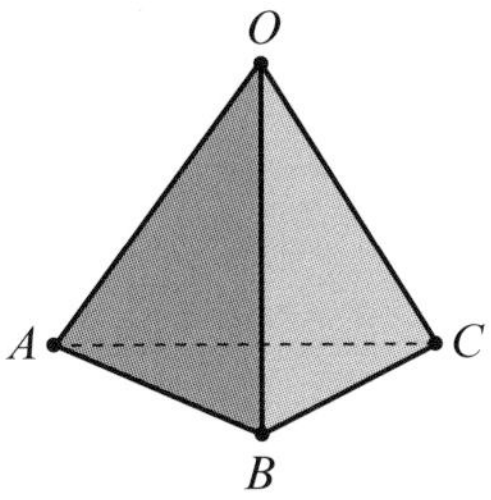

Points A, B and C have position vectors $6\mathbf{i} + 4\mathbf{j} + 2\mathbf{k}$, $-2\mathbf{i} + 2\mathbf{j} + 3\mathbf{k}$ and $-\mathbf{i} + 4\mathbf{j} - \mathbf{k}$ respectively and O is the origin.

a Find $\overrightarrow{OB} \times \overrightarrow{OC}$. **(3)**

b Find the area of the face OBC correct to three significant figures. **(2)**

The dice is to be 3D printed using a scale of 1 cm per unit and a plastic filament of density 1.35 g/cm³.

Given that the manufacturer has 1 kg of plastic filament,

c work out the number of dice that can be made. **(4)**

d Give a reason why your answer to part **c** might be an over-estimate. **(1)**

Exam-style practice

Further Mathematics A Level Further Pure 1

Time: 1 hour and 30 minutes
You must have: Mathematical Formulae and Statistical Tables, Calculator

1 A tetrahedron has vertices at $A(-1, 3, 2)$, $B(1, -4, 2)$, $C(-1, -5, 6)$ and $D(-7, -2, 2)$.
Find:

a The Cartesian equation of the plane ABC. **(3)**

b The volume of tetrahedron $ABCD$. **(3)**

The normal to the plane ABC through point D intersects the plane at point E.

c Find the angle DCE. Give your answer in radians correct to three significant figures. **(5)**

2

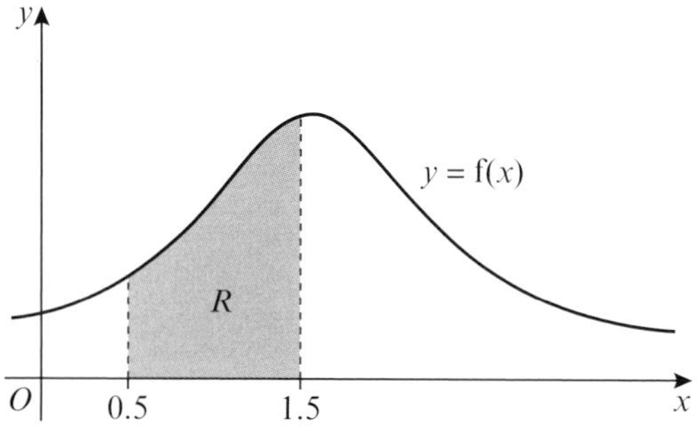

The diagram shows the graph of $y = f(x)$ where $f(x) = \dfrac{1}{4 - 3\sin x}$

The finite region R is bounded by the curve, the x-axis and the lines $x = 0.5$ and $x = 1.5$.

a Use Simpson's rule with 4 intervals to find an approximation for the area of R, giving your answer to 5 decimal places. **(5)**

b Use the substitution $t = \tan\frac{x}{2}$ to find the value of the integral $\int_{0.5}^{1.5} f(x)\,dx$ to 5 decimal places **(6)**

c Hence find, correct to 2 decimal places, the percentage error in using the method in part **a**, and suggest a way in which the approximation could be improved. **(2)**

3 An extreme sports enthusiast jumps from the top of a cliff attached to a parachute. Her velocity, $y\,\text{m}\,\text{s}^{-1}$, is related to the distance jumped, x, where x is measured in hundreds of metres, from the top of a cliff. She believes the differential equation used to model the relationship between x and y is

$$xy\frac{dy}{dx} + 3x^2 + y^2 = 0 \qquad (1)$$

a Show that the substitution $y = zx$ transforms (1) into the differential equation

$$x\frac{dv}{dx} + \frac{3 + 2z^2}{z} = 0 \qquad (2)$$ **(5)**

b By solving equation (2), find the particular solution to equation (1), given that her velocity is $5\,\text{m}\,\text{s}^{-1}$ when she is 100 metres from the top of the cliff. **(8)**

c Assuming that her velocity reaches zero as she lands, find, according to the model, the height of the cliff. **(2)**

d By considering your solution to part **b**, comment on the suitability of this model for small values of x. **(1)**

4 **a** Explain why you cannot use L'Hospital's rule to evaluate $\lim_{x\to 1}\frac{5x^4 - 3x^2 - 1}{11 - 2x - 9x^3}$ **(1)**

b Use L'Hospital's rule to find $\lim_{x\to 1}\frac{5x^4 - 3x^2 - 2}{11 - 2x - 9x^3}$ **(3)**

5 The line L has equation $y = mx + c$, where m and c are constants.

The hyperbola H has equation $\frac{x^2}{a^2} - \frac{y^2}{b^2} = 1$, where a and b are constants.

a Given that L is a tangent to H, show that $a^2m^2 = b^2 + c^2$. **(5)**

The hyperbola H' has equation $\frac{x^2}{26} - \frac{y^2}{25} = 1$

b Find the equations of the tangents to H' which pass through the point (2, 3). **(6)**

6 $\frac{d^2y}{dx^2} + \left(\frac{dy}{dx}\right)^2 + 2y = 0$

Given that when $x = 0$, $y = \frac{dy}{dx} = 1$, find a series solution for y in ascending powers of x, up to and including the term in x^3. **(9)**

7 Find the set of values of x such that

$$\left|\frac{x}{x + 3}\right| < 2 - x$$

expressing your answer in set notation. **(7)**

8 Given that

$$y = e^x \sin x$$

use Leibnitz's theorem to show that

$$\frac{d^6y}{dx^6} + 8\frac{dy}{dx} = 8y$$ **(4)**

Answers

CHAPTER 1

Prior knowledge check

1 -1

2 $\frac{x-1}{2} = \frac{y-4}{3} = \frac{x+2}{5}$ $(= \lambda)$

3 a 0.302 radians b $(6, 1, -7)$

Exercise 1A

1 a $5\mathbf{i}$ b $-3\mathbf{j}$
c $3\mathbf{j}$ d $-3\mathbf{j} - 3\mathbf{k}$
e $-2\mathbf{i} - 6\mathbf{k}$ f $2\mathbf{i} + 6\mathbf{k}$
g $\begin{pmatrix} 5 \\ -16 \\ -7 \end{pmatrix}$ h $\begin{pmatrix} 9 \\ 0 \\ -3 \end{pmatrix}$
i $\begin{pmatrix} -9 \\ -7 \\ -11 \end{pmatrix}$ j $\begin{pmatrix} 2 \\ -4 \\ -3 \end{pmatrix}$

2 a $-6\mathbf{i} + (3\lambda + 1)\mathbf{j} - 2\mathbf{k}$ b $(7\lambda - 3)\mathbf{i} + \mathbf{j} + (1 - 2\lambda)\mathbf{k}$

3 $-\frac{1}{3}\mathbf{i} - \frac{2}{3}\mathbf{j} + \frac{2}{3}\mathbf{k}$ or $\frac{1}{3}\mathbf{i} + \frac{2}{3}\mathbf{j} - \frac{2}{3}\mathbf{k}$

4 $\frac{1}{7}(-\mathbf{i} + 4\sqrt{2}\mathbf{j} + 4\mathbf{k})$

5 $\frac{1}{11}(6\mathbf{i} + 6\mathbf{j} + 7\mathbf{k})$

6 $\begin{pmatrix} \frac{4}{9} \\ \frac{4}{9} \\ -\frac{7}{9} \end{pmatrix}$

7 $-\mathbf{i} - 2\sqrt{2}\mathbf{j} + 4\mathbf{k}$

8 $\sqrt{8}$ or $2\sqrt{2}$ or 2.83 (to 3 s.f.)

9 a -14 b $-8\mathbf{i} - 24\mathbf{j} - 8\mathbf{k}$ c $\frac{1}{\sqrt{11}}(-\mathbf{i} - 3\mathbf{j} - \mathbf{k})$

10 a $\frac{\sqrt{221}}{15}$ b 1 c $\frac{\sqrt{21}}{11}$

11 Any multiple of $(\mathbf{i} + \mathbf{j} - \mathbf{k})$

12 $u = -1$, $v = 4$ and $w = 11$

13 a $a = 1$ and $b = -1$ b $-\frac{5}{6}$

14 $\lambda = \frac{3}{2}$ and $\mu = -\frac{3}{2}$

15 Given that $\mathbf{a} + \mathbf{b} + \mathbf{c} = \mathbf{0}$ *
Take the vector product of this with $\mathbf{a}$.
$\mathbf{a} \times (\mathbf{a} + \mathbf{b} + \mathbf{c}) = \mathbf{a} \times \mathbf{0}$
$\mathbf{a} \times \mathbf{a} + \mathbf{a} \times \mathbf{b} + \mathbf{a} \times \mathbf{c} = \mathbf{0}$
But $\mathbf{a} \times \mathbf{a} = \mathbf{0}$ and $\mathbf{a} \times \mathbf{c} = -\mathbf{c} \times \mathbf{a}$
Therefore $\mathbf{a} \times \mathbf{b} - \mathbf{c} \times \mathbf{a} = \mathbf{0}$
So $\mathbf{a} \times \mathbf{b} = \mathbf{c} \times \mathbf{a}$
Take the vector product of * with $\mathbf{b}$.
$\mathbf{b} \times (\mathbf{a} + \mathbf{b} + \mathbf{c}) = \mathbf{b} \times \mathbf{0}$
$\mathbf{b} \times \mathbf{a} + \mathbf{b} \times \mathbf{b} + \mathbf{b} \times \mathbf{c} = \mathbf{0}$
But $\mathbf{b} \times \mathbf{b} = \mathbf{0}$ and $\mathbf{b} \times \mathbf{a} = -\mathbf{a} \times \mathbf{b}$
Therefore $-\mathbf{a} \times \mathbf{b} + \mathbf{b} \times \mathbf{c} = \mathbf{0}$
So $\mathbf{b} \times \mathbf{c} = \mathbf{a} \times \mathbf{b}$
Therefore $\mathbf{a} \times \mathbf{b} = \mathbf{b} \times \mathbf{c} = \mathbf{c} \times \mathbf{a}$

Challenge

$\mathbf{a} \times \mathbf{b} = \mathbf{c} \times \mathbf{a}$
$\mathbf{a} \times \mathbf{b} - \mathbf{c} \times \mathbf{a} = \mathbf{0}$
$\mathbf{a} \times \mathbf{b} + \mathbf{a} \times \mathbf{c} = \mathbf{0}$
$\mathbf{a} \times (\mathbf{b} + \mathbf{c}) = \mathbf{0}$
As $\mathbf{a} \neq \mathbf{0}$ and $\mathbf{b}$ and $\mathbf{c}$ are non-parallel
$\mathbf{a}$ is parallel to $\mathbf{b} + \mathbf{c}$.

Exercise 1B

1 a 4.5 b $\frac{5\sqrt{2}}{2}$ c 16.5

2 a $2\sqrt{13}$ b 8.5

3 $\frac{3}{2}\sqrt{2}$

4 $\frac{5}{2}\sqrt{3}$

5 $5\sqrt{2}$

6 $10\sqrt{2}$

7 $3\sqrt{2}$

8 $\frac{\sqrt{3}}{2}a^2$

9 a Area of parallelogram $ABCD = 2 \times$ area of triangle ABC
$= 2 \times \frac{1}{2}|\overrightarrow{AB} \times \overrightarrow{AC}|$
$= |\overrightarrow{AB} \times \overrightarrow{AC}|$
As $\overrightarrow{AB} = (\mathbf{b} - \mathbf{a})$ and $\overrightarrow{AC} = (\mathbf{c} - \mathbf{a})$
Area $= |(\mathbf{b} - \mathbf{a}) \times (\mathbf{c} - \mathbf{a})|$
b $(\mathbf{b} - \mathbf{a}) \times (\mathbf{c} - \mathbf{a}) - (\mathbf{b} - \mathbf{a}) \times (\mathbf{d} - \mathbf{a}) = \mathbf{0}$
$\Rightarrow (\mathbf{b} - \mathbf{a}) \times \mathbf{c} + (\mathbf{b} - \mathbf{a}) \times (\mathbf{a} - \mathbf{a}) - (\mathbf{b} - \mathbf{a}) \times \mathbf{d} = \mathbf{0}$
$\Rightarrow (\mathbf{b} - \mathbf{a}) \times (\mathbf{c} - \mathbf{d}) = \mathbf{0}$
$\overrightarrow{AB}$ is parallel to $\overrightarrow{DC}$.

10 a $5\mathbf{i} - 5\mathbf{j} + 15\mathbf{k}$ b $\frac{5\sqrt{11}}{2}$

11 a $-15\mathbf{i} + 17\mathbf{j} + 20\mathbf{k}$
b $15.12\,\text{m}^2$
c The area of fabric needed will be larger as there will need to be excess fabric to attach to the masts and some slack in the sail to fill with air.

12 a $(2, -5, 1)$ b £4480

Challenge

$|\mathbf{p} \times (\mathbf{q} + \mathbf{r})| = ABFE$ as $BF = \mathbf{q} + \mathbf{r}$
$|\mathbf{p} \times \mathbf{q}| = ABCD$
$|\mathbf{p} \times \mathbf{r}| = CDEF$
$|\mathbf{p} \times \mathbf{q}| + |\mathbf{p} \times \mathbf{r}| = ABFE + BCF - ADE$
By definition, $AD = \mathbf{q}$ and $DE = \mathbf{r}$
$|\mathbf{p} \times \mathbf{q}| + |\mathbf{p} \times \mathbf{r}| = ABFE + \frac{1}{2}|\mathbf{q} \times \mathbf{r}| - \frac{1}{2}|\mathbf{q} \times \mathbf{r}| = |\mathbf{p} \times (\mathbf{q} + \mathbf{r})|$

Exercise 1C

1 a 21 b 21 c 21

2 0; $\mathbf{a}$ is parallel to the plane containing $\mathbf{b}$ and $\mathbf{c}$

3 17

4 18

5 $\frac{3}{2}$

6 a 3 b $\pm\frac{1}{3}(\mathbf{i} + 2\mathbf{j} - 2\mathbf{k})$ c $\frac{7}{3}$

7 a The distance between any two vertices is 2.
b $\frac{2}{3}\sqrt{2}$

8 a $\overrightarrow{AB} = -2\mathbf{i} - \mathbf{j} + 3\mathbf{k}$
$\overrightarrow{AC} = \mathbf{i} - 3\mathbf{j} + 2\mathbf{k}$
$\overrightarrow{AB} \times \overrightarrow{AC} = 7\mathbf{i} + 7\mathbf{j} + 7\mathbf{k}$
b $\frac{7\sqrt{3}}{2}$
c $\frac{7}{3}$

9 a $\overrightarrow{AB} \times \overrightarrow{BC} = 5\mathbf{i} - \mathbf{j} - 7\mathbf{k}$
$\overrightarrow{BD} \times \overrightarrow{DC} = 2\mathbf{i} - 8\mathbf{j} + \mathbf{k}$
b i $\frac{5}{2}\sqrt{3}$ ii $\frac{19}{6}$

10 a $\mathbf{i} + 2\mathbf{j}$
b $\overrightarrow{OP} = 2\sqrt{5}$
Area of $OQR = \frac{\sqrt{5}}{2}$
Volume of tetrahedron $= \frac{5}{3}$

c $\mathbf{a}.(\mathbf{b} \times \mathbf{c}) = 10$
This is $6 \times$ volume of tetrahedron so verified.

11 a $18\mathbf{i} + 12\mathbf{j} + 6\mathbf{k}$ b 127 g

12 1400 cubic angstroms

13 a 12:1 b Ratio will be unchanged as N moves.

14 $\frac{14}{3}$ units^3

Challenge

a Let: $\mathbf{a} = a_1\mathbf{i} + a_2\mathbf{j} + a_3\mathbf{k}$
$\mathbf{b} = b_1\mathbf{i} + b_2\mathbf{j} + b_3\mathbf{k}$
$\mathbf{c} = c_1\mathbf{i} + c_2\mathbf{j} + c_3\mathbf{k}$
$\mathbf{a}.(\mathbf{b} \times \mathbf{c}) = a_1(b_2c_3 - b_3c_2) + a_2(b_3c_1 - b_1c_3) + a_3(b_1c_2 - b_2c_1)$
$\mathbf{a} \times \mathbf{b} = (a_2b_3 - a_3b_2)\mathbf{i} + (a_3b_1 - a_1b_3)\mathbf{j} + (a_1b_2 - a_2b_1)\mathbf{k}$
$(\mathbf{a} \times \mathbf{b}).\mathbf{c} = (a_2b_3 - a_3b_2)c_1 + (a_3b_1 - a_1b_3)c_2 + (a_1b_2 - a_2b_1)c_3$
$= a_2b_3c_1 - a_3b_2c_1 + a_3b_1c_2 - a_1b_3c_2 + a_1b_2c_3 - a_2b_1c_3$
$= a_1(b_2c_3 - b_3c_2) + a_2(b_3c_1 - b_1c_3) + a_3(b_1c_2 - b_2c_1)$
Therefore, $\mathbf{a}.(\mathbf{b} \times \mathbf{c}) = (\mathbf{a} \times \mathbf{b}).\mathbf{c}$

b $\mathbf{d}.(\mathbf{a} \times \mathbf{b} + \mathbf{a} \times \mathbf{c}) = \mathbf{d}.(\mathbf{a} \times \mathbf{b}) + \mathbf{d}.(\mathbf{a} \times \mathbf{c})$
$= (\mathbf{d} \times \mathbf{a}).\mathbf{b} + (\mathbf{d} \times \mathbf{a}).\mathbf{c}$
$= (\mathbf{d} \times \mathbf{a}).(\mathbf{b} + \mathbf{c})$
$= \mathbf{d}.(\mathbf{a} \times (\mathbf{b} + \mathbf{c}))$

c As $\mathbf{d}$ can be any vector, if $\mathbf{d}.(\mathbf{a} \times \mathbf{b} + \mathbf{a} \times \mathbf{c}) = \mathbf{d}.(\mathbf{a} \times (\mathbf{b} + \mathbf{c}))$, then it follows that $\mathbf{a} \times \mathbf{b} + \mathbf{a} \times \mathbf{c} = \mathbf{a} \times (\mathbf{b} + \mathbf{c})$

Exercise 1D

1 a $\mathbf{r} \times (3\mathbf{i} + \mathbf{j} - 2\mathbf{k}) = -4\mathbf{i} + 10\mathbf{j} - \mathbf{k}$
b $\mathbf{r} \times (\mathbf{i} + \mathbf{j} + 5\mathbf{k}) = 3\mathbf{i} - 13\mathbf{j} + 2\mathbf{k}$
c $\mathbf{r} \times (-\mathbf{i} - 2\mathbf{j} + 3\mathbf{k}) = -4\mathbf{i} - 13\mathbf{j} - 10\mathbf{k}$

2 a $\frac{x-2}{3} = \frac{y-1}{1} = \frac{z-2}{-2} = \lambda$
b $\frac{x-2}{1} = \frac{y}{1} = \frac{z+3}{5} = \lambda$
c $\frac{x-4}{-1} = \frac{y+2}{-2} = \frac{z-1}{3} = \lambda$

3 a $\left(\mathbf{r} - \begin{pmatrix}1\\3\\5\end{pmatrix}\right) \times \begin{pmatrix}5\\1\\-3\end{pmatrix} = \mathbf{0}$
b $\left(\mathbf{r} - \begin{pmatrix}3\\4\\12\end{pmatrix}\right) \times \begin{pmatrix}1\\-1\\-7\end{pmatrix} = \mathbf{0}$
c $\left(\mathbf{r} - \begin{pmatrix}-2\\2\\6\end{pmatrix}\right) \times \begin{pmatrix}5\\5\\5\end{pmatrix} = \mathbf{0}$
d $\left(\mathbf{r} - \begin{pmatrix}1\\1\\1\end{pmatrix}\right) \times \begin{pmatrix}-3\\-1\\5\end{pmatrix} = \mathbf{0}$

4 a $\frac{x-1}{5} = \frac{y-3}{1} = \frac{z-5}{-3} = \lambda$
b $\frac{x-3}{1} = \frac{y-4}{-1} = \frac{z-12}{-7} = \lambda$
c $\frac{x+2}{5} = \frac{y-2}{5} = \frac{z-6}{5} = \lambda$ or $x + 2 = y - 2 = z - 6 = \mu$
d $\frac{x-4}{-3} = \frac{y-2}{-1} = \frac{z+4}{5} = \lambda$

5 a $(\mathbf{r} - (\mathbf{i} + \mathbf{j} - 2\mathbf{k})) \times (2\mathbf{i} - \mathbf{k}) = \mathbf{0}$
b $(\mathbf{r} - (\mathbf{i} + 4\mathbf{j})) \times (3\mathbf{i} + \mathbf{j} - 5\mathbf{k}) = \mathbf{0}$
c $(\mathbf{r} - (3\mathbf{i} + 4\mathbf{j} - 4\mathbf{k})) \times (2\mathbf{i} - 2\mathbf{j} - 3\mathbf{k}) = \mathbf{0}$

6 a $\mathbf{r} \times (2\mathbf{i} + 5\mathbf{j} + \frac{3}{2}\mathbf{k}) = -9\mathbf{i} - \frac{3}{2}\mathbf{j} + 17\mathbf{k}$
b $\mathbf{r} = 3\mathbf{i} - \mathbf{j} + \frac{3}{2}\mathbf{k} + t(2\mathbf{i} + 5\mathbf{j} + \frac{3}{2}\mathbf{k})$
or $\mathbf{r} = 3\mathbf{i} - \mathbf{j} + \frac{3}{2}\mathbf{k} + s(4\mathbf{i} + 10\mathbf{j} + 3\mathbf{k})$

7 $p = 3$ and $q = 3$

8 $\mathbf{r} = -\mathbf{j} + 2\mathbf{k} + t(\mathbf{i} + \mathbf{j} - \mathbf{k})$ (other answers are possible)

9 a $\frac{6}{\sqrt{184}}, \frac{2}{\sqrt{184}}, -\frac{12}{\sqrt{184}}$
b $\frac{x+3}{\frac{6}{\sqrt{184}}} = \frac{y-2}{\frac{2}{\sqrt{184}}} = \frac{z-7}{\frac{-12}{\sqrt{184}}}$

10 a 1, 0, 0 b 0, 1, 0 c 0, 0, 1 d $\frac{1}{\sqrt{3}}, \frac{1}{\sqrt{3}}, \frac{1}{\sqrt{3}}$

11 a $\frac{1}{\sqrt{14}}, \frac{2}{\sqrt{14}}, \frac{3}{\sqrt{14}}$ b $\frac{3}{\sqrt{14}}, \frac{2}{\sqrt{14}}, \frac{1}{\sqrt{14}}$
c $l_1l_2 + m_1m_2 + n_1n_2 =$
$\left(\frac{1}{\sqrt{14}} \times \frac{3}{\sqrt{14}}\right) + \left(\frac{2}{\sqrt{14}} \times \frac{2}{\sqrt{14}}\right) + \left(\frac{3}{\sqrt{14}} \times \frac{1}{\sqrt{14}}\right) = \frac{5}{7}$
$\cos\theta = \frac{L_1.L_2}{|L_1||L_2|} = \frac{1 \times 3 + 2 \times 2 + 3 \times 1}{\sqrt{14} \times \sqrt{14}} = \frac{5}{7}$

12 1.41

13 Use of formula: $\cos 2x \equiv 2\cos^2 x - 1$
$\text{LHS} = 2\left(\frac{x}{|L|}\right)^2 - 1 + 2\left(\frac{y}{|L|}\right)^2 - 1 + 2\left(\frac{z}{|L|}\right)^2 - 1$
$= 2\frac{x^2}{L} + 2\frac{y^2}{L} + 2\frac{z^2}{L} - 3$
$= \frac{2(x^2 + y^2 + z^2)}{L} - 3$
$L = x^2 + y^2 + z^2 \Rightarrow 2 - 3 = -1 = \text{RHS}$

14 68.2°, 56.1°, 42.0°

15 $\mathbf{r} \times \begin{pmatrix}1\\\frac{1}{\sqrt{2}}\\\frac{1}{\sqrt{2}}\end{pmatrix} = \mathbf{0}, \mathbf{r} \times \begin{pmatrix}1\\-\frac{1}{\sqrt{2}}\\\frac{1}{\sqrt{2}}\end{pmatrix} = \mathbf{0}$

16 a $\frac{1}{\sqrt{3}}$ in each direction
b $\frac{x-1}{\frac{1}{\sqrt{3}}} = \frac{y-2}{\frac{1}{\sqrt{3}}} = \frac{z+1}{\frac{1}{\sqrt{3}}}$

17 a $\left(\mathbf{r} - \begin{pmatrix}0\\0\\6\end{pmatrix}\right) \times \begin{pmatrix}\sqrt{6} - \sqrt{2}\\\sqrt{6} + \sqrt{2}\\0\end{pmatrix} = \mathbf{0}$ (or equivalent)
b If the wires intersect, then:
$\begin{pmatrix}0\\0\\6\end{pmatrix} + \lambda\begin{pmatrix}\sqrt{6} - \sqrt{2}\\\sqrt{6} + \sqrt{2}\\0\end{pmatrix} = \begin{pmatrix}5\\2\\1\end{pmatrix} + \mu\begin{pmatrix}5 - 2(\sqrt{6} - \sqrt{2})\\2 - 2(\sqrt{6} + \sqrt{2})\\-5\end{pmatrix}$
$\mathbf{k}$: $6 = 1 - 5\mu \Rightarrow \mu = -1$
$\mathbf{i}$: $\lambda(\sqrt{6} - \sqrt{2}) = 5 + (-1)(5 - 2(\sqrt{6} - \sqrt{2})) \Rightarrow \lambda = 2$
$\mathbf{j}$: $2(\sqrt{6} + \sqrt{2}) = 2 + (-1)(2 - 2(\sqrt{6} + \sqrt{2})) = 2(\sqrt{6} + \sqrt{2})$
Therefore the wires intersect.
c The cable may not be completely horizontal (it may 'droop').

Challenge

a $l = m = \frac{\sqrt{6}}{4}, n = \frac{1}{2}$

b $l = \cos\theta\sin\varphi, m = \sin\theta\sin\varphi, n = \cos\varphi$

Exercise 1E

1 a $3x + y - z = 2$
b $7x - 2y + z = 5$
c $x + 2y - z = 3$
d $2x - 6y - z = 2$

2 a $\mathbf{r}.(2\mathbf{i} - 9\mathbf{j} + 4\mathbf{k}) = -15$
b $\mathbf{r}.(2\mathbf{i} - \mathbf{j} + \mathbf{k}) = 2$
c $\mathbf{r}.(8\mathbf{i} - 5\mathbf{j} + \mathbf{k}) = 22$

3 a $\mathbf{r} = \frac{5}{2}\mathbf{i} + \frac{5}{2}\mathbf{k} + \lambda(3\mathbf{i} + 2\mathbf{j} + 5\mathbf{k})$
b $\mathbf{r} = 3\mathbf{i} - \mathbf{j} + \lambda(2\mathbf{i} + 4\mathbf{j} + 3\mathbf{k})$
c $\mathbf{r} = -3\mathbf{i} - \frac{13}{3}\mathbf{j} + \lambda(3\mathbf{i} + 2\mathbf{j} + 3\mathbf{k})$

4 $a = 21.7°$ (3 s.f.)

5 $\frac{13}{11}$

6 a The line $\mathbf{r} = 2\mathbf{i} + 3\mathbf{j} + \mathbf{k} + \lambda(-\mathbf{i} + 2\mathbf{j} + \mathbf{k})$ passes through the point (2, 3, 1).
The point (2, 3, 1) also lies on the plane $\mathbf{r}.(\mathbf{i} + \mathbf{j} + \mathbf{k}) = 4$ as $2 \times 1 + 3 \times 1 - 1 = 4$.
So the line and plane have a point in common.
The line is in the direction $-\mathbf{i} + 2\mathbf{j} + \mathbf{k}$.
This direction is parallel to the plane as it is perpendicular to the normal $\mathbf{i} + \mathbf{j} - \mathbf{k}$, as $-1 \times 1 + 2 \times 1 + 1 \times -1 = 0$.
As the line also has a common point with the plane it lies in the plane.
b Line is parallel to the other line, which is in the plane.
$\frac{7\sqrt{3}}{3} = 4.04$ (3 s.f.)

7 a $-11x + 6y + z = 4$ b $\frac{67}{3}$ c 0.918

8 a 4 b $-7x + 5y - 3z - 4 = 0$ c 2.31 (3 s.f.)

9 a $\begin{pmatrix} -\frac{2}{3} \\ \frac{1}{3} \\ \frac{2}{3} \end{pmatrix}$ b 19° c 1.67

10 a 3.74 (3 s.f.) b 0.201 c $\mathbf{r} = \begin{pmatrix} -1 \\ 0 \\ 1 \end{pmatrix} + t\begin{pmatrix} 1 \\ 2 \\ 0 \end{pmatrix}$

11 $\mathbf{r}.\begin{pmatrix} -42 \\ -41 \\ 31 \end{pmatrix} = 147$

12 $\frac{25}{2268}$

Challenge

a $x + y + z = 0 \Rightarrow z = -x - y$
Applying the transformation to a general point on the plane gives:

$$\begin{pmatrix} 2 & -1 & 2 \\ 2 & 2 & -1 \\ -1 & 2 & 2 \end{pmatrix}\begin{pmatrix} x \\ y \\ -x - y \end{pmatrix} = \begin{pmatrix} 2x - y - 2x - 2y \\ 2x + 2y + x + y \\ -x + 2y - 2x - 2y \end{pmatrix} = \begin{pmatrix} -3y \\ 3x + 3y \\ -3x \end{pmatrix}$$

$$-3y + 3x + 3y - 3x = 0$$

Therefore, the image also lies on the plane.
Hence the plane is invariant under the linear transformation.

b To be invariant the point must map to itself.

$$\begin{pmatrix} x \\ y \\ z \end{pmatrix} = \begin{pmatrix} 2 & -1 & 2 \\ 2 & 2 & -1 \\ -1 & 2 & 2 \end{pmatrix}\begin{pmatrix} x \\ y \\ z \end{pmatrix} = \begin{pmatrix} 2x - y - 2z \\ 2x + 2y - z \\ -x + 2y + 2z \end{pmatrix}$$

$x = y + 2z$ (1)
$x = \frac{1}{2}(z - y)$ (2)
$x = 2y + z$ (3)
Equating (1) and (3): $y + 2z = 2y + z \Rightarrow y = z$
Substituting into (2): $x = 0$
Substituting into (1): $0 = 3y \Rightarrow y = 0$ and $z = 0$
Therefore, the only invariant point is the origin.

Mixed exercise 1

1 a $4\mathbf{i} + 10\mathbf{j} + 8\mathbf{k}$ b 38 c $3\sqrt{5}$ d $\frac{19}{3}$

2 a $13\mathbf{i} + 4\mathbf{j} - \mathbf{k}$ b 288 cm²

3 Volume of parallelepiped $= \overrightarrow{EA}.(\overrightarrow{EC} \times \overrightarrow{EF})$
$\overrightarrow{EA} = -3\mathbf{i} + \mathbf{j} - 2\mathbf{k}$
$\overrightarrow{EC} = \mathbf{i} + 2\mathbf{j} - 4\mathbf{k}$
$\overrightarrow{EF} = -\mathbf{i} - 4\mathbf{j} - \mathbf{k}$

$$\text{Volume} = \begin{vmatrix} -3 & 1 & -2 \\ 1 & 2 & -4 \\ -1 & -4 & -1 \end{vmatrix} = 63$$

Volume of tetrahedron $= \frac{1}{6}\overrightarrow{EA}.(\overrightarrow{EC} \times \overrightarrow{EM})$
$\overrightarrow{EA} = -3\mathbf{i} + \mathbf{j} - 2\mathbf{k}$
$\overrightarrow{EC} = \mathbf{i} + 2\mathbf{j} - 4\mathbf{k}$
$\overrightarrow{EM} = -\frac{2}{3}\mathrm{i} - \frac{8}{3}\mathrm{j} - \frac{2}{3}\mathrm{k}$

$$\text{Volume} = \frac{1}{6}\begin{vmatrix} -3 & 1 & -2 \\ 1 & 2 & -4 \\ -\frac{2}{3} & -\frac{8}{3} & -\frac{2}{3} \end{vmatrix} = \frac{63}{9}$$

4 $2x - 5y + 3z + 10 = 0$ and $2x - 5y + 3z - 10 = 0$

5 a Equation of L_1 is $\mathbf{r} = 3\mathbf{i} - 3\mathbf{j} - 2\mathbf{k} + \lambda(\mathbf{j} + 2\mathbf{k})$
When $s = 2$: $\mathbf{r} = 3\mathbf{i} - \mathbf{j} + 2\mathbf{k}$, so P lies on L_1.
Equation of L_2 is $\mathbf{r} = 8\mathbf{i} + 3\mathbf{j} + \mu(5\mathbf{i} + 4\mathbf{j} - 2\mathbf{k})$
When $t = -1$: $\mathbf{r} = 3\mathbf{i} - \mathbf{j} + 2\mathbf{k}$, so P lies on L_2.
b $-10\mathbf{i} + 10\mathbf{j} - 5\mathbf{k}$
c $2x - 2y + z = 10$
d 15

6 a $\mathbf{i} - \mathbf{j} - \mathbf{k}$ b -2 c $\frac{2}{3}\sqrt{3}$

7 a $\mathbf{r} = 2\mathbf{i} - 3\mathbf{j} + \mathbf{k} + t(-4\mathbf{i} + \mathbf{j} - 2\mathbf{k})$
b $\frac{5}{2}\sqrt{5}$ or 5.59 (3 s.f.)

8 a $\frac{1}{\sqrt{50}}(3\mathbf{i} + 5\mathbf{j} + 4\mathbf{k})$ b $3x + 5y + 4z = 30$ c $3\sqrt{2}$

9 a $\mathbf{j} \times (\mathbf{i} - \mathbf{k}) = -\mathbf{i} - \mathbf{k}$, which is parallel to $\mathbf{i} + \mathbf{k}$
b $\frac{\sqrt{2}}{2}$ or 0.707 (to 3 s.f.) c $x + z = 1$

10 a $-15\mathbf{i} - 20\mathbf{j} + 10\mathbf{k}$ or a multiple of $(3\mathbf{i} + 4\mathbf{j} - 2\mathbf{k})$
b $3x + 4y - 2z - 5 = 0$ c 5

11 a $-6\mathbf{i} - 4\mathbf{j} + 2\mathbf{k}$ b $\mathbf{r}.(3\mathbf{i} + 2\mathbf{j} - \mathbf{k}) = 0$ c $(-1, 1, -1)$

12 a $-\mathbf{i} + 7\mathbf{j} + 5\mathbf{k}$ b $-x + 7y + 5z = 0$ c $(1, -2, 3)$

13 a 73° (nearest degree)
b $\mathbf{r} \times (2\mathbf{i} + \mathbf{j} + 6\mathbf{k}) = (-5\mathbf{i} - 32\mathbf{j} + 7\mathbf{k})$

14 a The normal to the plane Π is in the direction $(4\mathbf{i} + \mathbf{j} + 2\mathbf{k}) \times (3\mathbf{i} + 2\mathbf{j} - \mathbf{k})$

$$\begin{vmatrix} \mathbf{i} & \mathbf{j} & \mathbf{k} \\ 4 & 1 & 2 \\ 3 & 2 & -1 \end{vmatrix} = -5\mathbf{i} + 10\mathbf{j} + 5\mathbf{k}$$

The line L is in the direction $2\mathbf{i} + 3\mathbf{j} - 4\mathbf{k}$
As $(-5\mathbf{i} + 10\mathbf{j} + 5\mathbf{k}).(2\mathbf{i} + 3\mathbf{j} - 4\mathbf{k}) = 0$
the line L is perpendicular to the normal to the plane.
Thus L is parallel to the plane Π.
b $2\sqrt{6} = 4.90$

15 a $\mathbf{r} = \mathbf{i} + 2\mathbf{j} + \mathbf{k} + \lambda(2\mathbf{i} + \mathbf{j} + 3\mathbf{k})$
b (3, 3, 4) c $5\mathbf{i} - \mathbf{j} - 3\mathbf{k}$ d $\frac{\sqrt{35}}{\sqrt{34}}$ e (5, 4, 7)

16 a $x - y - 2z + 7 = 0$ b $\frac{3}{\sqrt{2}\sqrt{14}} = 0.567$ (3s.f.)
c (0, 5, 7) or (4, 1, −1)

17 a 6:1:−8 b $\frac{6}{\sqrt{101}}, \frac{1}{\sqrt{101}}, -\frac{8}{\sqrt{101}}$
c $\frac{x + 2}{\frac{6}{\sqrt{101}}} = \frac{y - 1}{\frac{1}{\sqrt{101}}} = \frac{z - 5}{-\frac{8}{\sqrt{101}}}$

18 Use trigonometric identity $\sin^2\theta \equiv 1 - \cos^2\theta$
$\Rightarrow 1 - \frac{x^2}{|a|^2} + 1 - \frac{y^2}{|a|^2} + 1 - \frac{z^2}{|a|^2} = 3 - \left(\frac{x^2 + y^2 + z^2}{L}\right) = 2$

19 If L_1 and L_2 are parallel then $l_1 = l_2$; $m_1 = m_2$ and $n_1 = n_2$, therefore $\frac{l_1}{l_2} = \frac{m_1}{m_2} = \frac{n_1}{n_2} = 1$

20 **a** $(2, \sqrt{2}, \sqrt{2})$ **b** 24.5 m
c Guide wires likely not to be perfectly straight.

21 $d = -20$

Challenge

Find the equation of the plane passing through $A(p, 0, 0)$, $B(0, q, 0)$ and $C(0, 0, r)$:

$\overrightarrow{AB} = \overrightarrow{OB} - \overrightarrow{OA} = -p\mathbf{i} + q\mathbf{j}$

$\overrightarrow{AC} = \overrightarrow{OC} - \overrightarrow{OA} = -p\mathbf{i} + r\mathbf{k}$

$$\overrightarrow{AB} \times \overrightarrow{AC} = \begin{vmatrix} \mathbf{i} & \mathbf{j} & \mathbf{k} \\ -p & q & 0 \\ -p & 0 & r \end{vmatrix} = qr\mathbf{i} + pr\mathbf{j} + pq\mathbf{k}$$

$\mathbf{r}.(qr\mathbf{i} + pr\mathbf{j} + pq\mathbf{k}) = p\mathbf{i}.(qr\mathbf{i} + pr\mathbf{j} + pq\mathbf{k})$

$qrx + pry + pqz = pqr$

Distance between plane and origin:

$$d = \frac{|pqr|}{\sqrt{(qr)^2 + (pr)^2 + (pq)^2}}$$

$$d^2 = \frac{(pqr)^2}{(qr)^2 + (pr)^2 + (pq)^2}$$

$$\frac{1}{d^2} = \frac{(qr)^2 + (pr)^2 + (pq)^2}{(pqr)^2} = \frac{1}{p^2} + \frac{1}{q^2} + \frac{1}{r^2}$$

CHAPTER 2

Prior knowledge check

1

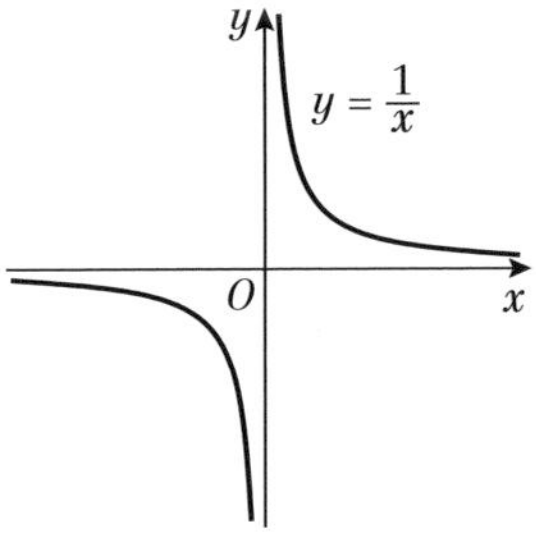

2 (1, 14) and (3, 10)

3 $\frac{dy}{dx} = 4x + 6 \Rightarrow \left.\frac{dy}{dx}\right|_{x=1} = 10$

$y - 0 = 10(x - 1)$

$y = 10x - 10$

Exercise 2A

1 **a** $y^2 = 20x$ **b** $y^2 = 2x$ **c** $y^2 = 200x$ **d** $y^2 = \frac{4}{5}x$
e $y^2 = 10x$ **f** $y^2 = 4\sqrt{3}x$ **g** $x^2 = 8y$ **h** $x^2 = 12y$

2 **a** $xy = 1$ **b** $xy = 49$ **c** $xy = 45$ **d** $xy = \frac{1}{25}$

3 **a** $xy = 9$

b

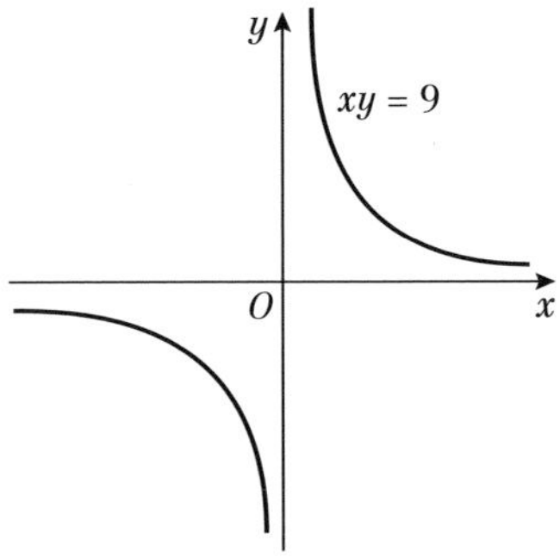

4 **a** $xy = 2$

b

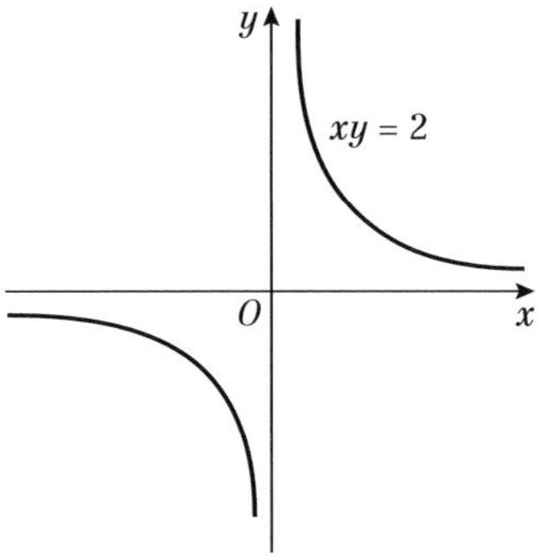

Exercise 2B

1 **a** $y^2 = 20x$ **b** $y^2 = 32x$ **c** $y^2 = 4x$
d $y^2 = 6x$ **e** $y^2 = 2\sqrt{3}x$

2 **a** $(3, 0)$; $x + 3 = 0$ **b** $(5, 0)$; $x + 5 = 0$
c $\left(\frac{5}{2}, 0\right)$; $x + \frac{5}{2} = 0$ **d** $(\sqrt{3}, 0)$; $x + \sqrt{3} = 0$
e $\left(\frac{\sqrt{2}}{4}, 0\right)$; $x + \frac{\sqrt{2}}{4} = 0$ **f** $\left(\frac{5\sqrt{2}}{4}, 0\right)$; $x + \frac{5\sqrt{2}}{4} = 0$

3 **a** $(6, 0)$; $y^2 = 24x$ **b** $(3\sqrt{2}, 0)$; $y^2 = 12\sqrt{2}x$

Challenge

1 **a** $x^2 = 16y$ **b** $(x - 3)^2 = 6y - 9$
c $y^2 = 12x - 60$

2 This is a parabola of the form $y^2 = 4ax$, rotated by $\frac{\pi}{4}$ anticlockwise about the origin. The distance between the origin and (2, 2) is $2\sqrt{2}$. Use $\begin{pmatrix} \cos\theta & -\sin\theta \\ \sin\theta & \cos\theta \end{pmatrix}$ with $\theta = \frac{\pi}{4}$ to obtain $\begin{pmatrix} \frac{1}{\sqrt{2}} & -\frac{1}{\sqrt{2}} \\ \frac{1}{\sqrt{2}} & \frac{1}{\sqrt{2}} \end{pmatrix}$.

Let $a = 2\sqrt{2}$. Calculate

$$\begin{pmatrix} \frac{1}{\sqrt{2}} & -\frac{1}{\sqrt{2}} \\ \frac{1}{\sqrt{2}} & \frac{1}{\sqrt{2}} \end{pmatrix}\begin{pmatrix} 2\sqrt{2}t^2 \\ 4\sqrt{2}t^2 \end{pmatrix} = \begin{pmatrix} 2t^2 - 4t \\ 2t^2 + 4t \end{pmatrix}.$$

Substitute $x = 2t^2 - 4t$ and $y = 2t^2 + 4t$ into the given equation.

Exercise 2C

1 (3, 3) and $\left(\frac{3}{4}, -\frac{3}{2}\right)$

2 $16\sqrt{2}$

3 $M(25, 5)$

4 **a** $(6, 0)$; $x + 6 = 0$

b

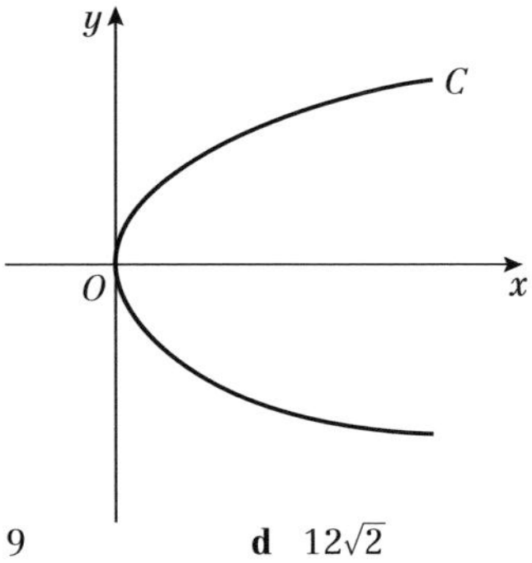

c 9 **d** $12\sqrt{2}$ **e** $18\sqrt{2}$

5 a $y^2 = 5x$ **b** 5
c $\left(-\frac{5}{4}, 3\right)$ **d** $8x - 25y + 85 = 0$

6 a $(1, 0)$ **b** 4 **c** $4x - 3y - 4 = 0$
d $\left(\frac{1}{4}, -1\right)$ **e** $\frac{5}{4}$

7 a $R(-3, 0), S(3, 0)$
b $P(9, 6\sqrt{3}), Q(-3, 6\sqrt{3})$
c $54\sqrt{3}$

8 a $a = 1, b = -4$ **b** $y = x - 8$ **c** $(10, 2)$
d $y = -x + 12$ **e** $x = 14 \pm 2\sqrt{13}$

9 a $\dfrac{2t}{t^2 - 1}$ **b** $\left(\dfrac{a}{t^2}, -\dfrac{2a}{t}\right)$

10 $40\sqrt{10}$

11 $\frac{64}{3}$

12 a $a = 4, b = -4$ **b** $y = x + 2$ **c** $\frac{16}{3}$

13 a $S(4, 0)$ **b** $x = 1$ **c** $y = -\frac{4}{3}x + \frac{16}{3}$
d $\frac{224}{3}$

Exercise 2D

1 a

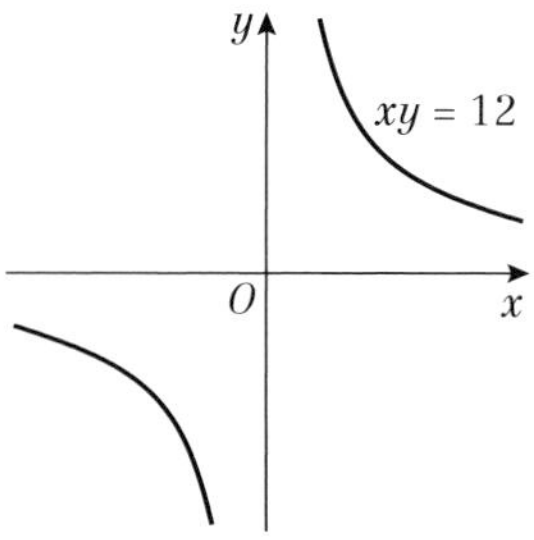

b $P(1, 12)$ and $Q(4, 3)$ **c** $y = \frac{1}{3}x + \frac{20}{3}$
d $x = -10 \pm 2\sqrt{34}$

2 a $P(-3, -3)$ and $Q(3, 3)$
b $S(1, 9)$ and $T(9, 1)$, so $ST = 8\sqrt{2}$
c $(5, 5)$ has $y = x$.

3 $\frac{1075}{12}$

4 Gradient of PQ is $-\dfrac{1}{pq}$. Use $y - y_1 = m(x - x_1)$ with either set of coordinates.

5 Solve the equations simultaneously to find single solution, $x = \dfrac{c^{\frac{4}{3}}}{(4a)^{\frac{1}{3}}}$ and $y = (4a)^{\frac{1}{3}}c^{\frac{2}{3}}$.

6 $\dfrac{9\sqrt{5}}{4}c$

7 a Substitute $x = 9t$ and $y = \dfrac{9}{t}$ into $4x - 3y + 69 = 0$ and simplify.
b $t = \frac{1}{3} \Rightarrow (3, 27)$ and $t = -\frac{9}{4} \Rightarrow \left(-\frac{81}{4}, -4\right)$

8 a $xy = 144$ **b** $22\sqrt{10}$ **c** $y = 3x - 104$

9 a $P(2, 4)$ and $Q(8, 1)$ **b** $15 - 8\ln 4$

Challenge

Use $\begin{pmatrix} \cos\theta & -\sin\theta \\ \sin\theta & \cos\theta \end{pmatrix}$ with $\theta = \dfrac{\pi}{4}$ to obtain $\begin{pmatrix} \frac{1}{\sqrt{2}} & -\frac{1}{\sqrt{2}} \\ \frac{1}{\sqrt{2}} & \frac{1}{\sqrt{2}} \end{pmatrix}$.

Use general point $\left(cp, \dfrac{c}{p}\right)$: $\begin{pmatrix} \frac{1}{\sqrt{2}} & -\frac{1}{\sqrt{2}} \\ \frac{1}{\sqrt{2}} & \frac{1}{\sqrt{2}} \end{pmatrix}\begin{pmatrix} cp \\ \frac{c}{p} \end{pmatrix} = \begin{pmatrix} \frac{cp}{\sqrt{2}} - \frac{c}{p\sqrt{2}} \\ \frac{cp}{\sqrt{2}} + \frac{c}{p\sqrt{2}} \end{pmatrix}$

So $x^2 = \dfrac{c^2p^2}{2} - c^2 + \dfrac{c^2}{2p^2}$ and $y^2 = \dfrac{c^2p^2}{2} + c^2 + \dfrac{c^2}{2p^2}$

So $y^2 - x^2 = 2c^2$. So $k^2 = 2c^2$ and therefore $k = c\sqrt{2}$.

Exercise 2E

1 a $x - 4y + 16 = 0$ **b** $\sqrt{2}x - 2y + 4\sqrt{2} = 0$
c $x + y - 10 = 0$ **d** $16x + y - 16 = 0$
e $x + 2y + 7 = 0$ **f** $2x + y - 8\sqrt{2} = 0$

2 a $x + y - 15 = 0$ **b** $2x - 8y - 45 = 0$

3 a $x - 8y - 126 = 0$ **b** $\left(128, \frac{1}{4}\right)$

4 a Gradient of PQ is $\frac{3}{2}$. Use $y - y_1 = m(x - x_1)$ with either set of coordinates.
b $(6\sqrt{2}, 4\sqrt{2})$ or $(-6\sqrt{2}, -4\sqrt{2})$

5 a $A(3, 9)$ and $B(3, -9)$
b

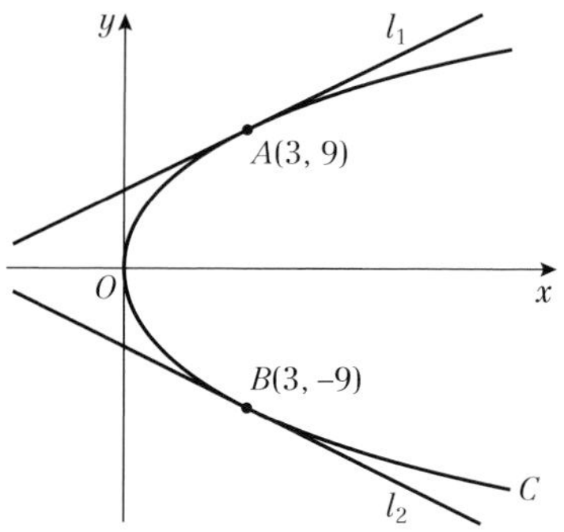

c i $l_1: 3x - 2y + 9 = 0$ **ii** $l_2: 3x + 2y + 9 = 0$

6 a $y = \dfrac{3}{x}$ **b** $8x - 2y - 15\sqrt{3} = 0$ **c** $\left(\dfrac{-\sqrt{3}}{8}, -8\sqrt{3}\right)$

7 a $t = \frac{1}{2}, P(1, 4)$ **b** $(-15, 0)$
c $(-1, 0)$ **d** 28

Exercise 2F

1 a Gradient of tangent is $\dfrac{1}{t}$. Use $y - y_1 = m(x - x_1)$ with given coordinates.
b Gradient of tangent is $-t$. Use $y - y_1 = m(x - x_1)$ with given coordinates.

2 a Gradient of tangent is $-\dfrac{1}{t^2}$. Use $y - y_1 = m(x - x_1)$ with given coordinates.
b Gradient of tangent is t^2. Use $y - y_1 = m(x - x_1)$ with given coordinates.

3 a 5
b Gradient of tangent is $\dfrac{1}{t}$. Use $y - y_1 = m(x - x_1)$ with given coordinates.
c $\frac{25}{2}t^3$

4 a Gradient of tangent is $\dfrac{1}{t}$. Use $y - y_1 = m(x - x_1)$ with given coordinates.
b $(a, -2a)$ and $(16a, 8a)$

5 a Gradient of tangent is $-\dfrac{1}{t^2}$. Use $y - y_1 = m(x - x_1)$ with given coordinates.
b $(-4, 5)$
c $(8, 2)$ and $\left(-\frac{8}{5}, -10\right)$
d $x + 4y - 16 = 0$; $25x + 4y + 80 = 0$

6 a $(-at^2, 0)$
b $(2a + at^2, 0)$
c $2a^2t(1 + t^2)$

7 a Gradient of tangent is $-t$. Use $y - y_1 = m(x - x_1)$ with given coordinates.
b $(0, 0)$, $(8, 8)$ and $(8, -8)$
c $y = 0$, $2x + y - 24 = 0$ and $2x - y - 24 = 0$

8 a $(0, at)$ **b** $(a, 0)$
c Show that the gradient of $SQ = -t$, gradient of $PQ = \dfrac{1}{t}$

9 a Gradient of tangent is $\frac{1}{t}$. Use $y - y_1 = m(x - x_1)$ with given coordinates

b -6

c $(24, 24)$ and $\left(\frac{3}{2}, -6\right)$

10 Normal at P: $y = -px + 8p + 4p^3$. Normal at Q: $y = -qx + 8q + 4q^3$. Equate ys and solve for x. Substitute to find y.

11 a Find $\frac{dy}{dx} = -\frac{1}{p^2}$ and substitute $\frac{dy}{dx} = -\frac{1}{p^2}$ and $(x_1, y_1) = \left(8p, \frac{8}{p}\right)$ into $y - y_1 = m(x - x_1)$.

Expand and simplify.

b $p^2y + x = 16p$ and $q^2y + x = 16q$ intersect at $R\left(\frac{16pq}{p+q}, \frac{16}{p+q}\right)$. Gradient of $OR = \frac{1}{pq}$

Gradient of $PQ = -\frac{1}{pq}$

Perpendicular gradients multiply to -1:

$\frac{1}{pq} \times -\frac{1}{pq} = -1 \Rightarrow p^2q^2 = 1$

12 a $\frac{dy}{dx} = \frac{1}{t}$ and use $y - 2at = \frac{1}{t}(x - at^2)$.

b Substitute $y = 0$ into $ty = x + at^2$

c Gradient of $PT = \frac{1}{t}$. Gradient of $PS = \frac{2t}{t^2 - 1}$

$\frac{1}{t} \times \frac{2t}{t^2 - 1} = -1 \Rightarrow t^2 = -1$. But $t^2 \neq -1$, so lines can never be perpendicular.

13 a Use $y - y_1 = m(x - x_1)$ with $m_T = \frac{1}{p}$ and $(x_1, y_1) = (p^2, 2p)$.

b $\frac{2p^3}{3}$

Exercise 2G

1
$$(x - 7)^2 + y^2 = (x + 7)^2$$
$$x^2 - 14x + 49 + y^2 = x^2 + 14x + 49$$
$$y^2 = 28x$$
$$a = 7$$

2 $(x - 2\sqrt{5})^2 + y^2 = (x + 2\sqrt{5})^2$

$x^2 - 4\sqrt{5}x + 20 + y^2 = x^2 + 4\sqrt{5}x + 20$

$y^2 = 8\sqrt{5}x$

$a = 2\sqrt{5}$

3 a $(y - 2)^2 + x^2 = (y + 2)^2$

$y^2 - 4y + 4 + x^2 = y^2 + 4y + 4$

$y = \frac{1}{8}x^2$

$k = \frac{1}{8}$

b $(0, 2)$; $y + 2 = 0$

c

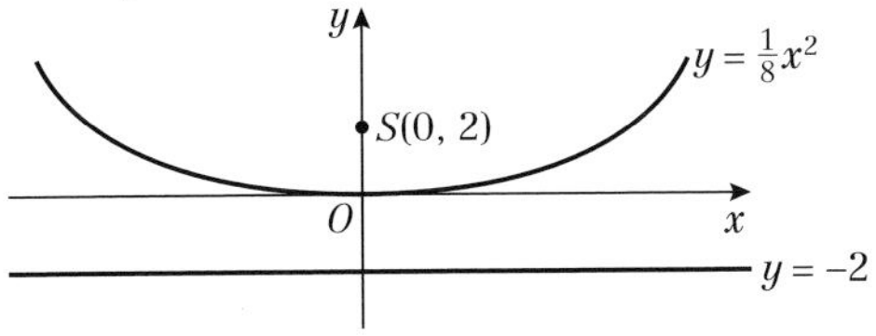

4
$$(x - a)^2 + y^2 = (x + a)^2$$
$$x^2 - 2ax + a^2 + y^2 = x^2 + 2ax + a^2$$
$$y^2 = 4ax$$

5 a $(x - 3)^2 + y^2 = (x + 3)^2$

$x^2 - 6x + 9 + y^2 = x^2 + 6x + 9$

$y^2 = 12x$

$k = 12$

b Find gradient of $PS = \frac{2\sqrt{6}}{5}$ and then use $y - y_1 = m(x - x_1)$ with $(x_1, y_1) = (18, 6\sqrt{6})$

c $R\left(\frac{1}{2}, -\sqrt{6}\right)$

d Area $= \frac{343\sqrt{6}}{4}$

6 Calculate xy, with $x = ct$ and $y = \frac{c}{2t}$:

$xy = \frac{1}{2}c^2$

7 a Let the coordinates of M be (x,y)

Area of triangle $= q$

$\frac{1}{2}(2x \times 2y) = q$

$2xy = q$

$xy = \frac{q}{2}$

Therefore the locus is a rectangular hyperbola

b $c = \sqrt{\frac{q}{2}}$

Challenge

Each crease line is formed of all the points that are equidistant from $(a, 0)$ and a particular point $(-a, y_1)$ on $x + a = 0$, so is the perpendicular bisector of these two points and has equation $y - \frac{y_1}{2} = \frac{2a}{y_1}x$. Consider the point (x_1, y_1) on the crease line.

Considering the distances from (x_1, y_1) to each of $(a, 0)$ and $(-a, y_1)$, $(x_1 + a)^2 = (x_1 - a)^2 + y_1^2 \Rightarrow y_1^2 = 4ax_1$

So all such points (x_1, y_1) form a parabola with equation $y^2 = 4ax$.

Solve this equation simultaneously with the equation of the crease line to see that the crease line meets the parabola at only one point, and hence is tangent.

Mixed exercise 2

1 a $(3, 0)$ **b** $\left(\frac{4}{3}, 4\right)$ **c** 6

2 a $\frac{3}{2}$

b $(6, 0)$

c Gradient of line through S and P is $-\frac{4}{3}$

Use $y - y_1 = m(x - x_1)$ with coordinates of either S or P.

d 30

3 a $y^2 = 48x$ **b** $x + 12 = 0$

c $(16, 16\sqrt{3})$ **d** $96\sqrt{3}$

4 a $(1, 4)$ and $(64, 32)$

b Gradient of normal is t. Use $y - y_1 = m(x - x_1)$ with given coordinates.

c $x + 2y - 9 = 0$ and $4x + y - 288 = 0$

d Coordinates are $(81, -36)$ so are in the form $(4t^2, 8t)$ where $t = -\frac{9}{2}$.

e $9\sqrt{97}$

5 a Focus of $C(a, 0)$, $Q(-a, 0)$

b $(a, 2a)$ or $(a, -2a)$

6 a Gradient of tangent is t^2. Use $y - y_1 = m(x - x_1)$ with coordinate $(ct, \frac{c}{t})$.

b $4x - y = 45$ **c** $\left(-\frac{3}{4}, -48\right)$

7 $x + 4y - 12 = 0$ and $x + 4y + 12 = 0$

8 a $X(2ct, 0)$ and $Y\left(0, \frac{2c}{t}\right)$ **b** $6\sqrt{2}$

9 a Gradient of tangent is $-\frac{1}{2t}$. Use $y - y_1 = m(x - x_1)$ with coordinates for P.

b $4ty = x + 16at^2$ **c** $(8at^2, 6at)$

10 a Gradient of tangent is $-\frac{1}{t^2}$. Use $y - y_1 = m(x - x_1)$ with coordinates for P.

b Substitute $(2a, 0)$ into equation for tangent to find t in terms of a. Then use expression for t in general point of H.

c $\frac{c^2}{2a}$ **d** $y = \frac{c^2x}{4a^2}$ **e** $\frac{8a}{5}$

f Gradient of OQ is $\frac{c^2}{4a^2}$. Gradient of XP is $-\frac{c^2}{a^2}$

Use the fact that the product of the gradients is -1 to find the required expression

g $\frac{4a}{5}$

11 a $P(3, -6)$ and $Q(12, 12)$

b Area $= 30$

12 a $2y\frac{dy}{dx} = 36 \Rightarrow \frac{dy}{dx} = \frac{18}{y}$

Gradient of normal $= -1 \div \left(\frac{18}{18p}\right) = -p$

Equation of normal:

$y - 18p = -p(x - 9p^2) \Rightarrow y + px = 18p + 9p^3$

b $(9, -18)$, $(0, 0)$ or $(9, 18)$

c $(81, 54)$

d 1458

13 a Find $m = \frac{2}{p+q}$ and use $y - 2ap = \frac{2}{p+q}(x - ap^2)$ to show $(p + q)y - 2x = 2apq$.

b Substitute $S(a, 0)$ to obtain $-2a = 2apq$ and conclude that $pq = -1$.

c $(apq, a(p + q))$

d $pq = -1$ implies $x = -a$, which is the equation of the directrix.

14 a Equation of tangent is $x + t^2y = 2ct$. At A, $y = 0$, so $x = 2ct$. So $A(2ct, 0)$.

At B, $x = 0$, so $y = \frac{2c}{t}$. So $B\left(0, \frac{2c}{t}\right)$.

$|PB|^2 = |AP|^2 = c^2\left(t^2 + \frac{1}{t^2}\right)$

b Area $= \frac{1}{2}(2ct)\left(\frac{2c}{t}\right) = 2c^2$, which is a constant.

15 a Gradient of $PQ = \frac{2}{p+q}$. Equation of line joining P and Q: $(p + q)y - 2x = 2apq$. PQ passes through $(a, 0) \Rightarrow 0 - 2a = 2apq \Rightarrow pq = -1$.

Equation of tangent at P: $y = \frac{x}{p} + ap$. Equation of tangent at Q: $y = \frac{x}{q} + aq$. Equate y terms and simplify to obtain $x = apq$. As $pq = -1$, $x = -a$, which is the equation of the directrix.

b Midpoint $= M\left(\frac{a(p^2 + q^2)}{2}, a(p + q)\right)$

Substitute $x = \frac{a(p^2 + q^2)}{2}$ and $y = a(p + q)$ into $y^2 = 2a(x - a)$ and simplify using $pq = -1$.

Challenge

a The gradient of the normal is $-t$. Since the ray is parallel to the x-axis, the right-angled triangle shown below gives $\tan\alpha = t$.

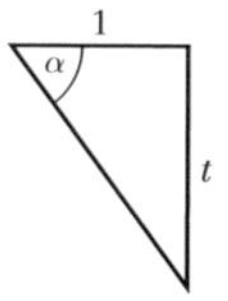

b Use the double-angle formula for tan:

$\tan 2\alpha \equiv \frac{2\tan\alpha}{1 - \tan^2\alpha} = \frac{2t}{1 - t^2}$

The gradient of the reflected ray is $\frac{2t}{t^2 - 1}$ using $\tan(\pi - \theta) = -\tan\theta$.

c Check the gradient of the reflected ray is also the gradient of the line PS.

Gradient of $PS = \frac{2at}{at^2 - a} = \frac{2at}{a(t^2 - 1)} = \frac{2t}{t^2 - 1}$

CHAPTER 3

Prior knowledge check

1 $\frac{1}{2}$

2 $x = \pm\frac{a}{\sqrt{1 + k^2}}$

3 $y = -3x + 10$

Exercise 3A

1 i a

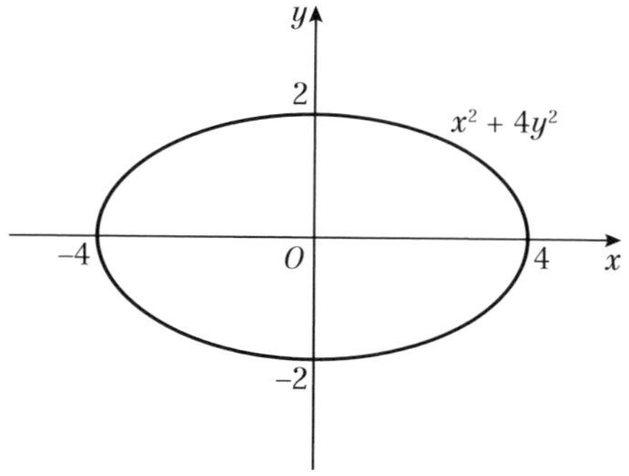

b $x = 4\cos\theta$, $y = 2\sin\theta$

ii a

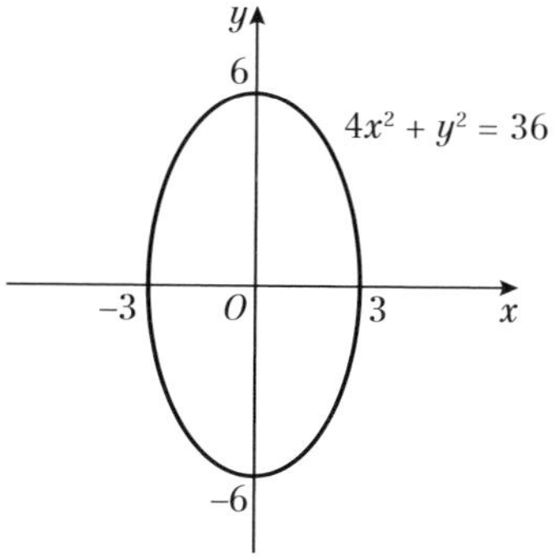

b $x = 3\cos\theta$, $y = 6\sin\theta$

iii a

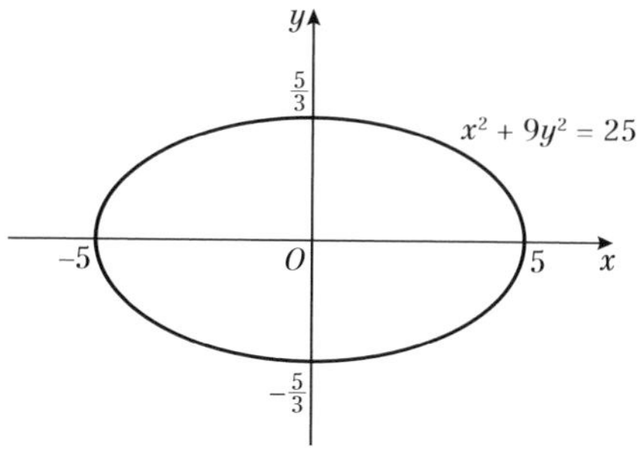

b $x = 5\cos\theta$, $y = \frac{5}{3}\sin\theta$

2 i a

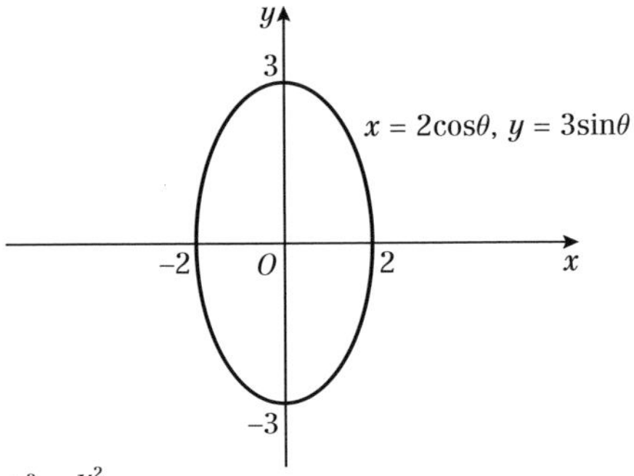

b $\frac{x^2}{4} + \frac{y^2}{9} = 1$

ii a

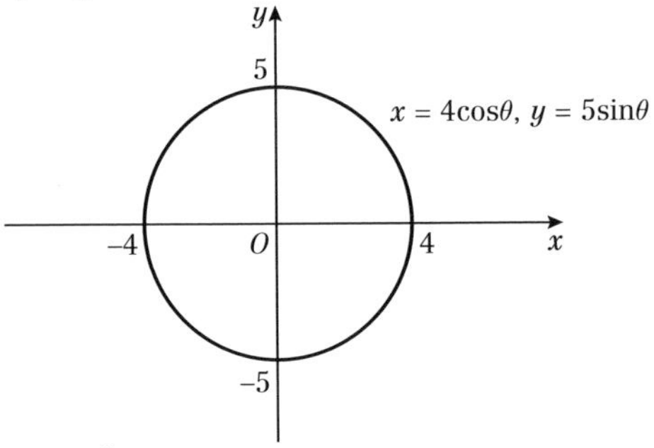

b $\frac{x^2}{16} + \frac{y^2}{25} = 1$

iii a

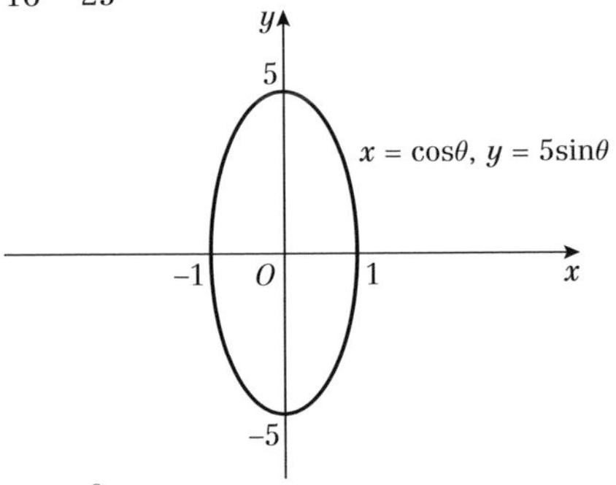

b $x^2 + \frac{y^2}{25} = 1$

iv a

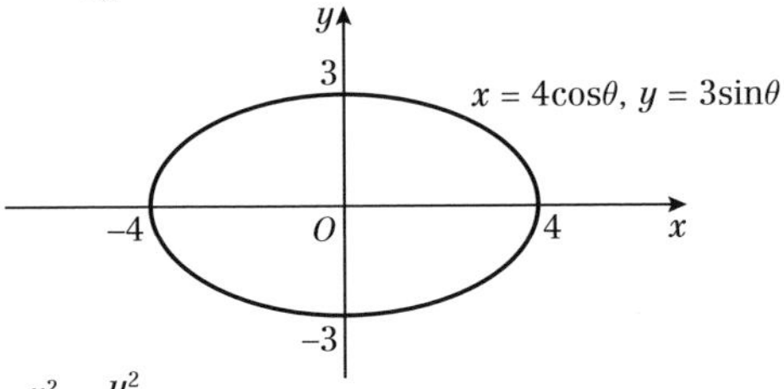

b $\frac{x^2}{16} + \frac{y^2}{9} = 1$

3 a $(b\cos\theta, a\sin\theta)$

b Ellipse $\frac{x^2}{b^2} + \frac{y^2}{a^2} = 1$

c

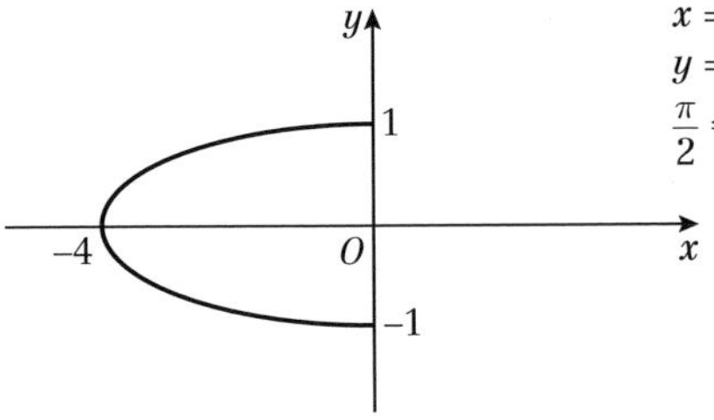

$x = 4\cos t$
$y = \sin t$
$\frac{\pi}{2} \leqslant t \leqslant \frac{3\pi}{2}$

Challenge

$$\begin{pmatrix} \frac{1}{\sqrt{2}} & -\frac{1}{\sqrt{2}} \\ \frac{1}{\sqrt{2}} & \frac{1}{\sqrt{2}} \end{pmatrix}\begin{pmatrix} a\cos t \\ b\sin t \end{pmatrix} = \begin{pmatrix} \frac{a}{\sqrt{2}\cos t} - \frac{b}{\sqrt{2}\sin t} \\ \frac{a}{\sqrt{2}\cos t} + \frac{b}{\sqrt{2}\sin t} \end{pmatrix}$$

Show $(x + y)^2 = 2a^2\cos^2 t$

Show $(x - y)^2 = 2b^2\sin^2 t$

Substitute into $\frac{(x + y)^2}{2a^2} + \frac{(x - y)^2}{2b^2} = 1$ and simplify.

Exercise 3B

1 a

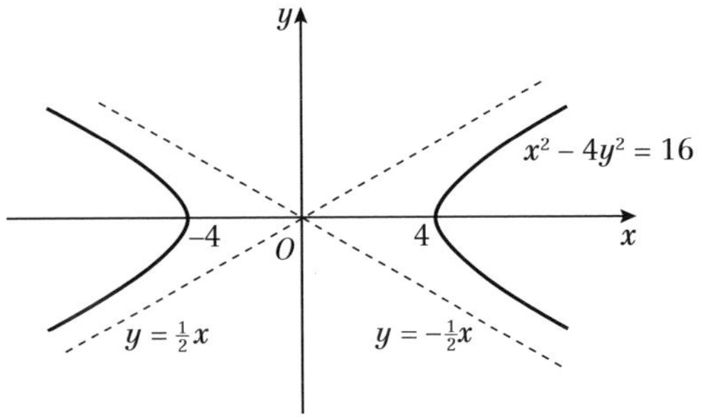

b

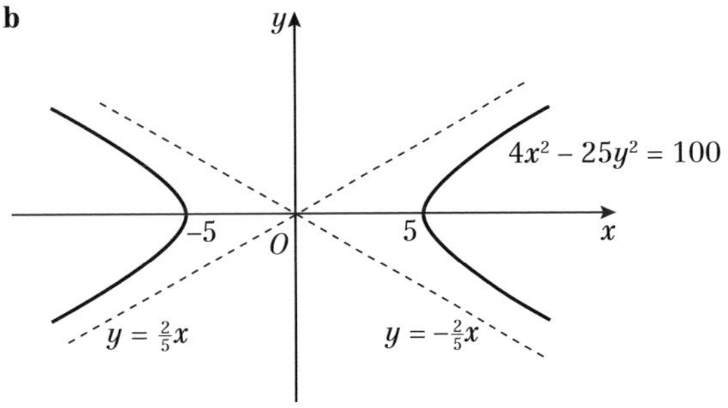

c

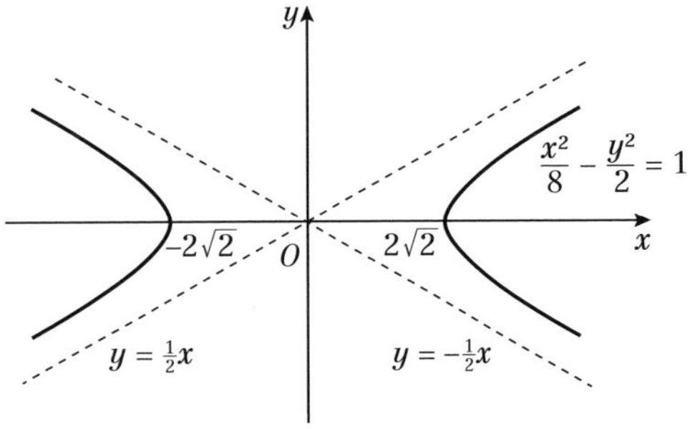

2 i a

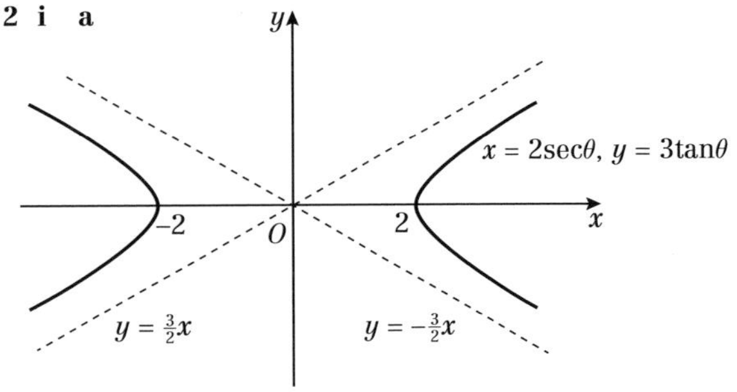

b $\frac{x^2}{4} - \frac{y^2}{9} = 1$

ii a

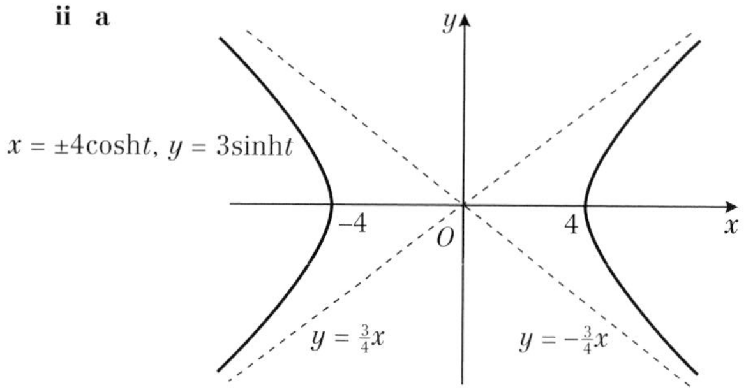

b $\dfrac{x^2}{16} - \dfrac{y^2}{9} = 1$

iii a

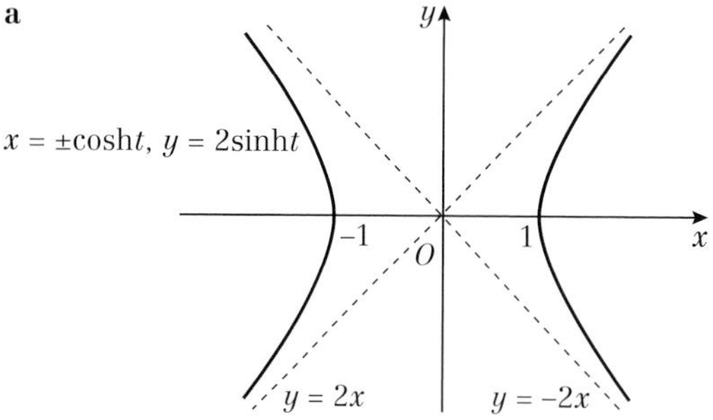

b $x^2 - \dfrac{y^2}{4} = 1$

iv a

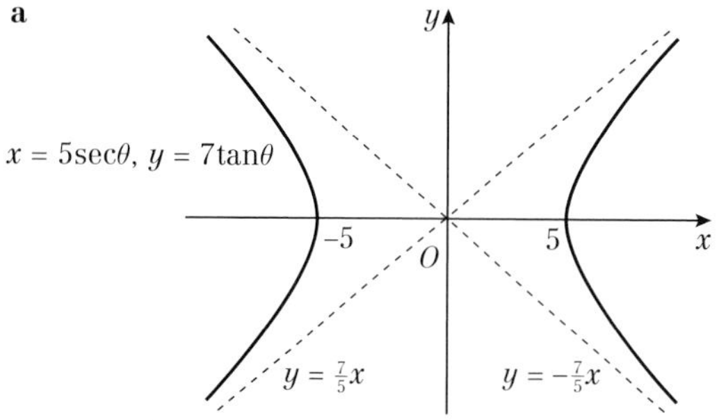

b $\dfrac{x^2}{25} - \dfrac{y^2}{49} = 1$

Challenge

$\begin{pmatrix} \frac{1}{\sqrt{2}} & -\frac{1}{\sqrt{2}} \\ \frac{1}{\sqrt{2}} & \frac{1}{\sqrt{2}} \end{pmatrix}\begin{pmatrix} ct \\ \frac{c}{t} \end{pmatrix} = \begin{pmatrix} \frac{ct}{\sqrt{2}} - \frac{c}{t\sqrt{2}} \\ \frac{ct}{\sqrt{2}} + \frac{c}{t\sqrt{2}} \end{pmatrix}$, so $x^2 = \dfrac{c^2t^2}{2} - c^2 + \dfrac{c^2}{2t^2}$ and

$y^2 = \dfrac{c^2t^2}{2} + c^2 + \dfrac{c^2}{2t^2}$. Therefore $y^2 - x^2 = 2c^2$,

so $a^2 = 2c^2 \Rightarrow a = \pm c\sqrt{2}$, so $y^2 - x^2 = a^2$

Exercise 3C

1 a $e = \frac{2}{3}$

b $e = \dfrac{\sqrt{7}}{4}$

c $e = \dfrac{1}{\sqrt{2}}$

2 a Foci = $(\pm 1, 0)$; directrices $x = \pm 4$

b Foci = $(\pm 3, 0)$; directrices $x = \pm\frac{16}{3}$

c Foci = $(0, \pm 2)$; directrices $y = \pm\frac{9}{2}$

3 a The foci are on the x-axis, so $a > b$.

b i $e = \frac{1}{2}$ **ii** $a = 6$, $b = 3\sqrt{3}$

c

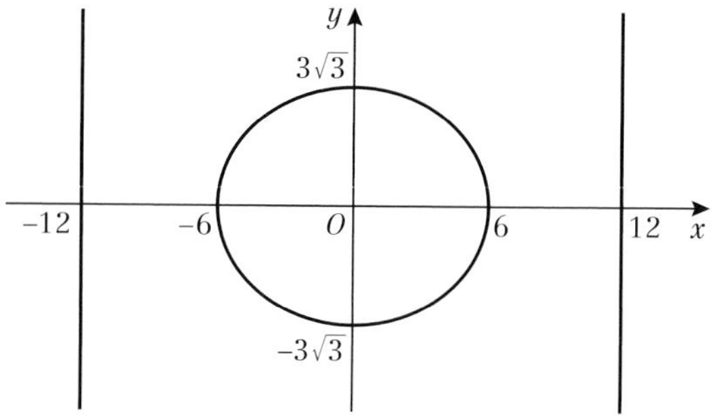

4 a The foci are on the y-axis, so $b > a$.

b i $e = \frac{1}{2}$ **ii** $a = 2\sqrt{3}$, $b = 4$

c

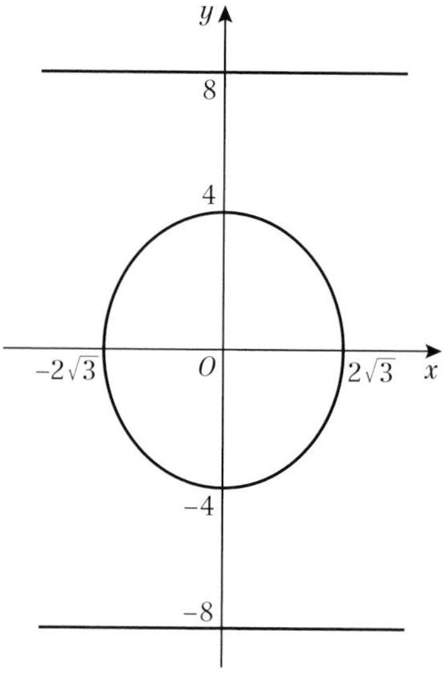

5 a $e = \dfrac{2\sqrt{10}}{5}$ **b** $e = \frac{4}{3}$ **c** $e = \frac{5}{3}$

6 a

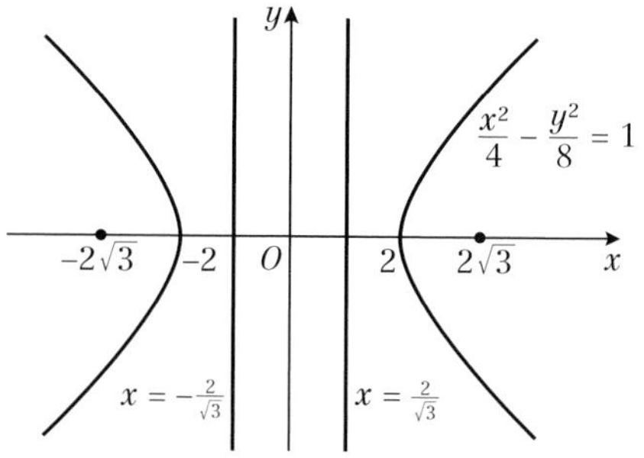

b

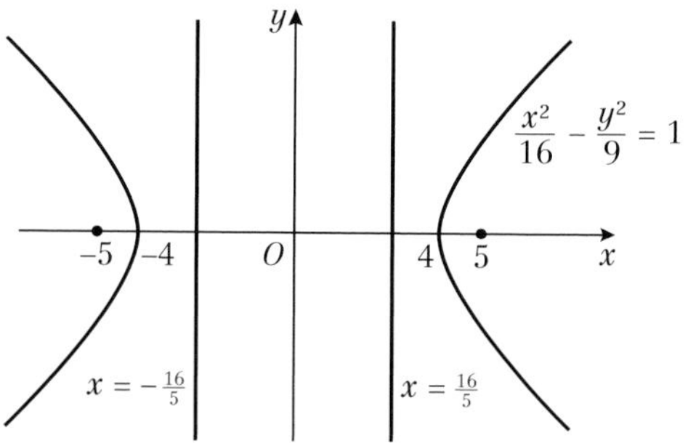

c

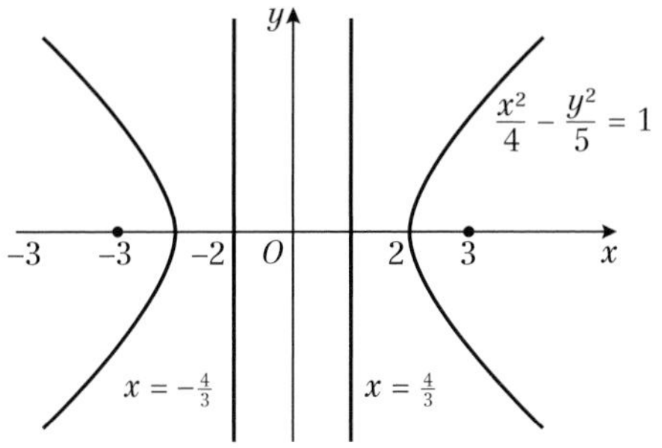

7 a i $e = \frac{5}{\sqrt{24}}$; foci are $\left(\pm\frac{5}{\sqrt{24}} \times \sqrt{24}, 0\right) = (\pm 5, 0)$

ii $e = 5$; foci are $(\pm 5 \times 1, 0) = (\pm 5, 0)$

iii $e = \frac{5}{4}$; foci are $\left(\frac{5}{4} \times 4, 0\right) = (\pm 5, 0)$

iv $e = \frac{5}{3}$; foci are $\left(\frac{5}{3} \times 3, 0\right) = (\pm 5, 0)$

b

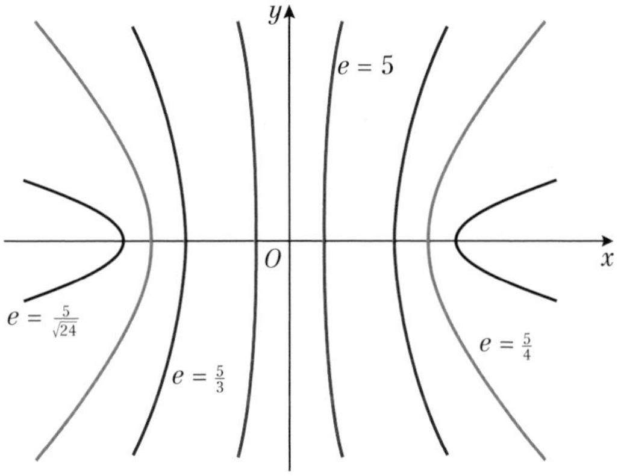

8 Use the fact that $(ae, 0)$ is on the chord.

$\frac{(ae)^2}{a^2} + \frac{y^2}{b^2} = 1$ simplifies to $y = \pm\frac{b^2}{a}$.

Therefore the length of the chord is $\frac{2b^2}{a}$

9 a $\frac{4}{5}$ **b** $\frac{x^2}{36} + \frac{y^2}{100} = 1$

10 Let P have coordinates (x, y).

$PA^2 = (x + 3\sqrt{3})^2 + y^2 = x^2 + 6x\sqrt{3} + 27 + \left(9 - \frac{x^2}{4}\right)$

$= \frac{3}{4}(x + 4\sqrt{3})^2$

$x + 4\sqrt{3} > 0 \Rightarrow PA = \frac{\sqrt{3}}{2}x + 6$

Similarly, $PB = 6 - \frac{\sqrt{3}}{2}x$

So $PA + PB = 12$.

11

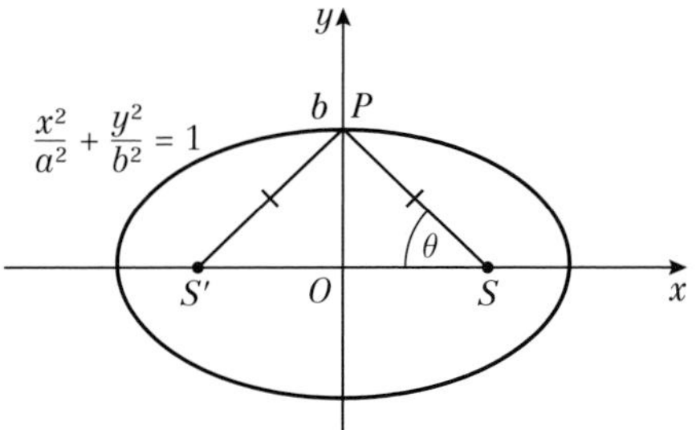

Consider $\triangle POS$

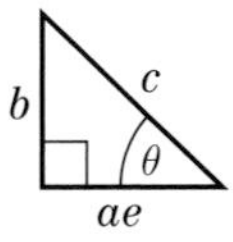

$c^2 = b^2 + a^2e^2$, but $b^2 = a^2(1 - e^2)$

$\Rightarrow c^2 = a^2 - a^2e^2 + a^2e^2 = a^2$

$\Rightarrow c = a$

So $\cos\theta = \frac{ae}{a} = e$

If you use the result that $SP + S'P = 2a$ then since $S'P = SP$ it is clear $SP = a$

Hence $\cos\theta = \frac{ae}{a} = e$

12

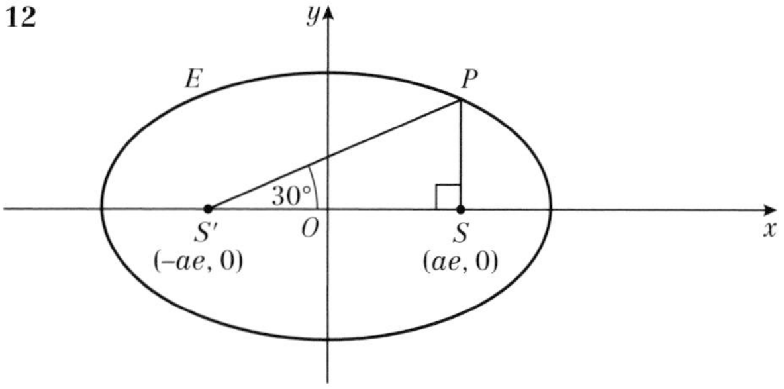

PS is y where $\frac{a^2e^2}{a^2} + \frac{y^2}{b^2} = 1$

$y^2 = b^2(1 - e^2)$

$y = b\sqrt{1 - e^2}$

$SS' = 2ae$

$\tan 30° = \frac{1}{\sqrt{3}} = \frac{y}{2ae} = \frac{b\sqrt{1 - e^2}}{2ae}$

But $b^2 = a^2(1 - e^2)$

$\Rightarrow \quad \frac{1}{\sqrt{3}} = \frac{a\sqrt{1 - e^2}\sqrt{1 - e^2}}{2ae}$

$\Rightarrow \quad \frac{2e}{\sqrt{3}} = 1 - e^2$

$\Rightarrow e^2 + \frac{2}{\sqrt{3}}e - 1 = 0$

$\Rightarrow e = \frac{1}{\sqrt{3}}, (e > 0)$

Exercise 3D

1 a Tangent: $x\cos\theta + 2y\sin\theta = 2$

Normal: $2x\sin\theta - y\cos\theta = 3\sin\theta\cos\theta$

b Tangent: $3x\cos\theta + 5y\sin\theta = 15$

Normal: $5x\sin\theta - 3y\cos\theta = 16\sin\theta\cos\theta$

2 a Tangent: $6y + \sqrt{5}x = 9$

Normal: $3\sqrt{5}y = 18x - 16\sqrt{5}$

b Tangent: $2\sqrt{3}y - x = 8$

Normal: $y + 2\sqrt{3}x = -3\sqrt{3}$

3 $\frac{dy}{dx} = -\frac{b\cos t}{a\sin t}$

So tangent is $y - b\sin t = -\frac{b\cos t}{a\sin t}(x - a\cos t)$

$\Rightarrow bx\cos t + ay\sin t = ab(\sin^2 t + \cos^2 t) = ab$

4 a $y = x + \sqrt{5}$ meets the ellipse when

$\frac{x^2}{4} + (x + \sqrt{5})^2 = 1$

$\Rightarrow 5x^2 + 8\sqrt{5}x + 16 = 0$

This has discriminant $(8\sqrt{5})^2 - 4 \times 5 \times 16 = 0$.

So the line meets the ellipse at only one point, therefore is a tangent to the ellipse.

b $\left(-\frac{4}{5}\sqrt{5}, \frac{1}{5}\sqrt{5}\right)$

5 a $2y\cos\theta - 3x\sin\theta = -5\sin\theta\cos\theta$

b P is $(3, 0)$, $(-3, 0)$, $\left(-\frac{3}{2}, \sqrt{3}\right)$ or $\left(-\frac{3}{2}, -\sqrt{3}\right)$

6 $c = \pm 2\sqrt{2}$

7 $m = \pm 2$

8 a $m = 2$ **b** $\left(-\frac{3}{2}, 1\right)$ **c** $\left(0, \frac{1}{4}\right)$ **d** $\frac{45}{16}$

9 a $\frac{dy}{dx} = \frac{\frac{dy}{d\theta}}{\frac{dx}{d\theta}} = \frac{2\cos\theta}{-3\sin\theta} = -\frac{2}{3}\cot\theta$

b $\frac{\left(\frac{9}{5}\right)^2}{9} + \frac{\left(-\frac{8}{5}\right)^2}{4} = \frac{9}{25} + \frac{16}{25} = 1$, so $Q\left(\frac{9}{5}, -\frac{8}{5}\right)$ lies on E.

c $\frac{1}{2}$

d $\tan\theta = \frac{1}{3}$; $\left(\frac{9}{\sqrt{10}}, \frac{2}{\sqrt{10}}\right)$ and $\left(-\frac{9}{\sqrt{10}}, -\frac{2}{\sqrt{10}}\right)$

10 $m = \pm\sqrt{2}$, $c = \pm 8$

11 $3x\sin\theta\cos\theta - 2y = 24\sin\theta$

12 a $m = \frac{3\cos\theta}{-5\sin\theta}$, $(x_1, y_1) = (5\cos\theta, 3\sin\theta)$
Substitute into $y - y_1 = m(x - x_1)$ and simplify.

b $3y\sin\theta\cos\theta - 9\cos\theta = 5x\sin^2\theta$

c At $(-4, 0)$, $-9\cos\theta = -20\sin^2\theta$
Use $\sin^2\theta + \cos^2\theta \equiv 1$ to obtain
$20\cos^2\theta + 9\cos\theta - 20 = 0$ and therefore $\cos\theta = \frac{4}{5}$

13 a $\frac{dy}{dx} = \frac{-2\cos t}{\sin t}$ and substitute into $y - y_1 = m(x - x_1)$
using $m = \frac{-2\cos t}{\sin t}$ and $(x_1, y_1) = (2\cos t, 4\sin t)$

b Find $l_2: 2y\cos t = x\sin t$ and equate/substitute l_1 and l_2.

14 x-intercept is $x = \frac{a}{\cos t}$, y-intercept is $y = \frac{b}{\sin t}$
Area $= \frac{1}{2} \times \frac{a}{\cos t} \times \frac{b}{\sin t}$ and simplify using
$\sin 2t \equiv 2\sin t\cos t$ to obtain answer.

15 Rearrange to obtain $y = \frac{2\sqrt{36 - x^2}}{3}$ and then integrate using the substitution $6\sin u = x$. Simplify using $\sin^2 u + \cos^2 u \equiv 1$. Integrate between $x = 3$ and $x = 6$ and multiply the answer by 2 (for the area underneath the x-axis.

Challenge

Rearrange to obtain $y = b\sqrt{1 - \frac{x^2}{a^2}}$
Use the substitution $\sin u = \frac{x}{a}$
Integrate between $x = 0$ and $x = a$ and then multiply the final answer by 4.

Exercise 3E

1 a Tangent: $8y = 3x - 4$
Normal: $3y + 8x = 108$

b Tangent: $3y = 2x - 6$
Normal: $2y + 3x = 48$

c Tangent: $5y = 2x - 5$
Normal: $2y + 5x = 56$

2 a Tangent: $5y\sinh t + 10 = 2x\cosh t$
Normal: $2y\cosh t + 5x\sinh t = 29\cosh t\sinh t$

b Tangent: $y\tan t + 3 = 3x\sec t$
Normal: $3y\sec t + x\tan t = 10\sec t\tan t$

3 $\frac{dy}{dx} = -\frac{b\sec t}{a\tan t}$
So tangent is $y - b\tan t = -\frac{b\sec t}{a\tan t}(x - a\sec t)$
$\Rightarrow bx\sec t - ay\tan t = ab(\sec^2 t - \tan^2 t) = ab$

4 $\frac{dy}{dx} = \frac{b\cosh t}{a\sinh t}$, so gradient of normal is $-\frac{a\sinh t}{b\cosh t}$
So equation of normal is
$y - b\sinh t = -\frac{a\sinh t}{b\cosh t}(x - a\cosh t)$
$\Rightarrow ax\sinh t + by\cosh t = (a^2 + b^2)\sinh t\cosh t$

5 a $\left(0, -\frac{3}{\sinh t}\right)$

b $\left(0, \frac{25}{3}\sinh t\right)$

c $\frac{2}{3}|(25\sinh^2 t + 9)\coth t|$

6 P and Q are $(4, 3\sqrt{3})$ and $(4, -3\sqrt{3})$

7 $c = \pm 6$

8 $m = \pm\frac{13}{7}$

9 $m = \pm 4$ and $c = \pm 7$

10 a $c = 3$

b $\left(\frac{25}{3}, -\frac{16}{3}\right)$

11 a Find normal gradient $= -\frac{a\sinh t}{b\cosh t}$ and substitute into $y - y_1 = m(x - x_1)$

b $\left(\left(\frac{a^2 + b^2}{a}\right)\cosh t, 0\right)$

c $\left(a, \frac{(a^2 + b^2)\sinh t\cosh t - a^2\sinh t}{b\cosh t}\right)$

12 a Substitute $m = \frac{5}{7\sin\theta}$ and
$(x_1, y_1) = (7\sec\theta, 5\tan\theta)$ into $y - y_1 = m(x - x_1)$.

b l_1 has gradient $\frac{5}{7\sin\theta}$, so equation of l_2 is $y = -\frac{7\sin\theta}{5}$
So at Q, $-\frac{49}{5}(\sin^2\theta)x = 5x - 35\cos\theta$
$\Rightarrow x = \frac{175\cos\theta}{25 + 49\sin^2\theta}$, $y = \frac{-245\sin\theta\cos\theta}{25 + 49\sin^2\theta}$

13 $\frac{dy}{dx} = \frac{x}{4y}$
$y - y_1 = m(x - x_1) \Rightarrow xx_1 - 4yy_1 = 16$, $xx_2 - 4yy_2 = 16$
Equate and substitute (m, n) to obtain $\frac{y_2 - y_1}{x_2 - x_1} = \frac{m}{4n}$
This is the gradient of the line joining (x_1, y_1) and (x_2, y_2)
$y - y_1 = \frac{m}{4n}(x - x_1) \Rightarrow mx - 4ny = 16$

14 Substitute (6, 4) into the general equation of the tangent to get $3\sec\theta - 4\tan\theta = 2 \Rightarrow 2\cos\theta + 4\sin\theta = 3$
$\Rightarrow \sqrt{20}\cos(\theta + 1.107...) = 3$
$\Rightarrow \theta + 1.107... = ..., 0.835..., 5.447..., 7.118..., ...$
This gives two values of θ in the range $[0, 2\pi)$, so there are two tangents to the hyperbola passing through (6, 4).

15 $P = (2, 2\sqrt{3})$
$\frac{dy}{dx} = \frac{4x}{y}$, so at P $\frac{dy}{dx} = \frac{4\sqrt{3}}{3}$
Line l has gradient $-\frac{\sqrt{3}}{4}$
$y - 2\sqrt{3} = -\frac{\sqrt{3}}{4}(x - 2) \Rightarrow y = -\frac{\sqrt{3}}{4}(x - 10)$
Line l cuts the x-axis at $x = 10$ so the right-angled triangle has area $= 8\sqrt{3}$
The remaining region has area $= \int_1^2 \sqrt{4(x^2 - 1)}\,dx$.
Substitute $x = \cosh u$ and $dx = \sinh u\,du$ so integral becomes
$2\int_0^{\text{arcosh } 2} \sqrt{\cosh^2 u - 1}\sinh u\,du = 2\int_0^{\text{arcosh } 2} \sinh^2 u\,du$
$= \frac{1}{2}\int_0^{\text{arcosh } 2} (e^u - e^{-u})(e^u - e^{-u})\,du$
$= \frac{1}{2}\int_0^{\text{arcosh } 2} (e^{2u} - 2 + e^{-2u})\,du$
$= \frac{1}{2}\left[\frac{1}{2}(e^{2u} - e^{-2u}) - 2u\right]_0^{\text{arcosh } 2} = 2\sqrt{3} - \text{arcosh } 2$
So total area $= 10\sqrt{3} - \text{arcosh } 2$

Online Full worked solutions are available in SolutionBank.

16 a The asymptotes of H are $y = x$ and $y = -x$.
Let $A = (a, a)$ and $B = (b, -b)$, so the midpoint of AB is $\left(\frac{a+b}{2}, \frac{a-b}{2}\right)$.
Now we compute a and b for the generic point P on H $P = (X, Y)$.
Differentiating H we get $2x - 2y\frac{dy}{dx} = 0 \Rightarrow \frac{dy}{dx} = \frac{x}{y}$
Gradient of the tangent at P is $\frac{X}{Y}$
So the tangent has equation $y - Y = \frac{X}{Y}(x - X)$
At A: $a - Y = \frac{X}{Y}(a - X) \Rightarrow a = X + Y$
At B: $b - Y = \frac{X}{Y}(-b - X) \Rightarrow b = Y - X$
So $X = \frac{a+b}{2}$ and $Y = \frac{a-b}{2}$

b $|OA| = \sqrt{2}|a|$ and $|OB| = \sqrt{2}|b|$
So $|OA| \times |OB| = 2|ab| = 2|X^2 - Y^2| = 2|1| = 2$
which is constant.

Exercise 3F

1 a $(apq, a(p + q))$

b Chord PQ has gradient
$$\frac{2ap - 2aq}{ap^2 - aq^2} = \frac{2a(p-q)}{a(p-q)(p+q)} = \frac{2}{(p+q)}$$
Equation of chord PQ is: $y - 2ap = \frac{2}{p+q}(x - ap^2)$
$\Rightarrow y(p + q) - 2ap^2 - 2apq = 2x - 2ap^2$
$\Rightarrow y(p + q) = 2x + 2apq$
Chord passes through $(a, 0) \Rightarrow 0 = 2a + 2apq$
or $pq = -1$
Locus of R is $x = -a$

c $y = a$

2 a $\frac{2x}{a^2} - \frac{2y}{b^2} \times \frac{dy}{dx} = 0 \Rightarrow \frac{dy}{dx} = \frac{b^2x}{a^2y}$
So gradient of tangent at P is $\frac{b^2 a\sec t}{a^2 b\tan t} = \frac{b}{a\sin t}$
Equation of tangent is $y - b\tan t = \frac{b}{a\sin t}(x - a\sec t)$
$\Rightarrow bx\sec t - ay\tan t = ab(\sec^2 t - \tan^2 t) = ab$

b A is where $y = 0 \Rightarrow x = \frac{ab}{b\sec t} = a\cos t$,
i.e. $A(a\cos t, 0)$.
B is where $x = 0 \Rightarrow y = \frac{ab}{-a\tan t} = -b\cot t$,
i.e. $B(0, -b\cot t)$.
Midpoint of AB is $\left(\frac{a}{2}\cos t, -\frac{b}{2}\cot t\right)$
$x = \frac{a}{2}\cos t \Rightarrow \sec t = \frac{a}{2x}$
$y = -\frac{b}{2}\cot t \Rightarrow \tan t = -\frac{b}{2y}$
Use $\sec^2 t \equiv 1 + \tan^2 t$
$\Rightarrow \frac{a^2}{4x^2} = 1 + \frac{b^2}{4y^2}$ which gives locus.

3 a $\frac{2x}{a^2} - \frac{2y}{b^2} \times \frac{dy}{dx} = 0 \Rightarrow \frac{dy}{dx} = \frac{b^2x}{a^2y}$
So gradient of normal at P is $-\frac{a^2 b\tan t}{b^2 a\sec t} = -\frac{a}{b}\sin t$
Equation of tangent is $y - b\tan t = -\frac{a}{b}\sin t(x - a\sec t)$
$\Rightarrow ax\sin t + by = (a^2 + b^2)\tan t$

b $y = 0 \Rightarrow x = \left(\frac{a^2+b^2}{a}\right)\sec t \Rightarrow A$ is $\left(\frac{a^2+b^2}{a}\sec t, 0\right)$
$x = 0 \Rightarrow y = \left(\frac{a^2+b^2}{b}\right)\tan t \Rightarrow B$ is $\left(0, \frac{a^2+b^2}{b}\tan t\right)$
Midpoint of AB is $\left(\frac{a^2+b^2}{2a}\sec t, \frac{a^2+b^2}{2b}\tan t\right)$
$x = \frac{a^2+b^2}{2a}\sec t \Rightarrow \sec t = \frac{2ax}{a^2+b^2}$
$y = \frac{a^2+b^2}{2b}\tan t \Rightarrow \tan t = \frac{2by}{a^2+b^2}$
Use $\sec^2 t \equiv 1 + \tan^2 t$:
$4a^2x^2 = (a^2 + b^2)^2 + 4b^2y^2$

4 a Find $\frac{dy}{dx} = \frac{3\cos t}{-5\sin t}$ and substitute into
$y - y_1 = m(x - x_1)$ using $m = \frac{5\sin\theta}{3\cos\theta}$ and
$(x_1, y_1) = (5\cos\theta, 3\sin\theta)$

b Find midpoint $\left(\frac{8}{5}\cos\theta, -\frac{8}{3}\sin\theta\right)$ and use
$\sin^2\theta + \cos^2\theta \equiv 1$

5 a Tangent at P is $x + p^2y = 2cp$
Tangent at Q is $x + q^2y = 2cq$
Tangents intersect when $(p^2 - q^2)y = 2c(p - q)$
So $R = \left(\frac{2cpq}{p+q}, \frac{2c}{p+q}\right)$

b Gradient of PQ is $\frac{\frac{c}{q} - \frac{c}{p}}{cq - cp} = -\frac{1}{pq}$
So equation of PQ is
$y - \frac{c}{p} = -\frac{1}{pq}(x - cp) \Rightarrow ypq + x = c(p + q)$

c **i** $y = -2x, x \neq 0$ **ii** $y = 2c^2, x < 0$ **iii** $x = 2c^2$

6 a $\frac{1}{t}$

b $y - 2at = \frac{1}{t}(x - at^2) \Rightarrow ty - 2at^2 = x - at^2$
$\Rightarrow x - ty + at^2 = 0$

c T is $(0, at)$. Perpendicular bisector of OT is $y = \frac{at}{2}$
Perpendicular bisector of OP is $y - at = -\frac{t}{2}\left(x - \frac{at^2}{2}\right)$
Centre of circle is where perpendicular bisectors intersect: $\frac{at}{2} - at = -\frac{t}{2}\left(x - \frac{at^2}{2}\right)$
Therefore centre of circle is $\left(\frac{at^2}{2} + a, \frac{at}{2}\right)$.

d $X = a + \frac{at^2}{2} \Rightarrow at^2 = 2(X - a)$
$Y = \frac{at}{2} \Rightarrow 2at = 4Y$
So $(4Y)^2 = 4a \times 2(X - a)$ or $2Y^2 = a(X - a)$

7 $y = \frac{1}{2}, x < 0$

8 a $\frac{x^2}{2^2} + \frac{(2y-6)^2}{4^2} = 1$

b Simplifies to $x^2 + (y - 3)^2 = 4$ which is a circle of centre $(0, 3)$ and radius 2.

Challenge

Let $A = (x_1, y_1)$ and $B = (x_2, y_2)$.

Then $k = \dfrac{y_2 - y_1}{x_2 - x_1}$ and the midpoint is $(x, y) = \left(\dfrac{x_1 + x_2}{2}, \dfrac{y_1 + y_2}{2}\right)$.

$$\left.\begin{array}{l} b^2x_1^2 + a^2y_1^2 = a^2b^2 \\ b^2x_2^2 + a^2y_2^2 = a^2b^2 \end{array}\right\} \Rightarrow a^2(y_2^2 - y_1^2) = -b^2(x_2^2 - x_1^2)$$

$\Rightarrow a^2(y_1 + y_2)(y_2 - y_1) = -b^2(x_1 + x_2)(x_2 - x_1)$

Substituting in k gives $-\dfrac{ka^2}{b^2}(y_1 + y_2) = (x_1 + x_2)$

Using the coordinates of the midpoint gives $-\dfrac{ka^2}{b^2}y = x$

$\Rightarrow ka^2y + b^2x = 0$

This is a line passing through the origin.

Mixed exercise 3

1 a $\dfrac{x^2}{16} + \dfrac{y^2}{81} = 1$

b

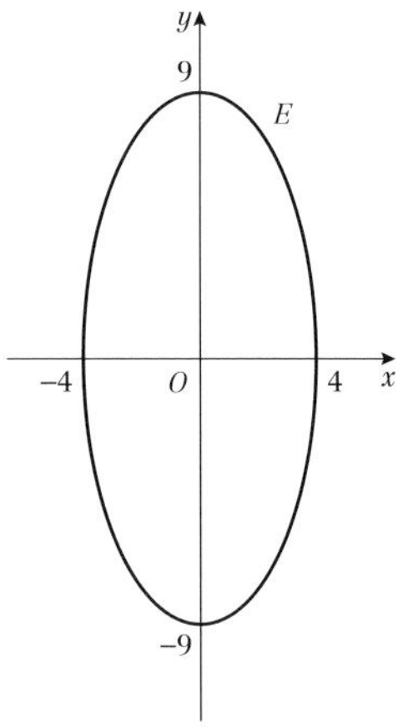

c $4x\sin\theta - 9y\cos\theta = -65\cos\theta\sin\theta$

2 a $\dfrac{x^2}{4} - \dfrac{y^2}{25} = 1$

b

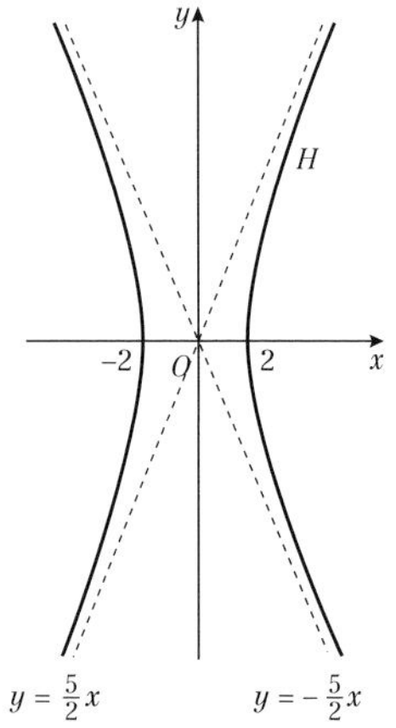

c $2y\sinh t + 10 = 5x\cosh t$

3 a $\dfrac{x^2}{a^2} - \dfrac{y^2}{a^2m^2} = 1$

b A is $\left(\dfrac{a^2 + b^2}{a}\sec t, 0\right)$ and B is $\left(0, \dfrac{a^2 + b^2}{b}\tan t\right)$.

So midpoint is $(x, y) = \left(\dfrac{a^2 + b^2}{2a}\sec t, \dfrac{a^2 + b^2}{2b}\tan t\right)$.

Using $\sec^2 t \equiv 1 + \tan^2 t$, $\dfrac{4a^2x^2}{(a^2 + b^2)^2} = \dfrac{4b^2y^2}{(a^2 + b^2)^2} + 1$

So the locus of the midpoint is $4a^2x^2 = (a^2 + b^2)^2 + 4b^2y^2$.

4 a Gradient of chord $= \dfrac{\frac{c}{p} - \frac{c}{q}}{cp - cq} = \dfrac{c(q - p)}{pq\,c(p - q)} = -\dfrac{1}{pq}$

b

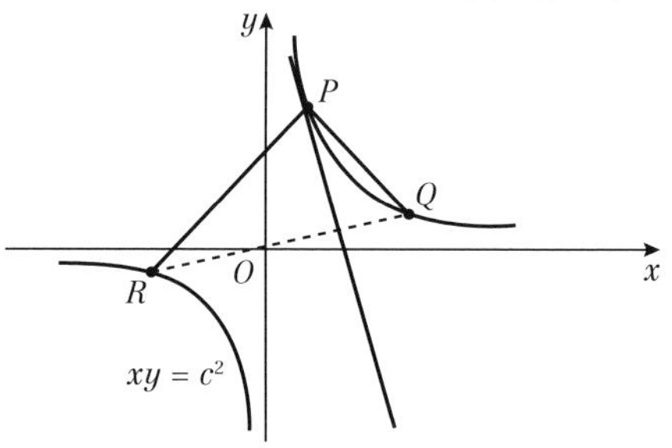

Gradient of $PQ = -\dfrac{1}{pq}$ and gradient of $PR = -\dfrac{1}{pr}$

So $-1 = p^2qr$ (1)

Gradient of tangent at P is $-\dfrac{1}{p^2}$ and gradient of chord $RQ = -\dfrac{1}{qr}$

So $\left(-\dfrac{1}{qr}\right)\left(-\dfrac{1}{p^2}\right) = \dfrac{1}{p^2qr}$

But from (1), $p^2qr = -1$

Therefore tangent at P is perpendicular to chord QR.

5 a $y = ct^{-1}$, $x = ct \Rightarrow \dfrac{\mathrm{d}y}{\mathrm{d}x} = \dfrac{-ct^{-2}}{c} = -\dfrac{1}{t^2}$

Equation of tangent is: $y - \dfrac{c}{t} = -\dfrac{1}{t^2}(x - ct)$

$\Rightarrow yt^2 - ct = -x + ct$ or $t^2y + x = 2ct$

b $\left(-\frac{4}{3}, -12\right)$ and $\left(12, \frac{4}{3}\right)$

6 a Let P have coordinates (x, y).

$PA^2 = (x + 4)^2 + y^2 = x^2 + 8x + 16 + \left(9 - \frac{9}{25}x^2\right)$

$= \left(\frac{4}{5}x + 5\right)^2$

$\frac{4}{5}x + 5 > 0 \Rightarrow PA = \frac{4}{5}x + 5$

Similarly, $PB = 5 - \frac{4}{5}x$, so $PA + PB = 10$.

b

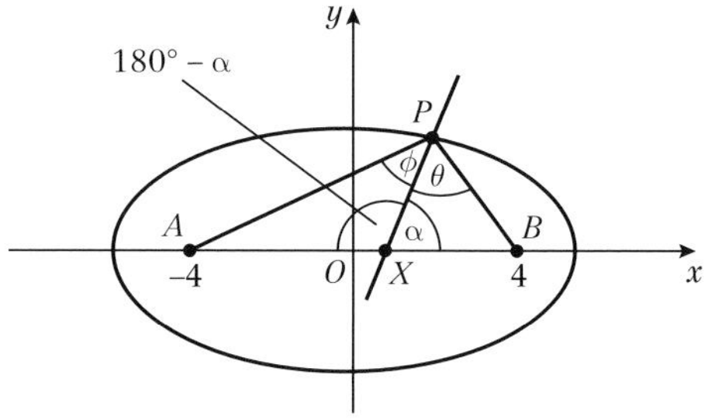

Normal at P is $5x\sin t - 3y\cos t = 16\cos t\sin t$

X is when $y = 0$, i.e. $x = \frac{16}{5}\cos t$

$PB^2 = (5\cos t - 4)^2 + (3\sin t)^2 = (4\cos t - 5)^2$

$\Rightarrow PB = 5 - 4\cos t$ and $PA = 10 - PB = 5 + 4\cos t$

$AX = 4 + \frac{16}{5}\cos t$, $BX = 4 - \frac{16}{5}\cos t$

Consider sine rule on $\triangle PAX$:

$\sin\phi = \dfrac{\sin(180° - \alpha)AX}{AP} = \dfrac{\sin\alpha\left(4 + \frac{16}{5}\cos t\right)}{5 + 4\cos t} = \frac{4}{5}\sin\alpha$

Online Full worked solutions are available in SolutionBank.

Consider sine rule on $\triangle PBX$:

$$\sin\theta = \frac{\sin\alpha BX}{PB} = \frac{\sin\alpha\left(4 - \frac{16}{5}\cos t\right)}{5 - 4\cos t} = \frac{4}{5}\sin\alpha$$

So $\sin\phi = \sin\theta$ and, since both angles are acute, $\theta = \phi$

Therefore normal bisects APB.

7 a $y = ct^{-1}$, $x = ct \Rightarrow \frac{dy}{dx} = \frac{-ct^2}{c} = -\frac{1}{t^2}$

Equation of tangent is: $y - \frac{c}{t} = -\frac{1}{t^2}(x - ct)$

$yt^2 - ct = -x + ct$ or $t^2y + x = 2ct$

b

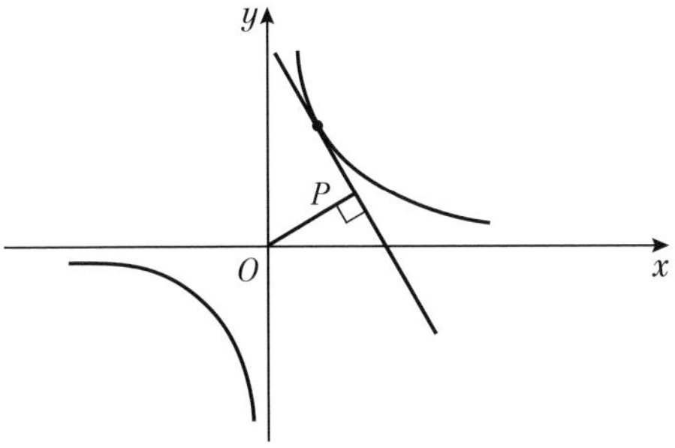

Gradient of tangent is $-\frac{1}{t^2}$

Gradient of OP is t^2

Equation of OP is $y = t^2x$

Equation of tangent is $t^2y = 2ct - x$

Solving, $t^4x = 2ct - x$

$$\Rightarrow \quad x = \frac{2ct}{1 + t^4},\ y = \frac{2ct^3}{1 + t^4}$$

$$x^2 + y^2 = \frac{4c^2t^2 + 4c^2t^6}{(1 + t^4)^2} = \frac{4c^2t^2}{1 + t^4}$$

$$\left.\begin{aligned}\Rightarrow (x^2 + y^2)^2 &= \frac{16c^4t^4}{(1 + t^4)^2}\\ xy &= \frac{4c^2t^4}{(1 + t^4)^2}\end{aligned}\right\} \Rightarrow (x^2 + y^2)^2 = 4c^2xy$$

8 a OP has gradient $\frac{2ap}{ap^2} = \frac{2}{p}$ and OQ has gradient $\frac{2}{q}$

Since OP and OQ are perpendicular, $\frac{2}{p} \times \frac{2}{q} = -1$, so $pq = -4$.

b $y + xq = aq^3 + 2aq$

c Normal at P is $y + xp = ap^3 + 2ap$

Solve equations simultaneously to get

$x = a(q^2 + p^2 + qp + 2)$, $y = apq(q + p)$

$pq = -4 \Rightarrow R$ is $(ap^2 + aq^2 - 2, -4pq(p + q))$

d $x = a((p + q)^2 - 2pq - 2) = a((p + q)^2 + 6)$

$$y = 4a(p + q) \Rightarrow p + q = \frac{y}{4a}$$

$$\Rightarrow \quad x = a\left(\frac{y^2}{16a^2} + 6\right)$$

$$\Rightarrow x - 6a = \frac{y^2}{16a} \Rightarrow y^2 = 16ax - 96a^2$$

9 $y = mx + c$ and $\frac{x^2}{a^2} + \frac{y^2}{b^2} = 1$

$\Rightarrow b^2x^2 + a^2(mx + c)^2 = a^2b^2$

$\Rightarrow x^2(b^2 + a^2m^2) + 2a^2mcx + a^2(c^2 - b^2) = 0$

For a tangent the discriminant is 0:

$4a^4m^2c^2 = 4(b^2 + a^2m^2)a^2(c^2 - b^2)$

$\Rightarrow \quad c = \pm\sqrt{a^2m^2 + b^2}$

So the lines $y = mx \pm \sqrt{a^2m^2 + b^2}$ are tangents.

10

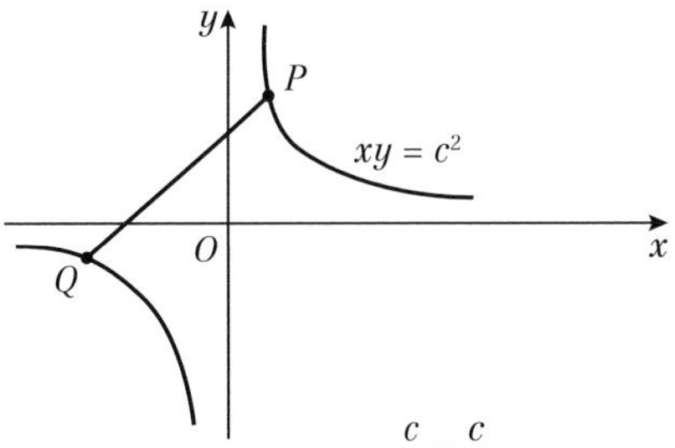

Chord PQ has gradient $\dfrac{\frac{c}{p} - \frac{c}{q}}{cp - cq} = \dfrac{c(q - p)}{pqc(p - q)} = -\dfrac{1}{pq}$

If gradient = 1, then $pq = -1$.

Tangent at P is $p^2y + x = 2cp$

Tangent at Q is $q^2y + x = 2cq$

Intersection: $(p^2 - q^2)y = 2c(p - q) \Rightarrow y = \frac{2c}{p + q}$

$\Rightarrow x = 2cp - \frac{2cp^2}{p + q} = \frac{2cpq}{p + q}$

So R is $\left(\frac{2cpq}{p + q}, \frac{2c}{p + q}\right)$

But $pq = -1$ so R is $\left(x = \frac{-2c}{p + q}, y = \frac{2c}{p + q}\right)$

The locus of R is the line $y = -x$

11 a Find $\frac{dy}{dx} = \frac{4\cos\theta}{-6\sin\theta} = \frac{2\cos\theta}{-3\sin\theta}$ and substitute into $y - y_1 = m(x - x_1)$ using $m = \frac{2\cos\theta}{-3\sin\theta}$ and $(x_1, y_1) = (6\cos\theta, 4\sin\theta)$

b Find midpoint $\left(\frac{3}{\cos\theta}, \frac{2}{\sin\theta}\right)$ and use $\sin^2\theta + \cos^2\theta \equiv 1$

12 a $m = \frac{5\cos\theta}{-13\sin\theta}$, $(x_1, y_1) = (13\cos\theta, 5\sin\theta)$

Substitute into $y - y_1 = m(x - x_1)$ and simplify.

b $5y\sin\theta\cos\theta - 25\cos\theta = 13x\sin^2\theta$

c $(-ae, 0) = (-12, 0)$ as $a = 13$, $b = 5$ and $e = \frac{12}{13}$

Given line passes through this point,

$-25\cos\theta = -156\sin^2\theta$

Use $\sin^2\theta + \cos^2\theta \equiv 1$ to obtain

$156\cos^2\theta + 25\cos\theta - 156 = 0$ and therefore

$\cos\theta = \frac{12}{13} = e$

13 a Find normal gradient $= -\frac{\sin\theta}{2}$ and substitute into $y - y_1 = m(x - x_1)$

b $A(20\sec\theta, 0)$, $B(1, 10\tan\theta)$ and midpoint of AB is $(10\sec\theta, 5\tan\theta)$

Use $\tan^2\theta + 1 \equiv \sec^2\theta$ to obtain $\frac{x^2}{100} - \frac{y^2}{25} = 1$

14 a $\frac{dy}{dx} = \frac{b\cos t}{-a\sin t}$

So gradient of normal at $(a\cos t, b\sin t)$ is $\frac{a\sin t}{b\cos t}$

Equation of normal is $y - b\sin t = \frac{a\sin t}{b\cos t}(x - a\cos t)$

$\Rightarrow ax\sin t - by\cos t = (a^2 - b^2)\cos t\sin t$

b $y = 0 \Rightarrow x = \left(\frac{a^2 - b^2}{a}\right)\cos t \Rightarrow M$ is $\left(\frac{a^2 - b^2}{a}\cos t, 0\right)$

$x = 0 \Rightarrow y = -\left(\frac{a^2 - b^2}{b}\right)\sin t \Rightarrow N$ is $\left(0, -\frac{a^2 - b^2}{b}\sin t\right)$

Midpoint of MN is $\left(\frac{a^2 - b^2}{2a}\cos t, -\frac{a^2 - b^2}{2b}\sin t\right)$

$x = \frac{a^2 - b^2}{2a}\cos t \Rightarrow \cos t = \frac{2ax}{a^2 - b^2}$

$y = -\frac{a^2 - b^2}{2b}\sin t \Rightarrow \sin t = \frac{2by}{a^2 - b^2}$

$\sin^2 t + \cos^2 t \equiv 1 \Rightarrow 4b^2y^2 + 4a^2x^2 = (a^2 - b^2)^2$

15 $a = 5, b = 3 \Rightarrow e = \frac{4}{5}$, so foci are $(\pm 4, 0)$.
Let P have coordinates (x, y).
$PS^2 = (x + 4)^2 + y^2 = x^2 + 8x + 16 + \left(9 - \frac{9}{25}x^2\right)$
$= \left(\frac{4}{5}x + 5\right)^2$
$\frac{4}{5}x + 5 > 0 \Rightarrow PS = \frac{4}{5}x + 5$
Similarly, $PS' = 5 - \frac{4}{5}x$, so $PS + PS' = 10$.

16 x-intercept is $x = \dfrac{a}{\cos t}$, height is $b\sin t$.
Area $= \frac{1}{2} \times \dfrac{a}{\cos t} \times b\sin t$ which simplifies to $\frac{1}{2}ab\tan t$.

17 Area bounded by x-axis, $x = 3$ and ellipse is
$\frac{1}{2}\int_3^b \sqrt{36 - x^2}\,dx = 3\pi - \frac{9}{4}\sqrt{3}$
Area of triangle formed by $x = 3$, the tangent and the x-axis is
$\frac{1}{2} \times 9 \times \dfrac{3\sqrt{3}}{2} = \frac{27}{4}\sqrt{3}$
Shaded area $= \frac{27}{4}\sqrt{3} - (3\pi - \frac{9}{4}\sqrt{3}) = 9\sqrt{3} - 3\pi$

18 $\frac{108}{5}$

19 **a** $P(3, 2\sqrt{2})$ and $Q = (3, -2\sqrt{2})$

b The area of R is $2\left(\frac{1}{2}\right)\left(\frac{8}{3}\right)(2\sqrt{2}) - I$
where $I = \int_1^3 \sqrt{x^2 - 1}\,dx$
Substitute $x = \cosh u$ and $dx = \sinh u\,du$ so integral becomes
$\int_0^{\text{arcosh}\,3} \sqrt{\cosh^2 u - 1}\,\sinh u\,du = \int_0^{\text{arcosh}\,3} \sinh^2 u\,du$
$= \frac{1}{4}\int_0^{\text{arcosh}\,3} (e^u - e^{-u})(e^u - e^{-u})\,du$
$= \frac{1}{4}\int_0^{\text{arcosh}\,3} (e^{2u} - 2 + e^{-2u})\,du$
$= \frac{1}{4}\left[\frac{1}{2}(e^{2u} - e^{-2u}) - 2u\right]_0^{\text{arcosh}\,3} = 3\sqrt{2} - \frac{1}{2}\text{arcosh}\,3$
Area of $R = 2\left(\dfrac{8\sqrt{2}}{3} - I\right) = \text{arcosh}\,3 - \frac{2}{3}\sqrt{2}$

Challenge

$QS = ePS \Leftrightarrow QS^2 = e^2PS^2$
$QS^2 = a^2e^4\cos^2\theta - 2a^2e^3\cos\theta + a^2e^2$
$PS^2 = a^2\cos^2\theta - 2a^2e\cos\theta + a^2e^2 + b^2\sin^2\theta$
Use rearrangements of $b^2 = a^2(1 - e^2)$ to simplify.

CHAPTER 4

Prior knowledge check

1 **a** $x < -\frac{1}{3}$ or $x > 1$ **b** $-2 - \sqrt{6} < x < -2 + \sqrt{6}$

2 **a** $x > 2$ or $x < -\frac{4}{3}$ **b** $\frac{3}{2} < x < \frac{5}{2}$

Exercise 4A

1 **a** $-1 < x < 6$ **b** $x \leqslant -3$ or $x \geqslant 2$
c $-1 < x < 1$ **d** $-\sqrt{3} < x < -1$ or $1 < x < \sqrt{3}$
e $0 \leqslant x < 1$ or $x \geqslant \frac{3}{2}$ **f** $x < -1$ or $0 < x < 2$
g $x < -2$ or $-1 < x < 1$ or $x > 2$
h $-1 < x < 0$ or $0 < x < 2$
i $x < 4$ or $x > \frac{14}{3}$
j $-2 < x < 5$ or $x > \frac{17}{2}$

2 **a** $\{x : x > \frac{1}{3}\} \cup \{x : -5 < x < 0\}$
b $\{x : x < 0\} \cup \{x : 2 < x < 5\}$
c $\{x : x < -2\} \cup \{x : 0 < x < 1\}$
d $\{x : x < -3\} \cup \{x : -1 < x < 1\}$
e $\{x : -\frac{1}{3} < x < 0\} \cup \{x : 0 < x < \frac{1}{2}\}$
f $\{x : -1 < x < -\frac{1}{3}\} \cup \{x : x > \frac{1}{2}\}$

3 $-5 < x < -4$ and $-1 - \sqrt{7} < x < -1 + \sqrt{7}$

4 $\{x : -\frac{1}{2} < x < \dfrac{5 - \sqrt{29}}{2}\} \cup \{x : 3 < x < \dfrac{5 + \sqrt{29}}{2}\}$

5 **a** The student did not square the denominators before cross-multiplying. Multiplying by negative values does not preserve the inequality.
b $-\frac{4}{3} < x < -1$ or $0 < x < 4$

6 $\{x : -2 < x < -1\} \cup \{x : -\frac{1}{2} < x < 0\}$

Challenge

$x < \ln\frac{1}{2}$ or $x > \ln 1$

Exercise 4B

1 **a**

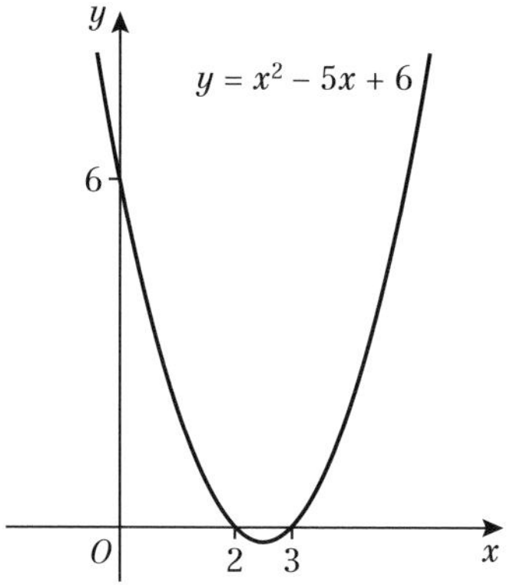

b

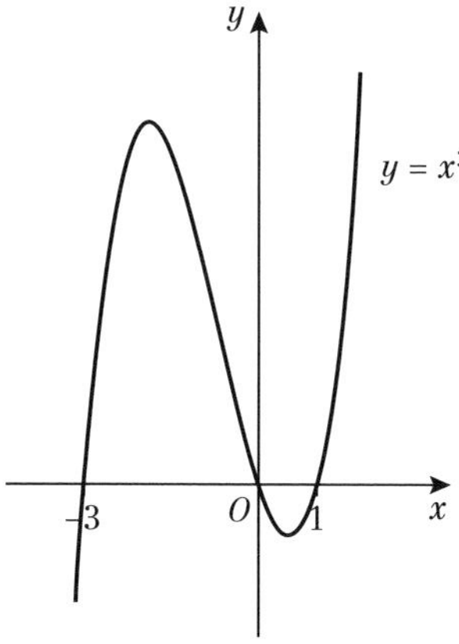

c

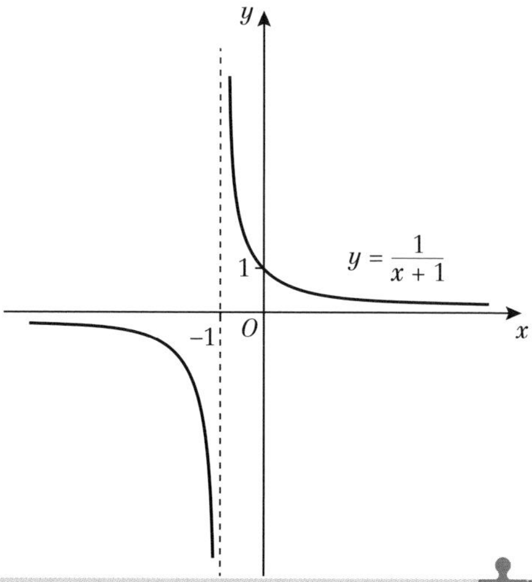

d

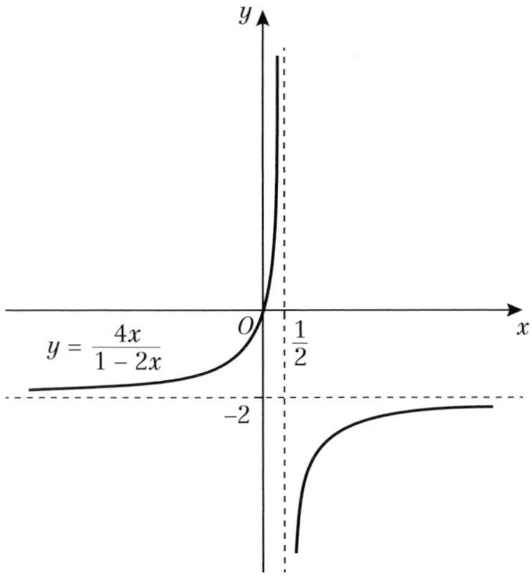

2 a

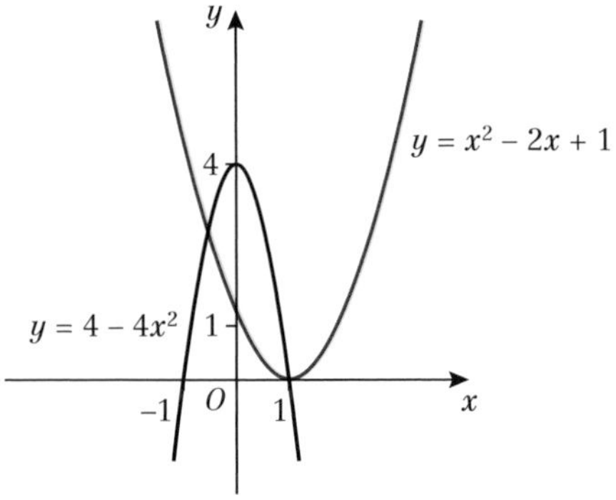

b

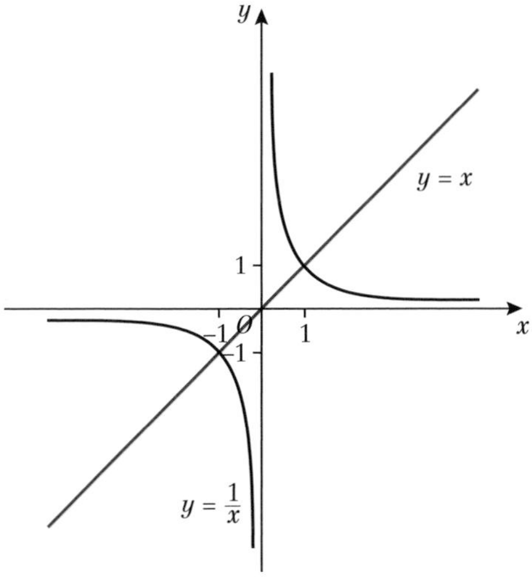

c

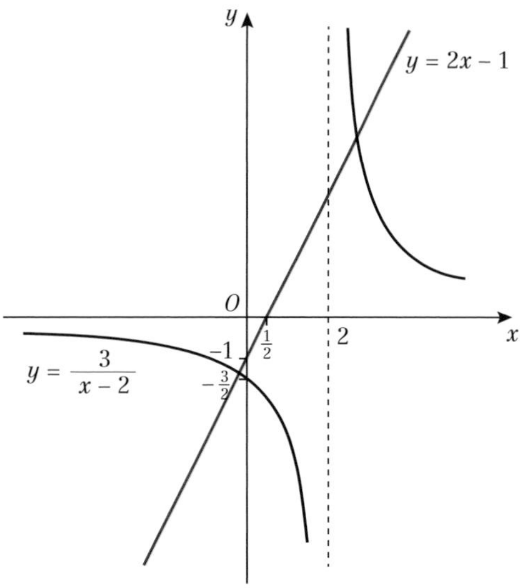

d

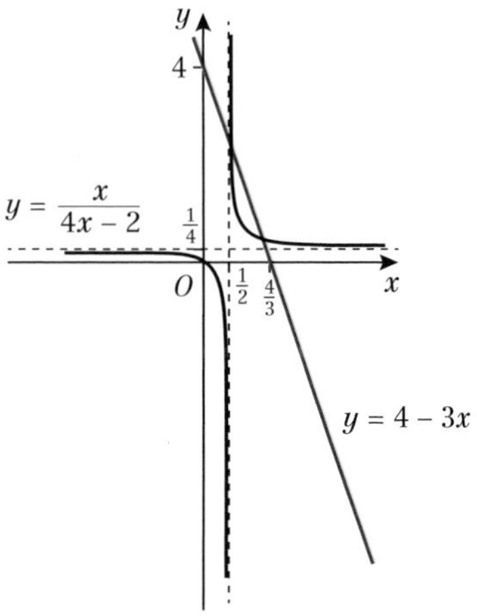

3 a $(7, \frac{1}{4})$

b (4, 2) and (–1, –3)

c (–2, 0), (0, –4), (4, 12)

4 a

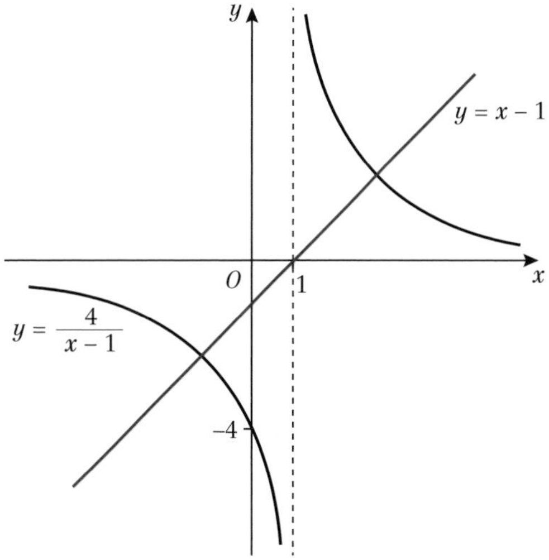

b (3, 2) and (–1, –2)

c $-1 < x < 1$ or $x > 3$

5 a

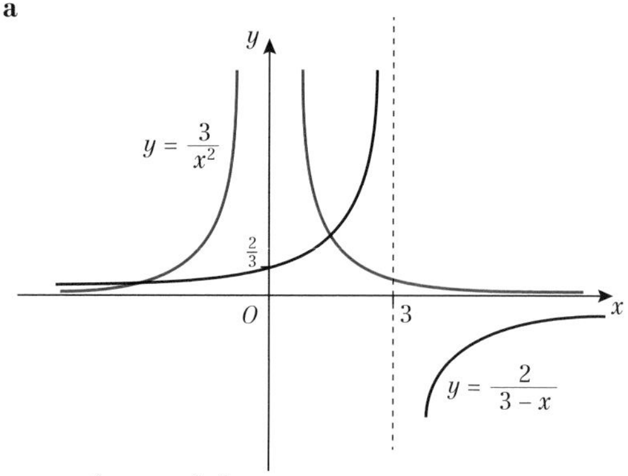

b $(-3, \frac{1}{3})$ and $(\frac{3}{2}, \frac{4}{3})$

c $\{x: -3 < x < 0\} \cup \{x: 0 < x < \frac{3}{2}\} \cup \{x: x > 3\}$

6 a

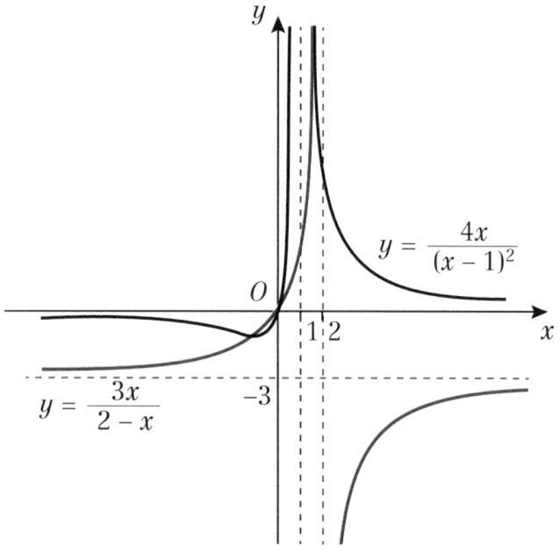

b (0, 0), $(\frac{5}{3}, 15)$ and (−1, −1)

c $x \leqslant -1$ or $0 \leqslant x < 1$ or $1 < x \leqslant \frac{5}{3}$ or $2 < x$

7 a

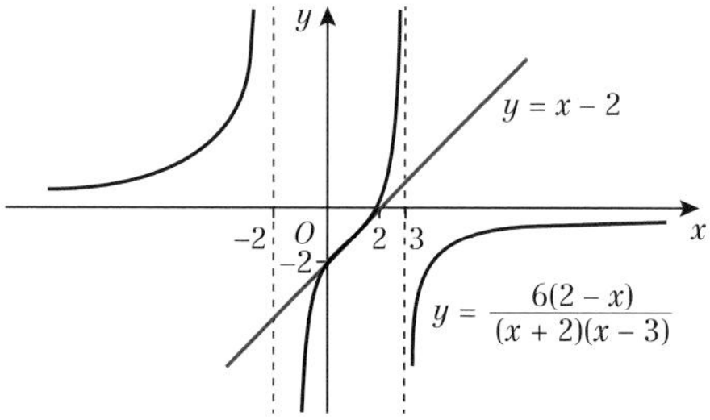

b (0, −2), (1, −1) and (2, 0)

c $x < -2$ or $0 \leqslant x \leqslant 1$ or $2 \leqslant x < 3$

8 a

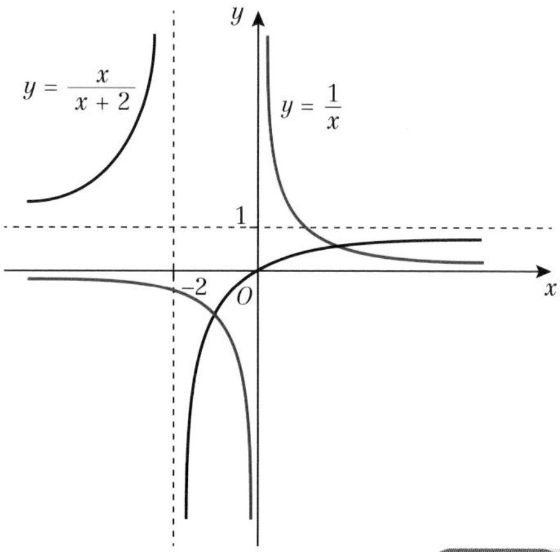

b $(-1, -1)$ and $(2, \frac{1}{2})$

c $-2 < x < -1$ or $0 < x < 2$

Challenge

a

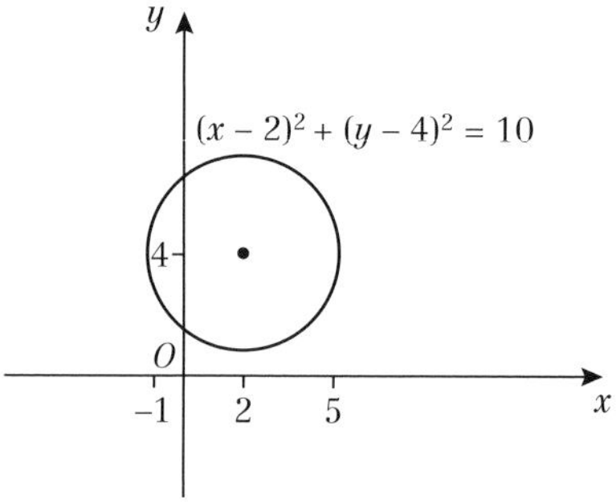

b (−1, 3), (1, 1), (3, 7), (5, 5),

c

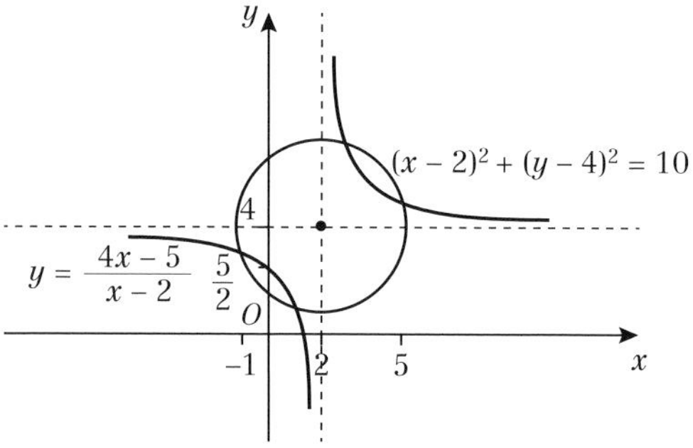

d $-1 < x < 1$ and $3 < x < 5$

Exercise 4C

1 a $x < \frac{6}{7}$

b $\frac{1}{2}(-\sqrt{13} - 1) < x < \frac{1}{2}(\sqrt{13} - 1)$

c $-7 < x < -2 - \sqrt{7}$ or $-2 + \sqrt{7} < x < 3$

d $x \geqslant 1$ or $x \leqslant -2$

e $x > 1$ or $x < -3$

f $x > 1$ or $x < -\frac{1}{3}$

2 a

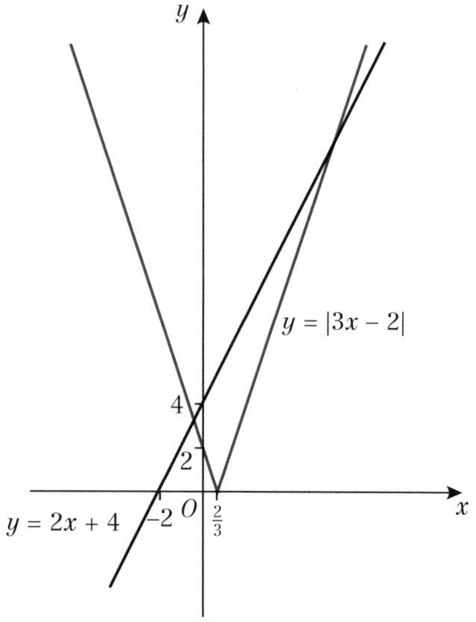

b $\{x: -\frac{2}{5} \leqslant x \leqslant 6\}$

3 a

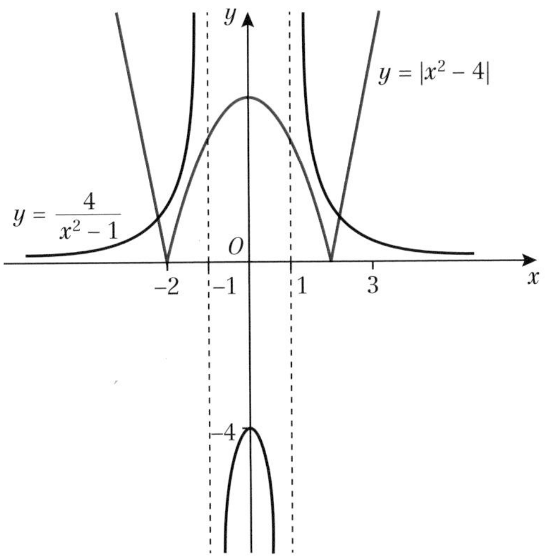

b $-\sqrt{5} \leqslant x < -1$ or $1 < x \leqslant \sqrt{5}$

4 $\{x: -1 < x < \frac{1}{3}\}$

5 $\{x:x < -1 - \sqrt{3}\} \cup \{x: -\sqrt{2} < x < \sqrt{3} - 1\}$

6 a

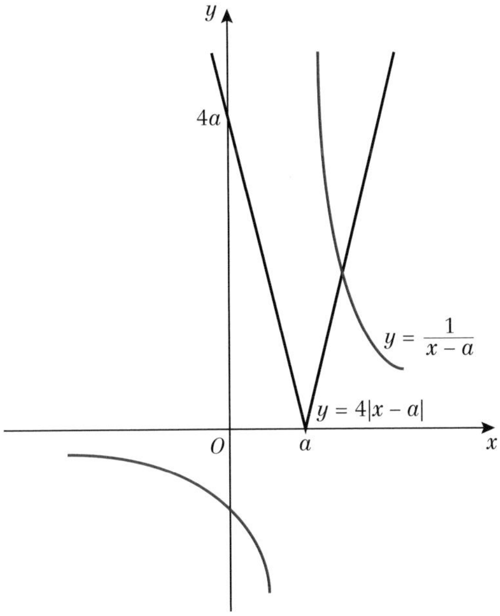

b $x < a$ or $x > a + \frac{1}{2}$

7 $-2 < x < 0$ or $x > 2$

8 a The student hasn't checked which critical values actually correspond to intersections of the graphs.

b $1 < x < 5$

Challenge

a $f(-1) = (-1)^3 + 3(-1)^2 - 13(-1) - 15 = -1 + 3 + 13 - 15 = 0$
So by the factor theorem $(x + 1)$ is a factor.

b $f(x) = (x + 1)(x + 5)(x - 3)$

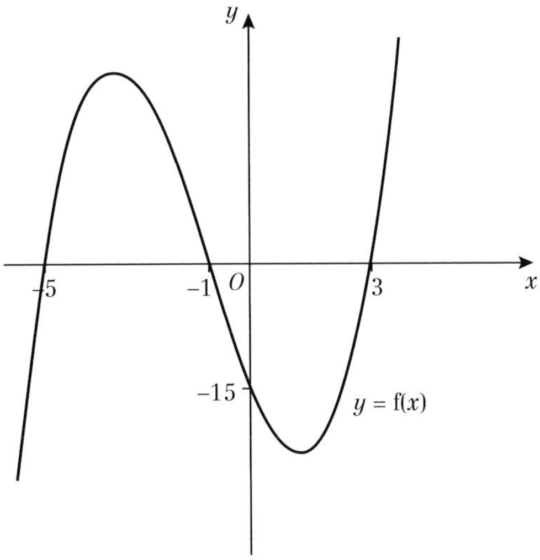

c $x = -5$, $1 - \sqrt{5} \leqslant x \leqslant 1 - \sqrt{3}$, $1 + \sqrt{3} \leqslant x \leqslant 1 + \sqrt{5}$

Mixed exercise 4

1 $0 < x < 2$ or $x \geqslant 4$

2 $-2 < x < 1 - \sqrt{6}$ or $x > 1 + \sqrt{6}$

3 $0 < x < 2$ or $x > \frac{7}{2}$

4 $\{x:0 < x < \frac{3}{2}\} \cup \{x:3 < x < 4\}$

5 $\{x:x < -1\} \cup \{x:1 < x < 11\}$

6 a

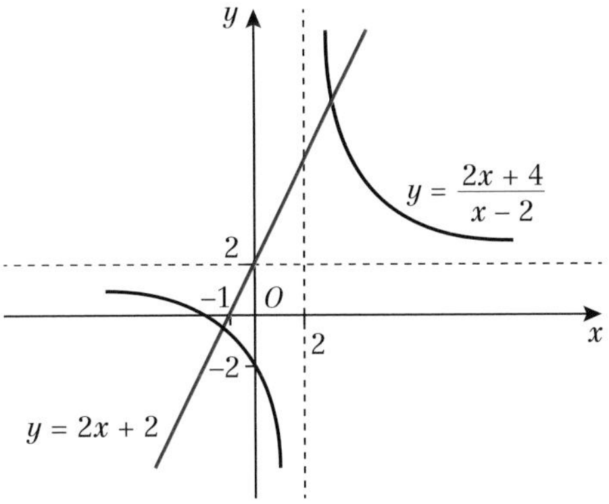

b $1 - \sqrt{5} < x < 2$ or $x > 1 + \sqrt{5}$

7 a

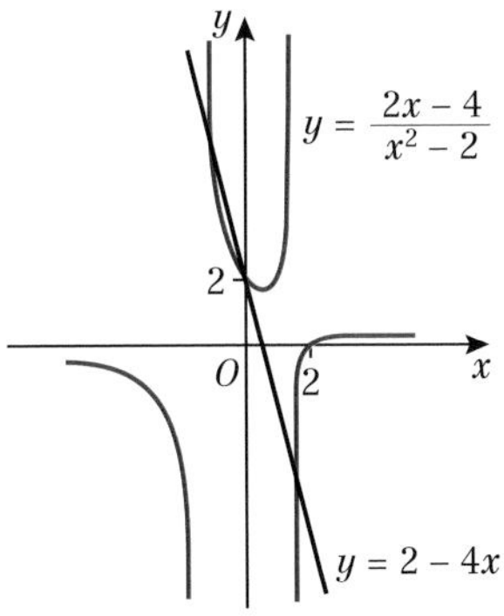

b $-\sqrt{2} < x < -1$ or $0 < x < \sqrt{2}$ or $x > \frac{3}{2}$

8 a

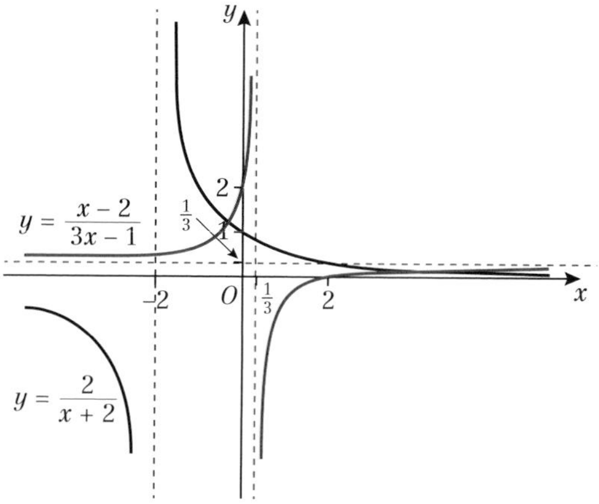

b $-2 < x < 3 - \sqrt{11}$ and $\frac{1}{3} < x < 3 + \sqrt{11}$

9 a

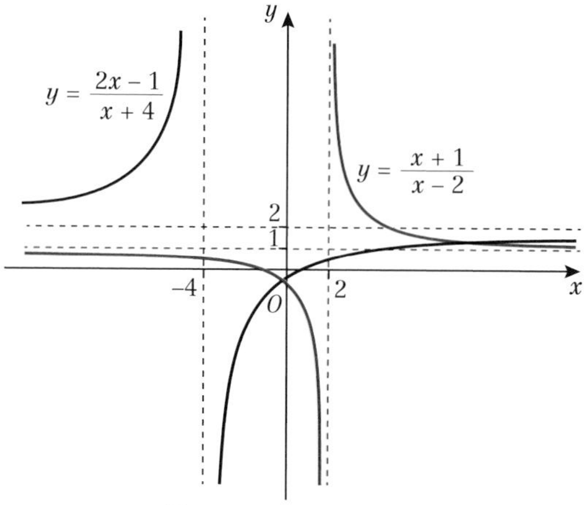

b $x < -4,\ 5 - 3\sqrt{3} < x < 2,\ 5 + 3\sqrt{3} < x$

10 $1 < x < 5$

11 $-3 < x < 3$

12 $x < \frac{2}{7}$

13 $x < \sqrt{3} - 1$ or $x > \sqrt{3} + 1$

14 a

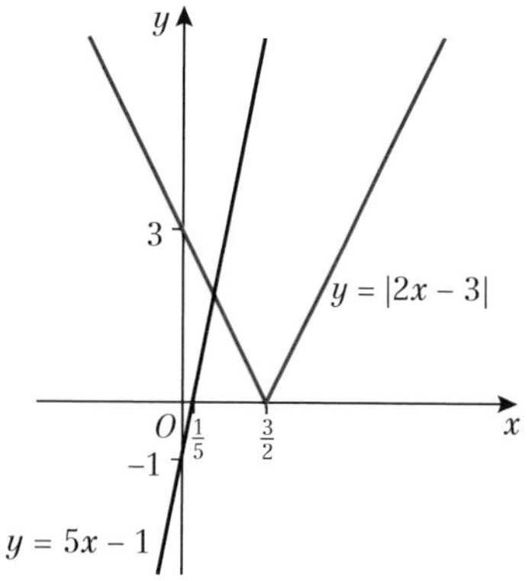

b $x > \frac{4}{7}$

15 a $x = -1 - \sqrt{7}, 0, 1, -1 + \sqrt{7}$

b

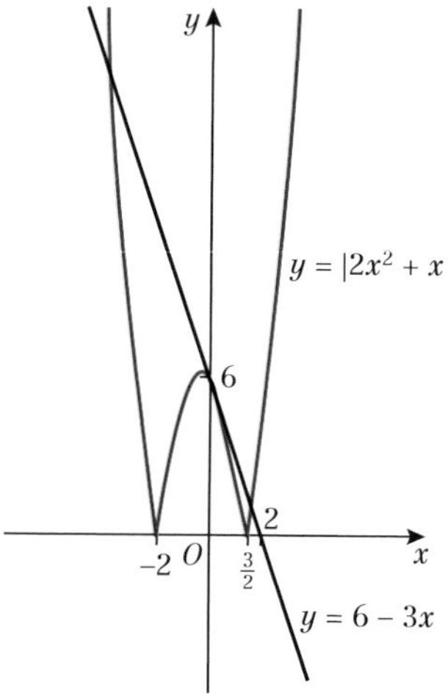

c $x < -1 - \sqrt{7}$ or $0 < x < 1$ or $x > -1 + \sqrt{7}$

16 a

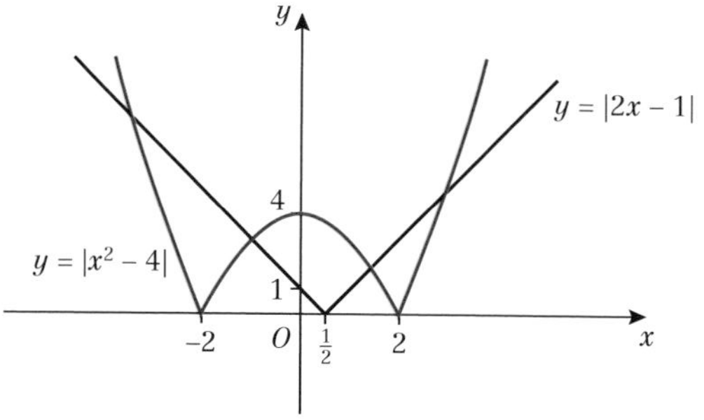

b $x = -1 - \sqrt{6}, -1, -1 + \sqrt{6}, 3$

c $x < -1 - \sqrt{6}$, or $-1 < x < -1 + \sqrt{6}$ or $x > 3$

17 a The student has correctly found critical values, but not checked which correspond to points of intersection.

b $\{x : x < -3 + \sqrt{6}\} \cup \{x : x > 1\}$

Challenge

Solving $x^2 - 5x + 2 = x - 3$ and $x^2 - 5x + 2 = 3 - x$ we find that the critical values are $x = 2 - \sqrt{5}, 1, 2 + \sqrt{5}, 5$

Sketching the graphs we have

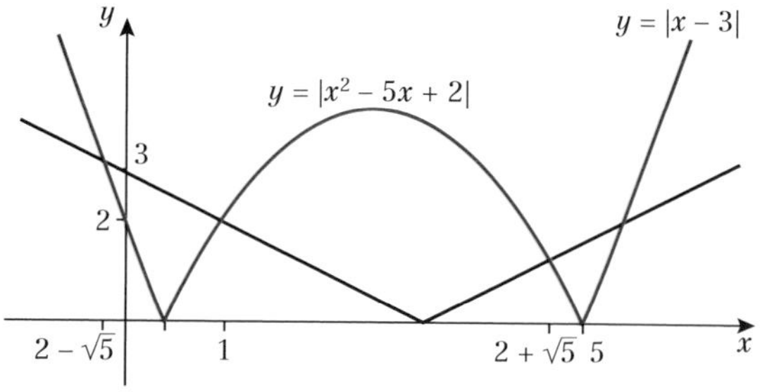

$\{x : x < 2 - \sqrt{5}\} \cup \{x : 1 < x < 2 + \sqrt{5}\} \cup \{x : x > 5\}$

Review exercise 1

1 $2\sqrt{2}$

2 a $\begin{pmatrix} -3 \\ 3k \\ 2 + k \end{pmatrix}$

b $\frac{3\sqrt{35}}{5}$, which occurs when $k = -0.2$

3 $\overrightarrow{AB} \times \overrightarrow{AC} = (\mathbf{b} - \mathbf{a}) \times (\mathbf{c} - \mathbf{a})$
$= \mathbf{b} \times \mathbf{c} - \mathbf{b} \times \mathbf{a} - \mathbf{a} \times \mathbf{c} + \mathbf{a} \times \mathbf{a}$
$= \mathbf{a} \times \mathbf{b} + \mathbf{b} \times \mathbf{c} + \mathbf{c} \times \mathbf{a}$

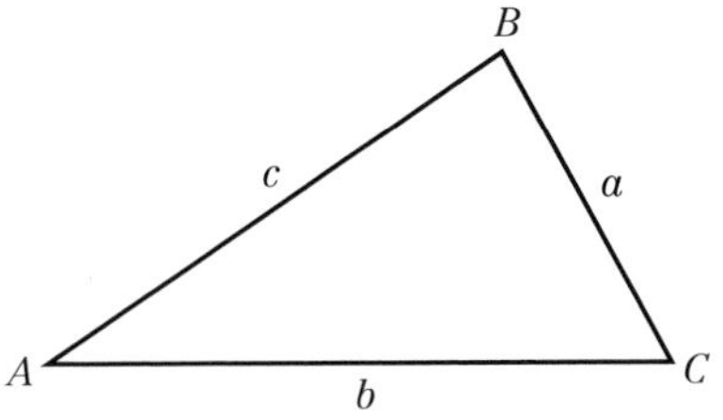

Area $= \frac{1}{2}AC \times AB \sin A = \frac{1}{2}|\overrightarrow{AB} \times \overrightarrow{AC}|$
$= \frac{1}{2}|\mathbf{a} \times \mathbf{b} + \mathbf{b} \times \mathbf{c} + \mathbf{c} \times \mathbf{a}|$, as required.

4 a $5\mathbf{i} - 3\mathbf{j} - 4\mathbf{k}$ **b** 100 **c** 50

5 a $\begin{pmatrix} -30 \\ -15 \\ 45 \end{pmatrix}$ **b** $\mathbf{r} = \begin{pmatrix} 3 \\ 1 \\ 2 \end{pmatrix} + \lambda\begin{pmatrix} -2 \\ -1 \\ 3 \end{pmatrix}$

c $\begin{pmatrix} 5 \\ 2 \\ -1 \end{pmatrix} = \begin{pmatrix} 3 \\ 1 \\ 2 \end{pmatrix} + \lambda\begin{pmatrix} -2 \\ -1 \\ 3 \end{pmatrix}$, i.e. B is the point with $\lambda = -1$.

d 35

6 a $\begin{pmatrix} -6 \\ 2 \\ -4 \end{pmatrix}$

b A vector equation of Π is $\mathbf{r}.\begin{pmatrix} -6 \\ 2 \\ -4 \end{pmatrix} = -14$

So $-6x + 2y - 4z = -14 \Rightarrow 3x - y + 2z = 7$

c $(-1, 8, 9)$

d $\mathbf{t} - \mathbf{a} = 3(\mathbf{b} - \mathbf{a})$, so $\overrightarrow{AT}$ and $\overrightarrow{AB}$ are in the same direction and have A as a common endpoint. Thus A, B and T lie on the same straight line.

7 a Equating the x- and y-components, and solving the resulting simultaneous equations gives $t = u = \frac{1}{4}$. Substituting these values into the z-components gives $\frac{11}{4}$ for l and $-\frac{3}{4}$ for m, which are not equal, so the lines do not intersect.

b $(1 - 2t_1 - 2u_1)\mathbf{i} + (-t_1 + u_1)\mathbf{j} + (-4 + t_1 + u_1)\mathbf{k}$

c $u_1 = \frac{3}{5}, t_1 = \frac{3}{5}$

8 a $\frac{-3}{\sqrt{17}}, \frac{-2}{\sqrt{17}}, \frac{2}{\sqrt{17}}$ **b** $\frac{x-2}{-3} = \frac{y-5}{-2} = \frac{z}{2}$

9 a $\mathbf{i} + 4\mathbf{j} + 2\mathbf{k}$ **b** $\mathbf{r}.(\mathbf{i} + 4\mathbf{j} + 2\mathbf{k}) = 7$ **c** 2

10 a $-\mathbf{i} + 8\mathbf{j} - 4\mathbf{k}$ **b** $3\mathbf{i} + \mathbf{j} - \mathbf{k}$

c $\mathbf{n}_1 \times \mathbf{n}_2 = \begin{vmatrix} \mathbf{i} & \mathbf{j} & \mathbf{k} \\ -1 & 8 & -4 \\ 3 & 1 & -1 \end{vmatrix} = \begin{vmatrix} 8 & -4 \\ 1 & -1 \end{vmatrix}\mathbf{i} - \begin{vmatrix} -1 & -4 \\ 3 & -1 \end{vmatrix}\mathbf{j} + \begin{vmatrix} -1 & 8 \\ 3 & 1 \end{vmatrix}\mathbf{k}$
$= -4\mathbf{i} - 13\mathbf{j} - 25\mathbf{k} = -1(4\mathbf{i} + 13\mathbf{j} + 25\mathbf{k})$

d $\mathbf{r} = \mathbf{i} + \mathbf{j} + \mathbf{k} + t(4\mathbf{i} + 13\mathbf{j} + 25\mathbf{k})$

11 a $a(4\mathbf{i} + \mathbf{j} + 2\mathbf{k}).(\mathbf{i} - 5\mathbf{j} + 3\mathbf{k}) = a(4 \times 1 + 1 \times (-5) + 2 \times 3)$
$= 5a$

b $\overrightarrow{BA} = a(4\mathbf{i} + \mathbf{j} + 2\mathbf{k}) - a(2\mathbf{i} + 11\mathbf{j} - 4\mathbf{k})$
$= 2a(\mathbf{i} - 5\mathbf{j} + 3\mathbf{k})$
$\overrightarrow{BA}$ is parallel to $\mathbf{i} - 5\mathbf{j} + 3\mathbf{k}$, which is perpendicular to Π. Hence $\overrightarrow{BA}$ is perpendicular to Π.

c 22.3° (1 d.p.)

12 a $6\mathbf{i} + \mathbf{j} - 4\mathbf{k}$

b The vector equation for Π_1 is
$\mathbf{r}.(6\mathbf{i} + \mathbf{j} - 4\mathbf{k}) = (\mathbf{i} + 6\mathbf{j} - \mathbf{k}).(6\mathbf{i} + \mathbf{j} - 4\mathbf{k}) = 16$
So $(x\mathbf{i} + y\mathbf{j} + z\mathbf{k}).(6\mathbf{i} + \mathbf{j} - 4\mathbf{k}) = 16$
$\Rightarrow 6x + y - 4z = 16$

c -2

d $(\mathbf{r} - (\frac{8}{3}\mathbf{j} - \frac{10}{3}\mathbf{k})) \times (-9\mathbf{i} + 10\mathbf{j} - 11\mathbf{k}) = \mathbf{0}$

13 a $(-5 - 4c)\mathbf{i} + (-6 - 5c)\mathbf{j} + \mathbf{k}$

b Equating coefficients of $\mathbf{i}$ and $\mathbf{j}$ of $\overrightarrow{RP} \times \overrightarrow{RQ}$, $-5 - 4c = 3 \Rightarrow c = -2$, and then $d = -6 + 10 = 4$.

c $\mathbf{r}.(3\mathbf{i} + 4\mathbf{j} + \mathbf{k}) = 7$

d $-5\mathbf{i} - 3\mathbf{j} + 8\mathbf{k}$

14 a $-15\mathbf{i} - 10\mathbf{j} - 10\mathbf{k}$ **b** $\mathbf{r}.(3\mathbf{i} + 2\mathbf{j} + 2\mathbf{k}) = 7$

c $(\mathbf{r} - (3\mathbf{i} - \mathbf{j})) \times (-2\mathbf{i} + \mathbf{j} + 2\mathbf{k}) = \mathbf{0}$

d $\left(\frac{13}{9}, -\frac{2}{9}, \frac{14}{9}\right)$

15 a $3\mathbf{i} - 6\mathbf{j} + 6\mathbf{k}$

b $(0, 0, 0)$, $(2, 0, -1)$ and $(4, 3, 1)$ all satisfy $x - 2y + 2z = 0$, so $x - 2y + 2z = 0$ is the equation of the plane through O, A and B.

c 14

d $\mathbf{r} = 4\mathbf{i} + 3\mathbf{j} + \mathbf{k} + t(\mathbf{j} + \mathbf{k})$

e $4\mathbf{i} + \mathbf{j} - \mathbf{k}$

16 a $2\mathbf{i} - 3\mathbf{j} - 2\mathbf{k}$ **b** $\frac{\sqrt{17}}{2}$

c $\mathbf{r}.(2\mathbf{i} - 3\mathbf{j} - 2\mathbf{k}) = -7$ **d** $2x - 3y - 2z = -7$

e $\frac{7}{\sqrt{17}}$ **f** 3.2° (1 d.p.)

17 a $\mathbf{r} = \begin{pmatrix} 1 \\ 2 \\ 3 \end{pmatrix} + t\begin{pmatrix} -3 \\ 5 \\ 1 \end{pmatrix}$ **b** $\frac{11}{6}$

c $\mathbf{r}.\begin{pmatrix} -3 \\ 5 \\ 1 \end{pmatrix} = -1$ **d** $\left(\frac{68}{35}, \frac{15}{35}, \frac{94}{35}\right)$

e $DE^2 = \left(1 - \frac{68}{35}\right)^2 + \left(2 - \frac{15}{35}\right)^2 + \left(3 - \frac{94}{35}\right)^2 = \frac{121}{35}$
$\Rightarrow DE = \frac{11\sqrt{35}}{35}$

f $\left(\frac{101}{35}, -\frac{40}{35}, \frac{83}{35}\right)$

18 a Equating the x- and y-components of $\mathbf{r}$ for l_1 and l_2 and solving the resulting simultaneous equations gives $s = 2$ and $t = 5$. Substituting these into the z-components gives 2 for both l_1 and l_2, so l_1 and l_2 intersect.
$(-2\mathbf{i} + \mathbf{j} + 3\mathbf{k}).(-\mathbf{i} + \mathbf{j} - \mathbf{k}) = 2 + 1 - 3 = 0$, so $l_1 \perp l_2$.

b $\mathbf{r} = -5\mathbf{i} + 4\mathbf{j} + 2\mathbf{k} + \mu(9\mathbf{i} + (\lambda - 4)\mathbf{j} - 5\mathbf{k})$

c $\frac{5\lambda + 11}{\sqrt{42}\sqrt{\lambda^2 - 8\lambda + 122}}$

d $-\frac{11}{5}$

19 a $\sqrt{10}$ **b** $\mathbf{r}.(\mathbf{i} - 2\mathbf{j} - 2\mathbf{k}) = -6$

c $2\mathbf{i} + \mathbf{j} + 3\mathbf{k}$

20 $(-8, 2)$

21 a

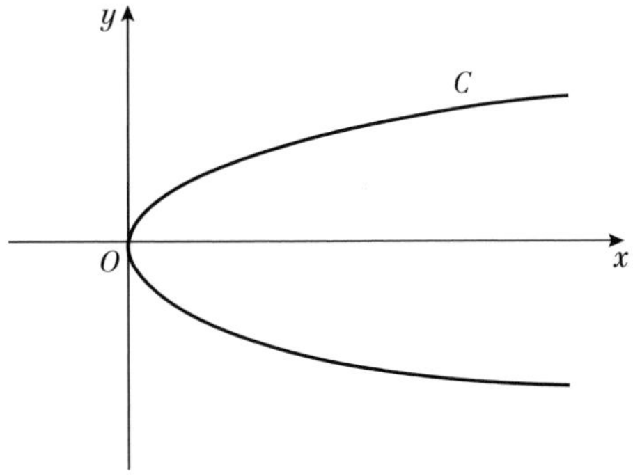

b $60\sqrt{2}$

22 $a = 2, b = 9$
Equation of perpendicular bisector of PQ is
$y = -2x + 14$
x-coordinates of M and N are $\frac{15}{2} + \frac{\sqrt{29}}{2}, \frac{15}{2} - \frac{\sqrt{29}}{2}$
$\lambda = \frac{15}{2}, \mu = \frac{1}{2}$

23 **a** $(4, 0)$ **b** $4x - 3y - 16 = 0$ **c** $(1, -4)$

24 $\frac{350}{3}$

25 $(2, 8), (-4, -4)$

26 $(-8, 2)$

27 **a** $P(2, 5)$ and $Q(10, 1)$ **b** $24 - 10\ln 5$

28 **a** $t = \frac{1}{2}, (6, 24)$ **b** $y = 2x + 12$
c $y = -4x + 48$

29 The equation of the tangent is $yt = x + at^2$ so T has coordinates $(-at^2, 0)$.
The equation of the normal is $y = -tx + at^3 + 2at$, so N has coordinates $(at^2 + 2a, 0)$.
$PT^2 = (2at^2)^2 + (2at)^2 = 4a^2t^2(1 + t^2) \Rightarrow PT = 2at\sqrt{1 + t^2}$
$PN^2 = (2a)^2 + (2at)^2 = (2a)^2(1 + t^2) \Rightarrow PN = 2a\sqrt{1 + t^2}$
$\Rightarrow \frac{PT}{PN} = \frac{2at\sqrt{1 + t^2}}{2a\sqrt{1 + t^2}} = t$

30 **a** $\frac{dy}{dx} = -\frac{9}{x^2}, x = 3t \Rightarrow \frac{dy}{dx} = -\frac{1}{t^2}$
So the tangent to H at $\left(3t, \frac{3}{t}\right)$ has equation
$y - \frac{3}{t} = -\frac{1}{t^2}(x - 3t)$, or $x + t^2y = 6t$.
b At A, $x = 6t \Rightarrow OA = 6t$.
At B, $t^2y = 6t \Rightarrow y = \frac{6}{t} \Rightarrow OB = \frac{6}{t}$
Area of triangle $OAB = \frac{1}{2} \times 6t \times \frac{6}{t} = 18$

31 **a** $\frac{dy}{dx} = -\frac{c^2}{x^2}. x = ct \Rightarrow \frac{dy}{dx} = -\frac{1}{t^2}$
So gradient of normal is t^2 and equation of normal is
$y - \frac{c}{t} = t^2(x - ct) \Rightarrow y = t^3x - ty - c(t^4 - 1) = 0$
b $y = x \Rightarrow t^3x - tx - c(t^4 - 1) = 0 \Rightarrow x = ct + \frac{c}{t}$
So G has coordinates $\left(ct + \frac{c}{t}, ct + \frac{c}{t}\right)$.
$PG^2 = \left(ct + \frac{c}{t} - ct\right)^2 + \left(ct + \frac{c}{t} - \frac{c}{t}\right)^2$
$= c^2\left(t^2 + \frac{1}{t^2}\right)$

32 **a** $(8, 0)$ **b** $x = -8$
c Line through PQ and Q has gradient $\frac{-40}{30} = -\frac{4}{3}$, so equation of this line is $y - 8 = -\frac{4}{3}(x - 2)$.
When $y = 0$, this gives $x = 8$, so the line goes through $S(8,0)$.
d $\frac{dy}{dx} = \frac{16}{y}$
Tangent to C at P is $y - 8 = \frac{16}{8}(x - 2)$, or $y = 2x + 4$,
and tangent at Q is $y + 32 = \frac{16}{-32}(x - 32)$,
or $y = -\frac{1}{2}x - 16$. Thus D is such that $2x + 4 = -\frac{1}{2}x - 16$
$\Rightarrow x = -8$, and hence lies on the directrix.

33 **a** $\frac{dy}{dx} = \frac{\sqrt{a}}{\sqrt{x}}. x = at^2 \Rightarrow \frac{dy}{dx} = \frac{1}{t}$
So gradient of normal is $-t$ and equation of normal is
$y - 2at = -t(x - at^2) \Rightarrow y + tx = 2at + at^3$
b $\left(a\left(\frac{t^2 + 2}{t}\right)^2, -2a\left(\frac{t^2 + 2}{t}\right)\right)$

34 **a** 8 **b** $y = 2x + 4$ **c** 4

35 **a** $\frac{dy}{dx} = -\frac{c^2}{x^2}. x = ct \Rightarrow \frac{dy}{dx} = -\frac{1}{t^2}$
So gradient of normal is t^2 and equation of normal is
$y - \frac{c}{t} = t^2(x - ct) \Rightarrow y = t^2x + \frac{c}{t} - ct^3$
b $\left(-\frac{c}{t^3}, -ct^3\right)$
c $(X, Y) = \frac{c}{2}\left(t - \frac{1}{t^3}, \frac{1}{t} - t^3\right)$
So $\frac{X}{Y} = \frac{t - \frac{1}{t^3}}{\frac{1}{t} - t^3} = -\frac{1}{t^2}\left(\frac{t - \frac{1}{t^3}}{t - \frac{1}{t^3}}\right) = -\frac{1}{t^2}$

36 **a** $\frac{dy}{dx} = -\frac{c^2}{x^2}. x = cp \Rightarrow \frac{dy}{dx} = -\frac{1}{p^2}$
So equation of tangent is
$y - \frac{c}{p} = -\frac{1}{p^2}(x - cp) \Rightarrow p^2y = -x + 2cp$ (1)
b The tangent at Q is $q^2y = -x + 2cq$ (2)
Subtracting (2) from (1) gives
$(p^2 - q^2)y = 2c(p - q) \Rightarrow y = \frac{2c}{p + q}$
c 1

37 **a** $\frac{dy}{dx} = -\frac{c^2}{x^2}. x = cp \Rightarrow \frac{dy}{dx} = -\frac{1}{p^2}$
So equation of tangent is
$y - \frac{c}{p} = -\frac{1}{p^2}(x - cp) \Rightarrow p^2y + x = 2cp$
b $(2cp, 0)$ **c** $\frac{c}{p}\sqrt{1 + p^4}$ **d** $\left(\frac{c}{3}, 3c\right)$

38 Midpoint of OP has general point $\left(\frac{ct}{2}, \frac{c}{2t}\right)$
$xy = \frac{c^2}{4}$, which is a hyperbola

39 Distance from P to line $= x + 5$
Distance from P to $(5, 0)$ is $\sqrt{(x - 5)^2 + y^2}$
So $(x + 5)^2 = (x - 5)^2 + y^2 \Rightarrow$ locus of P has equation $y^2 = 20x$, i.e. $a = 5$.

40 **a**

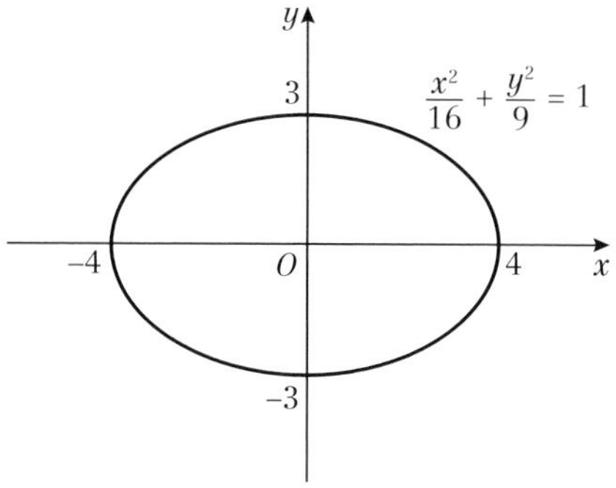

b $\frac{\sqrt{7}}{4}$ **c** $(\pm\sqrt{7}, 0)$

41 **a** $\frac{\sqrt{5}}{2}$ **b** $4\sqrt{5}$
c

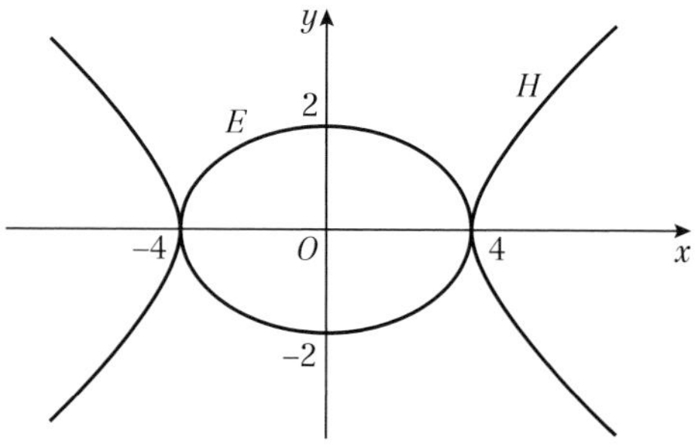

42 a $(\pm\sqrt{5}, 0)$

b The directrices are $x = \pm\frac{9}{\sqrt{5}}$

Let the line through P parallel to the x-axis intersect the directrices at N and N'.

Then $NN' = 2 \times \frac{9}{\sqrt{5}} = \frac{18}{\sqrt{5}}$

$SP = ePN$ and $S'P = ePN'$, so

$SP + S'P = ePN + ePN' = e(PN + PN') = eN'N$

$= \frac{\sqrt{5}}{3} \times \frac{18}{\sqrt{5}} = 6$

43 a $\frac{1}{2}$ **b** $y = -\frac{1}{2}x + 2$ **c** 2

44 a $x^2 + 4y^2 = a^2$ **b** $x = \pm\frac{2a}{\sqrt{3}}$ **c** $b = \frac{a}{2}$

d P is $\left(\frac{a}{\sqrt{2}}, \frac{a}{2\sqrt{2}}\right)$ and Q is $\left(0, \frac{a}{2}\right)$.

Gradient of PQ is $\dfrac{\frac{a}{2\sqrt{2}} - \frac{a}{2}}{\frac{a}{\sqrt{2}}} = \frac{1-\sqrt{2}}{2}$

So equation of line containing chord PQ is

$y = \frac{1-\sqrt{2}}{2}x + \frac{a}{2} \Rightarrow (\sqrt{2} - 1)x + 2y - a = 0$

45 a $\frac{\sqrt{5}}{3}$ **b** $(\pm\sqrt{5}, 0)$, $x = \pm\frac{9}{\sqrt{5}}$

c Gradient of ellipse at P is $\frac{2\cos\theta}{-3\sin\theta}$, so equation of tangent is $y - 2\sin\theta = -\frac{2\cos\theta}{3\sin\theta}(x - 3\cos\theta)$,

which can be rearranged to $\frac{x\cos\theta}{3} + \frac{y\sin\theta}{2} = 1$.

d Equation of perpendicular line is $y = \frac{3\sin\theta}{2\cos\theta}x$.

So foot of perpendicular, (x, y), satisfies

$2x\cos\theta + 3y\sin\theta = 6$

$2y\cos\theta - 3x\sin\theta = 0$

Solve these simultaneously to find

$\cos\theta = \frac{3x}{x^2 + y^2}$ and $\sin\theta = \frac{2y}{x^2 + y^2}$

Therefore $\left(\frac{3x}{x^2+y^2}\right)^2 + \left(\frac{2y}{x^2+y^2}\right)^2 = 1$

Rearranging, this gives that the locus of the foot of the perpendicular as $(x^2 + y^2)^2 = 9x^2 + 4y^2$.

46 a $\frac{dx}{d\theta} = -a\sin\theta$, $\frac{dy}{d\theta} = b\cos\theta \Rightarrow \frac{dy}{dx} = -\frac{b\cos\theta}{a\sin\theta}$

The gradient of the normal is $\frac{a\sin\theta}{b\cos\theta}$

So the equation of the normal is

$y - b\sin\theta = \frac{a\sin\theta}{b\cos\theta}(x - a\cos\theta)$

$\Rightarrow ax\sec\theta - by\operatorname{cosec}\theta = a^2 - b^2$

b $y = 0 \Rightarrow ax\sec\theta = a^2 - b^2 \Rightarrow x = \frac{a^2 - b^2}{a}\cos\theta$

So G has coordinates $\left(\frac{a^2 - b^2}{a}\cos\theta, 0\right)$

Midpoint has coordinates

$\left(\dfrac{a\cos\theta + \frac{a^2-b^2}{a}\cos\theta}{2}, \dfrac{b\sin\theta + 0}{2}\right)$

$= \left(\frac{2a^2 - b^2}{2a}\cos\theta, \frac{b}{2}\sin\theta\right)$

c $x = \frac{2a^2 - b^2}{2a}\cos\theta \Rightarrow \cos\theta = \frac{2ax}{2a^2 - b^2}$

$y = \frac{b}{2}\sin\theta \Rightarrow \sin\theta = \frac{2y}{b}$

So using $\cos^2\theta + \sin^2\theta \equiv 1$, M has locus

$\frac{4a^2x^2}{(2a^2 - b^2)^2} + \frac{4y^2}{b^2} = 1$, which is an ellipse.

d H has coordinates $\left(0, -\frac{a^2 - b^2}{b\operatorname{cosec}\theta}\right)$.

$A_1 = \text{Area of } \triangle OMG = \frac{1}{2} \times \frac{a^2 - b^2}{a\sec\theta} \times \frac{b}{2}\sin\theta$

$= \frac{b}{4a}(a^2 - b^2)\sin\theta\cos\theta$

$A_2 = \text{Area of } \triangle OGH = \frac{1}{2} \times \frac{a^2 - b^2}{b\operatorname{cosec}\theta} \times \frac{a^2 - b^2}{a\sec\theta}$

$= \frac{(a^2 - b^2)^2}{2ab}\sin\theta\cos\theta$

So $A_1 : A_2 = b^2 : 2(a^2 - b^2)$

47 Area of triangle to left of P is

$\frac{1}{2} \times 4 \times 2\sqrt{3} = 4\sqrt{3}$

Area to right of P is

$\int_4^8 \sqrt{16 - \left(\frac{x}{2}\right)^2}\,dx = \frac{16\pi}{3} - 4\sqrt{3}$

So total area is $4\sqrt{3} + \frac{16\pi}{3} - 4\sqrt{3} = \frac{16\pi}{3}$ and $a = \frac{16}{3}$

48 a Substitute $y = mx + c$ into $\frac{x^2}{a^2} + \frac{y^2}{b^2} = 1$:

$\frac{x^2}{a^2} + \frac{(mx + c)^2}{b^2} = 1$

$\Rightarrow (a^2m^2 + b^2)x^2 + 2a^2mcx + a^2(c^2 - b^2) = 0$

As the line is a tangent, need to have "$b^2 - 4ac = 0$".

$4a^4m^2c^2 - 4a^2(a^2m^2 + b^2)(c^2 - b^2) = 0$

$\Rightarrow 4(a^2m^2b^2 - b^2c^2 + b^4) = 0 \Rightarrow c^2 = a^2m^2 + b^2$

b $y = -3x + 13$ and $y = -\frac{3}{7}x + \frac{37}{7}$

49 a Substitute $y = mx + c$ into $\frac{x^2}{a^2} + \frac{y^2}{b^2} = 1$:

$\frac{x^2}{a^2} + \frac{(mx + c)^2}{b^2} = 1$

$\Rightarrow (a^2m^2 + b^2)x^2 + 2a^2mcx + a^2(c^2 - b^2) = 0$

b As the line is a tangent, need to have "$b^2 - 4ac = 0$".

$4a^4m^2c^2 - 4(a^2m^2 + b^2)(c^2 - b^2) = 0$

$\Rightarrow 4(a^2m^2b^2 - b^2c^2 + b^4) = 0 \Rightarrow c^2 = a^2m^2 + b^2$

c $\frac{b^2 + a^2m^2}{2m}$

d $T = \frac{b^2 + a^2m^2}{2m} = \frac{1}{2}b^2m^{-1} + \frac{1}{2}a^2m$

For a minimum, $\frac{dT}{dm} = -\frac{1}{2}b^2m^{-2} + \frac{1}{2}a^2 = 0$

$\frac{b^2}{m^2} = a^2 \Rightarrow m^2 = \frac{b^2}{a^2}$

As L has a positive gradient, $m = \frac{b}{a}$

At $m = \frac{b}{a}$, $\frac{d^2T}{dm^2} = \frac{b^2}{m^3} = \frac{a^3}{b} > 0$ and so this gives a minimum value of

$T = \dfrac{b^2 + a^2\left(\frac{b}{a}\right)^2}{2\left(\frac{b}{a}\right)} = \dfrac{2b^2}{2\left(\frac{b}{a}\right)} = ab$

e $-\frac{a}{\sqrt{2}}$

50 a Tangent at P: $x\cosh t - y\sinh t = 1$

Normal at P: $x\sinh t + y\cosh t = 2\sinh t\cosh t$

b Substitute $y = 0$ into the equation of the normal:

$x\sinh t = 2\sinh t\cosh t \Rightarrow x = 2\cosh t$, so G is $(2\cosh t, 0)$.

Q has $x = \cosh t$, and the asymptote in the first quadrant is $y = x$, so Q is $(\cosh t, \cosh t)$.

Gradient of GQ is $\frac{0 - \cosh t}{2\cosh t - \cosh t} = -1$

So GQ is perpendicular to the asymptote $y = x$.

c Substitute $y = 0$ into the equation of the tangent:

$x\cosh t = 1 \Rightarrow x = \dfrac{1}{\cosh t}$, so $T = \left(\dfrac{1}{\cosh t}, 0\right)$

Substitute $x = 0$ into the equation of the normal:
$y\cosh t = 2\sinh t\cosh t \Rightarrow y = 2\sinh t$, so R is $(0, 2\sinh t)$.

$TG = 2\cosh t - \dfrac{1}{\cosh t}$

$TR^2 = OR^2 + OT^2 = (2\sinh t)^2 + \left(\dfrac{1}{\cosh t}\right)^2$

$= 4(\cosh^2 t - 1) + \dfrac{1}{\cosh^2 t} = \left(2\cosh t - \dfrac{1}{\cosh t}\right)^2$

$= TG^2$

So $TR = TG$ and R lies on the circle with centre T and radius TG.

51 Let the point P have coordinates $(a\cosh t, b\sinh t)$

$\dfrac{dx}{dt} = a\sinh t, \dfrac{dy}{dt} = b\cosh t \Rightarrow \dfrac{dy}{dx} = \dfrac{b\cosh t}{a\sinh t}$

Equation of tangent is $y - b\sinh t = \dfrac{b\cosh t}{a\sinh t}(x - a\cosh t)$

$\Rightarrow ay\sinh t = bx\cosh t - ab(\cosh^2 t - \sinh^2 t)$
$= bx\cosh t - ab$

For T, $y = 0$, so $bx\cosh t = ab \Rightarrow x = \dfrac{a}{\cosh t}$

The coordinates of N are $(a\cosh t, 0)$

$OT \times ON = \dfrac{a}{\cosh t} \times a\cosh t = a^2$

52 a $\dfrac{dx}{dt} = a\sec t\tan t, \dfrac{dy}{dt} = b\sec^2$

$\dfrac{dy}{dx} = \dfrac{b\sec^2 t}{a\sec t\tan t} = \dfrac{b}{a\sin t}$

The gradient of the normal is $-\dfrac{a\sin t}{b}$

The equation of the normal is

$y - b\tan t = -\dfrac{a\sin t}{b}(x - a\sec t)$

$\Rightarrow ax\sin t + by = (a^2 + b^2)\tan t$

b $\dfrac{\pi}{3}, \dfrac{2\pi}{3}, \dfrac{4\pi}{3}, \dfrac{5\pi}{3}$

53 a $\dfrac{x^2}{a^2} - \dfrac{y^2}{a^2} = 1, b^2 = a^2(e^2 - 1)$

$b^2 = a^2 \Rightarrow a^2 = a^2(e^2 - 1)$

$\Rightarrow 1 = e^2 - 1 \Rightarrow e^2 = 2 \Rightarrow e = \sqrt{2}$

b $(a\sqrt{2}, 0), x = \dfrac{a\sqrt{2}}{2}$

c P is on the line with gradient -1 through $(a\sqrt{2}, 0)$,

$y = -x + a\sqrt{2}$, which intersects $y = x$ at $P\left(\dfrac{a\sqrt{2}}{2}, \dfrac{a\sqrt{2}}{2}\right)$.

Q is on the line with gradient 1 through $(a\sqrt{2}, 0)$,

$y = x - a\sqrt{2}$, which intersects $y = -x$ at $Q\left(\dfrac{a\sqrt{2}}{2}, -\dfrac{a\sqrt{2}}{2}\right)$.

P and Q both have $x = \dfrac{a\sqrt{2}}{2}$, so lie on directrix L.

d SP has equation $x + y = a\sqrt{2}$

So R is where $x^2 - (a\sqrt{2} - x)^2 = a^2$

$\Rightarrow x = \dfrac{3\sqrt{2}}{4}a, y = \dfrac{\sqrt{2}}{4}a$

$x^2 - y^2 = a^2 \Rightarrow \dfrac{dy}{dx} = \dfrac{x}{y}$, so at R, $\dfrac{dy}{dx} = 3$.

Therefore the tangent is

$y - \dfrac{\sqrt{2}}{4}a = 3\left(x - \dfrac{3\sqrt{2}}{4}a\right)$

$\Rightarrow y = 3x - 2a\sqrt{2}$

$x = \dfrac{a\sqrt{2}}{2} \Rightarrow y = -\dfrac{a\sqrt{2}}{2}$, which is the y-coordinate of Q, so the tangent passes through Q.

54 Let the equation of the tangent be $y = mx + c$.

$x^2 - 4(mx + c)^2 = 4 \Rightarrow (4m^2 - 1)x^2 + 8mcx + 4(c^2 + 1) = 0$

As the line is a tangent, this equation will have repeated roots, so '$b^2 - 4ac = 0$':

$64m^2c^2 - 16(4m^2 - 1)(c^2 + 1) = 0 \Rightarrow 16c^2 - 64m^2 + 16 = 0$

$\Rightarrow c = \pm\sqrt{4m^2 - 1}$, so the equations of the tangents are $y = mx \pm \sqrt{4m^2 - 1}$, where $|m| > \frac{1}{2}$

55 a $ay\sin t + bx\cos t = ab$

b $ax\sin t - by\cos t = (a^2 - b^2)\sin t\cos t$

c $\left(\dfrac{a^2 - b^2}{2a}\cos t, \dfrac{b}{2\sin t}\right)$

d $x = \dfrac{a^2 - b^2}{2a}\cos t \Rightarrow \cos t = \dfrac{2ax}{a^2 - b^2}$ and

$y = \dfrac{b}{2\sin t} \Rightarrow \sin t = \dfrac{b}{2y}$

So using $\cos^2 t + \sin^2 t \equiv 1$, M has locus

$\left(\dfrac{2ax}{a^2 - b^2}\right)^2 + \left(\dfrac{b}{2y}\right)^2 = 1$

56 a Tangent: $bx - ay\sin\theta = ab\cos\theta$

Normal: $ax\sin\theta + by = (a^2 + b^2)\tan\theta$

b Find the coordinates of P and Q by substituting $x = 0$ into the equations of the two lines.

$-ay\sin\theta = ab\cos\theta \Rightarrow y = -b\cot\theta$

$by = (a^2 + b^2)\tan\theta \Rightarrow y = \dfrac{a^2 + b^2}{b}\tan\theta$

So P is $(0, -b\cot\theta)$ and Q is $\left(0, \dfrac{a^2 + b^2}{b}\tan\theta\right)$.

The focus S with $x > 0$ is $(ae, 0)$.

PS has gradient $m = \dfrac{-b\cot\theta - 0}{0 - ae} = \dfrac{b}{ae}\cot\theta$

QS has gradient $m' = \dfrac{\dfrac{a^2 + b^2}{b}\tan\theta - 0}{0 - ae} = -\dfrac{a^2 + b^2}{abe}\tan\theta$

$mm' = -\dfrac{a^2 + b^2}{a^2e^2} = -1$, since $b^2 = a^2(e^2 - 1)$, so PS and QS are perpendicular. Thus PSQ is a right-angled triangle, and PQ is the diameter of a circle, C, through S. By symmetry, C also passes through the other focus, $(-ae, 0)$.

57 $x < -4, -1 < x < 2$

58 $\{x : x < 0\} \cup \{x : 2 < x < 4\}$

59 $\{x : -3 < x < 0\} \cup \{x : x > 4\}$

60 $\{x : -\frac{1}{2} < x < 0\} \cup \{x : x > 3\}$

61 $\{x : x < -4k\} \cup \{x : -2k < x < 0\} \cup \{x : x > 2k\}$

62 a

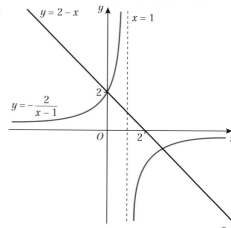

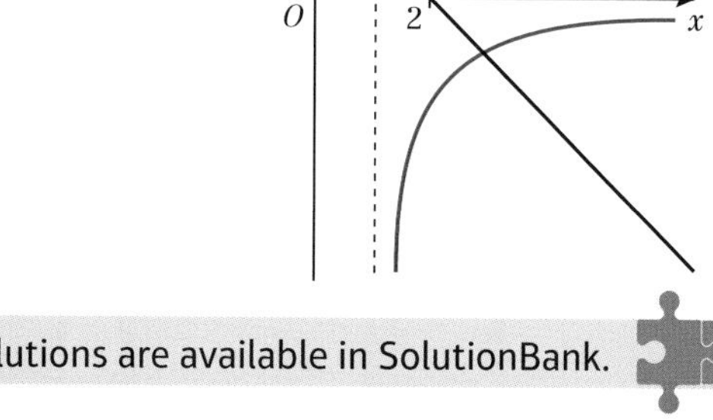

b (0, 2) and (3, –1) **c** $x < 0, 1 < x < 3$

63 a

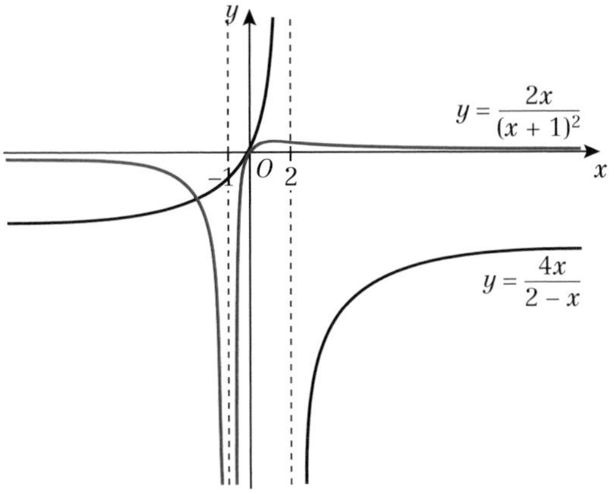

b (0, 0) and $\left(-\frac{5}{2}, -\frac{20}{9}\right)$

c $\left\{x : x \leqslant -\frac{5}{2}\right\} \cup \{x : x > 2\} \cup \{x : x = 0\}$

64 a

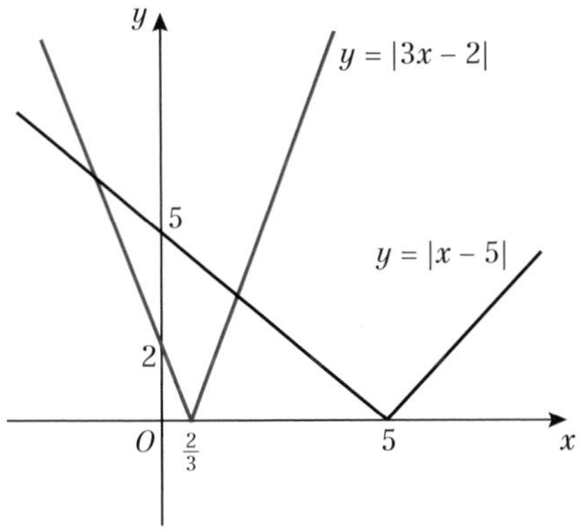

b $\left(-\frac{3}{2}, \frac{13}{2}\right), \left(\frac{7}{4}, \frac{13}{4}\right)$

c $x < -\frac{3}{2}, x > \frac{7}{4}$

65 a

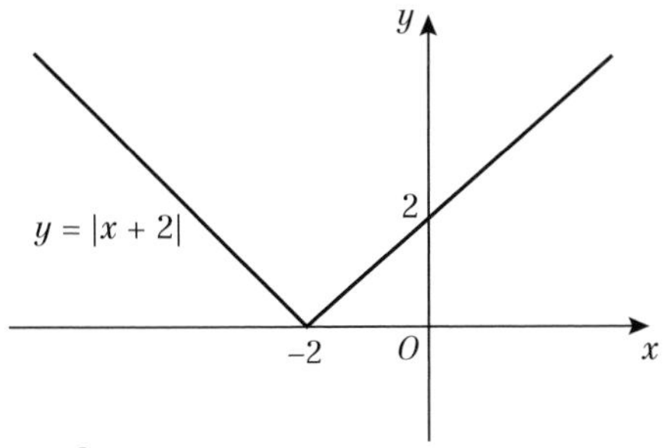

b $x > 2$

66 a

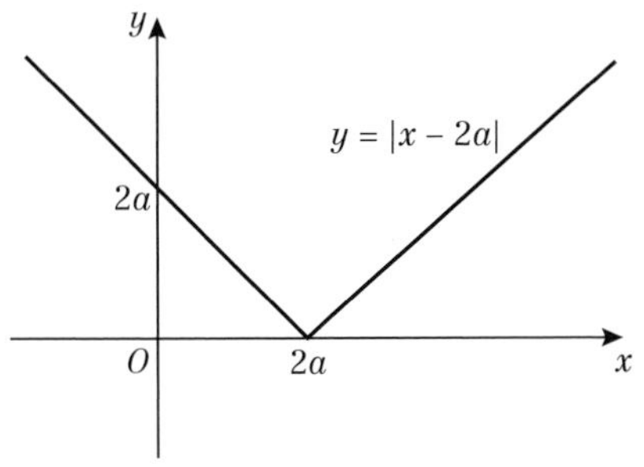

b $x < \frac{1}{3}a$

67 $\{x : x < 6 - 2\sqrt{3}\} \cup \{x : 4 < x < 6\}$

68 a

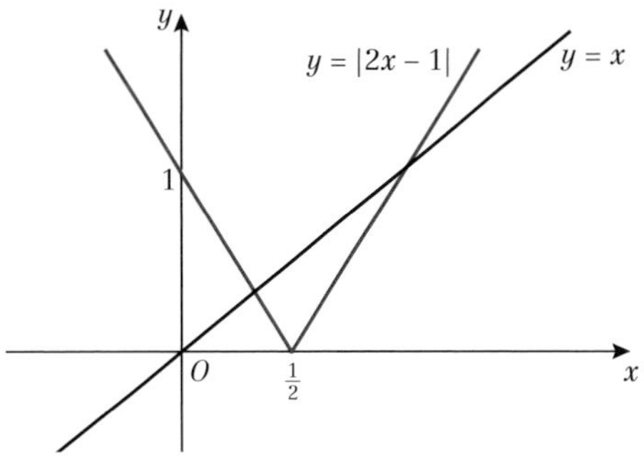

b $\left(\frac{1}{3}, \frac{1}{3}\right)$ and (1, 1) **c** $\{x : x < \frac{1}{3}\} \cup \{x : x > 1\}$

69 $\{x : -5 < x < \frac{1}{3}\}$

70 $-\frac{1}{3}a \leqslant x \leqslant -\frac{1}{7}a$

71 a

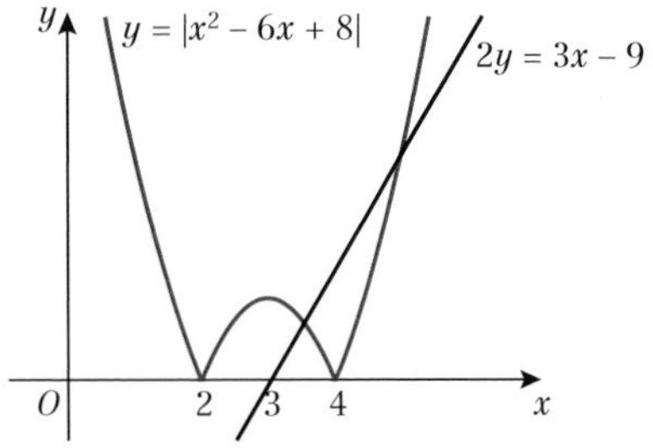

The curve meets the x-axis at (2, 0) and (4, 0).
The line meets the x-axis at (3, 0).

b $\left(\frac{7}{2}, \frac{3}{4}\right)$, (5, 3)

$x < \frac{7}{2}, x > 5$

72 a

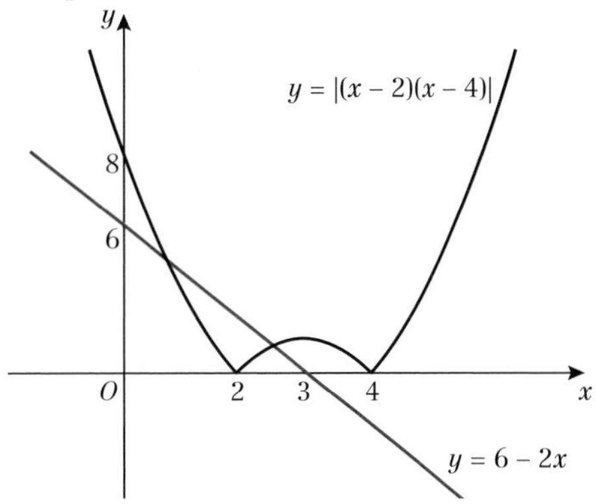

b $2 - \sqrt{2}$ and $4 - \sqrt{2}$ **c** $2 - \sqrt{2} < x < 4 - \sqrt{2}$

73 a $x = -\frac{5}{2}, x = -\frac{7}{4}$ or $x = 1$

b $\{x : x < -\frac{5}{2}\} \cup \{x : -\frac{7}{4} < x < 1\}$

Challenge

1 Use $\begin{pmatrix} \cos 135° & -\sin 135° \\ \sin 135° & \cos 135° \end{pmatrix} \begin{pmatrix} ct \\ \frac{c}{t} \end{pmatrix}$

Find x^2 and y^2 and show $y^2 - x^2 = 2c^2$

Then $k = \sqrt{2}c$

2 $0 < x < \frac{\pi}{6}, \frac{5\pi}{6} < x < \pi$

3 a

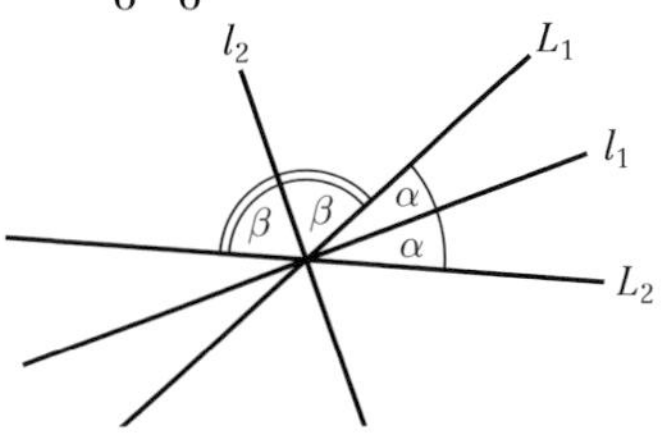

b L_1 has direction vector $\mathbf{v}_1 = \begin{pmatrix} l_1 \\ m_1 \\ n_1 \end{pmatrix}$ and L_2 has direction vector $\mathbf{v}_2 = \begin{pmatrix} l_2 \\ m_2 \\ n_2 \end{pmatrix}$. Then $\mathbf{v}_1$ and $\mathbf{v}_2$ form the rhombus with diagonal $\mathbf{v}_1 + \mathbf{v}_2$.

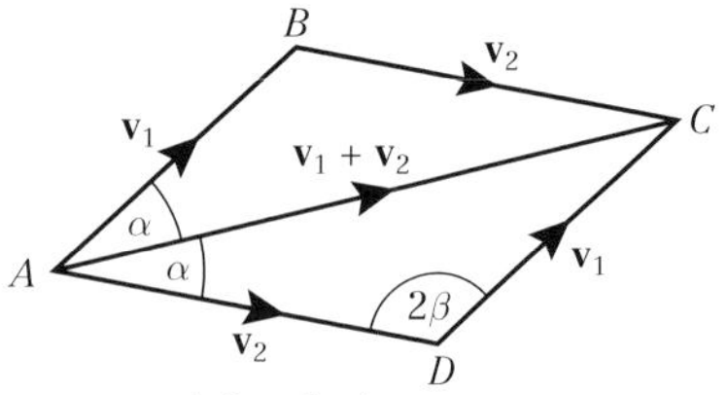

$\mathbf{v}_1 + \mathbf{v}_2 = \begin{pmatrix} l_1 + l_2 \\ m_1 + m_2 \\ n_1 + n_2 \end{pmatrix}$ bisects angle BAD so is parallel to l_1. Hence l_1 has direction ratios $l_1 + l_2 : m_1 + m_2 : n_1 + n_2$. The other diagonal of the rhombus is given by

$\mathbf{v}_1 - \mathbf{v}_2 = \begin{pmatrix} l_1 - l_2 \\ m_1 - m_2 \\ n_1 - n_2 \end{pmatrix}$ and bisects angle ADC so it is parallel to l_2. Hence l_2 has direction ratios $l_1 - l_2 : m_1 - m_2 : n_1 - n_2$ respectively.

In general, $|\mathbf{v}_1 + \mathbf{v}_2|$ and $|\mathbf{v}_1 - \mathbf{v}_2|$ are not equal to 1, so these values are not direction cosines.

4 a

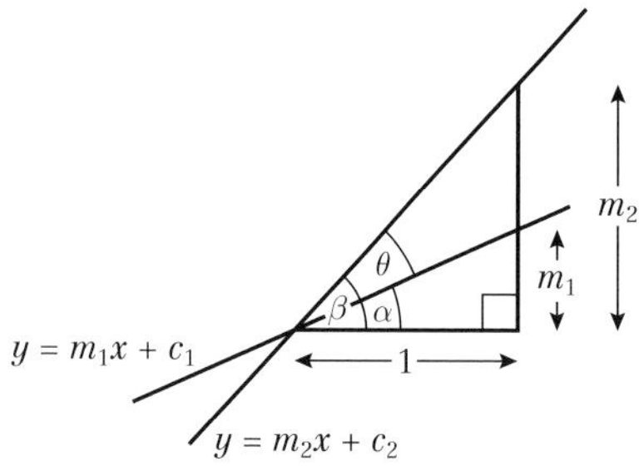

Using the identity for tan(A ± B):

$\tan\theta = \tan(\beta - \alpha) = \dfrac{\tan\beta - \tan\alpha}{1 + \tan\beta\tan\alpha} = \dfrac{m_2 - m_1}{1 + m_1 m_2}$

as required.

b

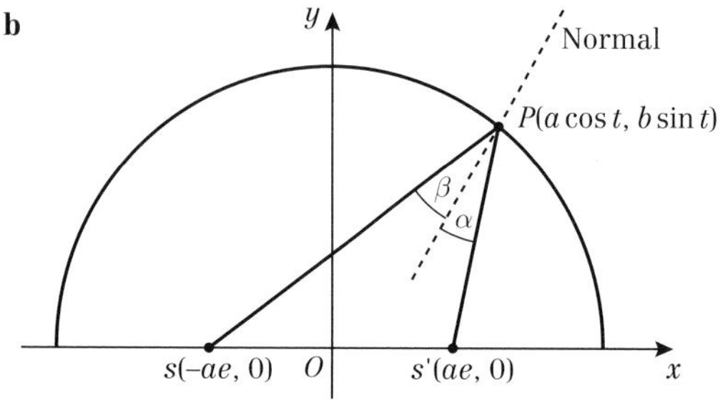

At P, $\dfrac{dy}{dx} = \dfrac{-b\cos t}{a\sin t}$, so gradient of normal is $\dfrac{a\sin t}{b\cos t}$

Gradient of PS' is $\dfrac{b\sin t}{a\cos t - ae}$ and gradient of PS is $\dfrac{b\sin t}{a\cos t + ae}$

So using the result from part **a**,

$$\tan\alpha = \frac{\dfrac{a\sin t}{b\cos t} - \dfrac{b\sin t}{a\cos t - ae}}{1 + \left(\dfrac{a\sin t}{b\cos t}\right)\left(\dfrac{b\sin t}{a\cos t - ae}\right)}$$

$$= \frac{a\sin t(a\cos t - ae) - b^2\sin t\cos t}{b\cos t(a\cos t - ae) + ab\sin^2 t}$$

$$= \frac{(a^2 - b^2)\sin t\cos t - a^2e\sin t}{ab(\cos^2 t + \sin^2 t) - abe\cos t}$$

$$= \frac{a^2e^2\sin t\cos t - a^2e\sin t}{ab - abe\cos t}$$

$$= \frac{a^2e\sin t(e\cos t - 1)}{ab(1 - e\cos t)} = \frac{-ae\sin t}{b}$$

Similarly, $\tan\beta = \dfrac{-ae\sin t}{b}$

So $\tan\alpha = \tan\beta$, and hence $\alpha = \beta$ as required.

CHAPTER 5

Prior knowledge check

1 $\cos^2\theta + \sin^2\theta \equiv 1 \Rightarrow 1 + \dfrac{\sin^2\theta}{\cos^2\theta} = \dfrac{1}{\cos^2\theta}$

$\Rightarrow 1 + \tan^2\theta \equiv \sec^2\theta$

2 $\sin 3\theta = \sin 2\theta\cos\theta + \sin\theta\cos 2\theta$
$= 2\sin\theta\cos^2\theta + (1 - 2\sin^2\theta)\sin\theta = 3\sin\theta - 4\sin^3\theta$

3 $\cos 2\theta = 1 - 2\sin^2\theta = \sin\theta$
$0 = 2\sin^2\theta + \sin\theta - 1 = (2\sin\theta - 1)(\sin\theta + 1)$
Hence $\sin\theta = -1, \frac{1}{2}$ which has solutions $\theta = \dfrac{\pi}{6}, \dfrac{5\pi}{6}, \dfrac{3\pi}{2}$

Exercise 5A

1	**a** $\frac{12}{13}$	**b** $\frac{5}{13}$	**c** $\frac{12}{5}$	
2	**a** $\frac{4}{5}$	**b** $-\frac{3}{5}$	**c** $-\frac{4}{3}$	**d** $-\frac{29}{12}$
3	**a** $\frac{24}{25}$	**b** $-\frac{7}{25}$	**c** $-\frac{25}{7}$	**d** $-\frac{7}{17}$
4	**a** $-\frac{119}{169}$	**b** $\frac{14400}{14161}$	**c** $-\frac{40391}{14280}$	**d** $\frac{3427320}{5098079}$
5	**a** $\frac{336}{527}$	**b** $\frac{354144}{390625}$	**c** $\frac{164833}{390625}$	**d** $\frac{164833}{354144}$

6 a $\cos^2\theta \equiv 1 - \sin^2\theta = 1 - \dfrac{4 - 2\sqrt{3}}{8} = \dfrac{4 + 2\sqrt{3}}{8}$

So $\cos\theta = \dfrac{\sqrt{3} + 1}{2\sqrt{2}}$, hence $\tan\theta = \dfrac{\sin\theta}{\cos\theta} = \dfrac{\sqrt{3} - 1}{\sqrt{3} + 1}$

b $\sin 2\theta = \frac{1}{2}$, $\cos 2\theta = \dfrac{\sqrt{3}}{2}$ **c** $\theta = \dfrac{\pi}{12}$

7 a $\sin^2 x \equiv 1 - \cos^2 x = 1 - \dfrac{2 + \sqrt{2}}{4} = \dfrac{2 - \sqrt{2}}{4}$

so $\sin x = \dfrac{\sqrt{2 - \sqrt{2}}}{2}$, hence $\tan x = \dfrac{\sin x}{\cos x} = -\dfrac{\sqrt{2 - \sqrt{2}}}{\sqrt{2 + \sqrt{2}}}$

b -1 **c** $x = \dfrac{7\pi}{8}$

8 a $\sin\dfrac{5\pi}{6} = \frac{1}{2} = \dfrac{2t}{1 + t^2} \Rightarrow 1 + t^2 = 4t \Rightarrow t^2 - 4t + 1 = 0$

b $\cos\dfrac{5\pi}{6} = -\dfrac{\sqrt{3}}{2} = \dfrac{1 - t^2}{1 + t^2} \Rightarrow -\sqrt{3} - \sqrt{3}\,t^2 = 2 - 2t^2$

$\Rightarrow t^2 = \dfrac{2 + \sqrt{3}}{2 - \sqrt{3}}$

c $2 + \sqrt{3}$

9 By considering angles and using Pythagoras' theorem, we can calculate

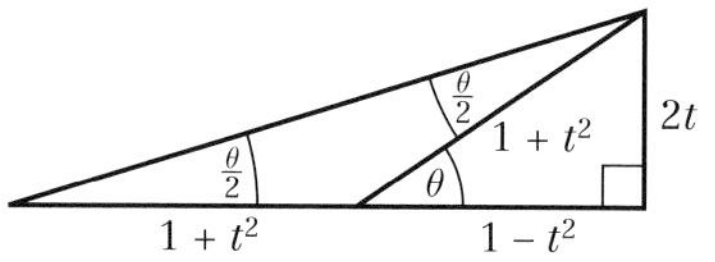

Hence $\tan\dfrac{\theta}{2} = \dfrac{2t}{1 + t^2 + 1 - t^2} = t$

Also, by considering the smaller triangle we see $\sin\theta = \dfrac{2t}{1 + t^2}$, $\cos\theta = \dfrac{1 - t^2}{1 + t^2}$ and $\tan\theta = \dfrac{2t}{1 - t^2}$

Online Full worked solutions are available in SolutionBank.

Exercise 5B

1 a $\sin^2\theta + \cos^2\theta \equiv \dfrac{4t^2 + 1 - 2t^2 + t^4}{(1+t^2)^2} \equiv \dfrac{(1+t^2)^2}{(1+t^2)^2} \equiv 1$

b $\dfrac{\tan^2\theta}{\tan^2\theta + 1} \equiv \dfrac{4t^2}{4t^2 + (1-t^2)^2} \equiv \dfrac{4t^2}{(1+t^2)^2} \equiv \sin^2\theta$

c $\dfrac{\operatorname{cosec}\theta}{\sin\theta} - \dfrac{\cot\theta}{\tan\theta} \equiv \dfrac{(1+t^2)^2}{4t^2} - \dfrac{(1-t^2)^2}{4t^2} \equiv \dfrac{4t^2}{4t^2} \equiv 1$

d $\cot 2\theta + \tan\theta \equiv \dfrac{1-t^2}{2t} + t \equiv \dfrac{1+t^2}{2t} \equiv \operatorname{cosec} 2\theta$

2 a $\tan\theta + \cot\theta \equiv \dfrac{2t}{1-t^2} + \dfrac{1-t^2}{2t}$

$\equiv \dfrac{(1+t^2)^2}{2t(1-t^2)} \equiv \sec\theta \operatorname{cosec}\theta$

b $\dfrac{1+\cos\theta}{\sin\theta} \equiv \dfrac{1+t^2+1-t^2}{2t} \equiv \dfrac{1}{t}$

$\equiv \dfrac{2t}{1+t^2-(1-t^2)} \equiv \dfrac{\sin\theta}{1-\cos\theta}$

c $\dfrac{1-\sin\theta}{\cos\theta} \equiv \dfrac{1-2t+t^2}{1-t^2} \equiv \dfrac{1-t}{1+t}$

$\equiv \dfrac{1-t^2}{(1+t)^2} \equiv \dfrac{\cos\theta}{1+\sin\theta}$

d $\tan\theta\sin\theta + \cos\theta \equiv \dfrac{4t^2}{1-t^4} + \dfrac{1-t^2}{1+t^2}$

$\equiv \dfrac{(1+t^2)^2}{1-t^4} \equiv \dfrac{1+t^2}{1-t^2} \equiv \sec\theta$

3 $\sin\theta + \sin\theta\cot^2\theta \equiv \dfrac{2t}{1+t^2} + \dfrac{2t}{1+t^2}\dfrac{(1-t^2)^2}{4t^2}$

$\equiv \dfrac{4t^2+(1-t^2)^2}{2t(1+t^2)} \equiv \dfrac{1+t^2}{2t} \equiv \operatorname{cosec}\theta$

4 $\dfrac{\cos\theta}{1-\sin\theta} - \dfrac{\cos\theta}{1+\sin\theta} \equiv \dfrac{1-t^2}{1+t^2-2t} - \dfrac{1-t^2}{1+t^2+2t}$

$\equiv \dfrac{1+t}{1-t} - \dfrac{1-t}{1+t} \equiv \dfrac{4t}{1-t^2} \equiv 2\tan\theta$

5 $\dfrac{\operatorname{cosec}x \cos x}{\tan x + \cot x} \equiv \dfrac{(1-t^2)^2}{4t^2+(1-t^2)^2}$

$\equiv \dfrac{(1-t^2)^2}{(1+t^2)^2} \equiv \cos^2 x$

6 $\dfrac{\cos\theta}{1+\sin\theta} + \dfrac{1+\sin\theta}{\cos\theta} \equiv \dfrac{1-t^2}{1+t^2+2t} + \dfrac{1+t^2+2t}{1-t^2}$

$\equiv \dfrac{1-t}{1+t} + \dfrac{1+t}{1-t} \equiv \dfrac{2(1+t^2)}{1-t^2}$

$\equiv 2\sec\theta$

7 $\sec\theta + \tan\theta \equiv \dfrac{1+t^2}{1-t^2} + \dfrac{2t}{1-t^2} \equiv \dfrac{1+t}{1-t}$

$\equiv \dfrac{1-t^2}{1+t^2-2t} \equiv \dfrac{\cos\theta}{1-\sin\theta}$

8 $\dfrac{1+\sin 2x - \cos 2x}{\sin 2x + \cos 2x - 1} \equiv \dfrac{1+t^2+2t-1+t^2}{2t+1-t^2-1-t^2}$

$\equiv \dfrac{1+t}{1-t} \equiv \dfrac{1+\tan x}{1-\tan x}$

9 $\dfrac{\cos\theta}{1-\sin\theta} - \tan\theta \equiv \dfrac{1-t^2}{1+t^2-2t} - \dfrac{2t}{1-t^2}$

$\equiv \dfrac{1+t}{1-t} - \dfrac{2t}{1-t^2} \equiv \dfrac{1+t^2}{1-t^2} \equiv \sec\theta$

10 $\tan^2\theta + \tan\theta\sec\theta + 1 \equiv \dfrac{4t^2}{(1-t^2)^2} + \dfrac{2t(1+t^2)}{(1-t^2)^2} + 1$

$\equiv \dfrac{1+2t+2t^2+2t^3+t^4}{(1-t^2)^2}$

$\equiv \dfrac{(1+t^2)^2+2t(1+t^2)}{(1-t^2)^2}$

$\equiv \dfrac{1+\sin\theta}{\cos^2\theta}$

11 LHS $\equiv \dfrac{\cos 2x}{1-\sin 2x} \equiv \dfrac{\frac{1-t^2}{1+t^2}}{1-\frac{2t}{1+t^2}} \equiv \dfrac{1-t^2}{1+t^2-2t}$

$\equiv \dfrac{(1-t)(1+t)}{(1-t)(1-t)} \equiv \dfrac{1+t}{1-t}$

RHS $\equiv \dfrac{\cot x + 1}{\cot x - 1} \equiv \dfrac{\frac{1}{\tan x}+1}{\frac{1}{\tan x}-1} \equiv \dfrac{\frac{1}{t}+1}{\frac{1}{t}-1} \equiv \dfrac{\frac{1}{t}(1+t)}{\frac{1}{t}(1-t)} \equiv \dfrac{1+t}{1-t}$

Hence $\dfrac{\cos 2x}{1-\sin 2x} \equiv \dfrac{\cot x + 1}{\cot x - 1}$

Challenge

$\dfrac{\sin^3\theta + \cos^3\theta}{\sin\theta + \cos\theta} \equiv \dfrac{8t^3 + (1-t^2)^3}{(1+t^2)^2(2t+1-t^2)} \equiv \dfrac{1-3t^2+8t^3+3t^4-t^6}{(1+t^2)^2(2t+1-t^2)}$

$1 - \sin\theta\cos\theta \equiv \dfrac{(1+t^2)^2 - 2t(1-t^2)}{(1+t^2)^2} \equiv \dfrac{1-2t+2t^2+2t^3+t^4}{(1+t^2)^2}$

and $1 - 3t^2 + 8t^3 + 3t^4 - t^6$

$\equiv (2t+1-t^2)(1-2t+2t^2+2t^3+t^4)$

Exercise 5C

1 a 1.57, 2.50 **b** 1.97, 4.71 **c** 2.21, 3.79

d 0.93, 3.54 **e** 1.05, 5.24

2 a $\sin 2\theta - 2\cos 2\theta = 1 - \sqrt{3}\cos 2\theta$

$\Rightarrow \dfrac{2t}{1+t^2} - \dfrac{2(1-t^2)}{1+t^2} = 1 - \dfrac{\sqrt{3}(1-t^2)}{1+t^2}$

$\Rightarrow (\sqrt{3}-1)t^2 - 2t - (\sqrt{3}-3) = 0$

b $t = 1, \sqrt{3}$. So $\theta = \dfrac{\pi}{4}, \dfrac{\pi}{3}, \dfrac{5\pi}{4}, \dfrac{4\pi}{3}$

3 a $16\cot x - 9\tan x = \dfrac{8(1-t^2)}{t} - \dfrac{18t}{1-t^2} = 0$

$\Rightarrow 8(1-t^2)^2 - 18t^2 = 8 - 34t^2 + 8t^4 = 0$

$\Rightarrow 4t^4 - 17t^2 + 4 = 0$

b $t = \pm\frac{1}{2}, \pm 2$, so $x \approx 0.93, 2.21, 4.07, 5.36$ (2 d.p.)

4 a $10\sin\theta\cos\theta - 3\cos\theta = -3$

$\Rightarrow \dfrac{20t(1-t^2)}{(1+t^2)^2} - \dfrac{3(1-t^2)}{1+t^2} = -3$

$\Rightarrow 20t(1-t^2) - 3(1-t^4) = -3(1+t^2)^2$

$\Rightarrow 3t^4 - 10t^3 + 3t^2 + 10t = 0$

$\Rightarrow t(3t^3 - 10t^2 + 3t + 10) = 0$

When $t = 2$, $3t^3 - 10t^2 + 3t + 10 = 0$ so $(t-2)$ is a factor.

$\Rightarrow t(t-2)(3t^2 - 4t - 5) = 0$

b $t = 0, 2, \dfrac{2 \pm \sqrt{19}}{3}$ so $\theta \approx 0, 2.21, 2.26, 4.95$ (2 d.p.)

Substituting 2π into the original equation also gives -3, so 6.28 (2 d.p.) is also a solution.

5 a $3\sin 2\theta + \cos 2\theta + 3\tan 2\theta = 1$

$\Rightarrow \dfrac{6t}{1+t^2} + \dfrac{1-t^2}{1+t^2} + \dfrac{6t}{1-t^2} = 1$

$\Rightarrow 6t(1-t^2) + (1-t^2)^2 + 6t(1+t^2) = 1 - t^4$

$\Rightarrow t^4 - t^2 + 6t = 0$

b $t = 0, -2$ so $\theta = 0, 2.03, 3.14, 5.18$ (2 d.p.)

6 a $\tan\theta + \cos 2\theta = 1 \Rightarrow t + \frac{1-t^2}{1+t^2} = 1 \Rightarrow t^3 - 2t^2 + t = 0$

b $t = 0, 1$ so $\theta = 0, \frac{\pi}{4}, \pi, \frac{5\pi}{4}, 2\pi$

7 a $2\sin 2\theta - \cos 4\theta - 4\tan\theta = -1$

$\Rightarrow \frac{4t}{1+t^2} - \frac{(1-t^2)^2}{(1+t^2)^2} + \frac{4t^2}{(1+t^2)^2} - 4t = -1$

$\Rightarrow 4t(1+t^2) - (1-t^2)^2 + 4t^2 - 4t(1+t^2)^2 = -(1+t^2)^2$

$\Rightarrow t^5 + t^3 - 2t^2 = 0$

b $t = 0, 1$ so $\theta = 0, \frac{\pi}{4}, \pi, \frac{5\pi}{4}, 2\pi$

8 $\theta = 4.07, 4.71$ (3 s.f.)

Challenge

$5\sin 2\theta + 12\cos\theta = -12$

$\Rightarrow \frac{20t(1-t^2)}{(1+t^2)^2} + \frac{12(1-t^2)}{1+t^2} = -12$

$\Rightarrow (t-2)(5t^2 + 4t + 3) = 0$

$\Rightarrow t = 2$ so $\theta = 2.21$ is a solution.

Check values of θ for which $\tan\left(\frac{\theta}{2}\right)$ is not defined:

$5 \times \sin 2\pi + 12\cos\pi = -12$, so $\theta = \pi$ is a solution.

Exercise 5D

1 a $\frac{ds}{dx} = -\cos x + 12\sin x = \frac{1}{1+t^2}(5t^2 + 24t - 5)$

b $x = 0.395$ (3 s.f.)

2 a $\frac{ds}{dx} = 2\cos x + 2\sin 2x = 2\cos x + 4\sin x\cos x$

$= \frac{2(1-t^2)(1+t^2)}{(1+t^2)^2} + \frac{8t(1-t^2)}{(1+t^2)^2}$

$= \frac{2}{(1+t^2)^2}(1-t^2)(t^2+4t+1)$

b $x = \frac{\pi}{2}, \frac{3\pi}{2}, 3.67, 5.76$ (3 s.f.)

3 a $v(x) = \frac{dh}{dx}(x) = 6\cos 2x + 8\sin 2x$

$= \frac{6(1-t^2)}{1+t^2} + \frac{16t}{1+t^2} = -\frac{2}{1+t^2}(3t^2 - 8t - 3)$

b Time between oscillations is π.

c $x = \arctan\left(-\frac{1}{3}\right) \approx 2.82$ (3 s.f.)

4 a $\frac{dy}{dx} = \frac{1}{10}\cos\frac{x}{5} + \frac{2}{5}\cos\frac{2x}{5} - \frac{1}{10}\sin\frac{x}{5}$

$= \frac{3t^4 - 2t^3 - 24t^2 - 2t + 5}{10(1+t^2)^2} = \frac{(3t^2 - 8t - 5)(t^2 + 2t - 1)}{10(1+t^2)^2}$

b i Comparing y-values on each graph $k \approx \frac{1}{10}$ would be sensible

ii The model is suitable for predicting times since both graphs have two distinct sets of peaks and similar periodicity.

Not suitable for predicting intensity since the peak height is constant for the model, but varies in the observed data.

c 98 milliseconds

Mixed exercise 5

1 a $\frac{3}{5}$	**b** $\frac{4}{5}$	**c** 3	**d** $\frac{25}{12}$
2 a $\frac{40}{9}$	**b** $\frac{41}{9}$	**c** $\frac{40}{41}$	**d** $\frac{5}{4}$
3 a $\frac{3}{5}$	**b** $-\frac{4}{5}$	**c** $\frac{9}{16}$	**d** $-\frac{15}{4}$
4 a $\frac{4}{3}$	**b** $\frac{25}{12}$	**c** $\frac{25}{9}$	**d** $\frac{25}{12}$

5 a $\sec^2\theta - 1 \equiv \frac{(1+t^2)^2}{(1-t^2)^2} - \frac{(1-t^2)^2}{(1-t^2)^2} \equiv \frac{4t^2}{(1-t^2)^2} \equiv \tan^2\theta$

b $\frac{\sqrt{3}-1}{\sqrt{3}+1}$ **c** $\sin 2\theta = \frac{1}{2}, \cos 2\theta = \frac{\sqrt{3}}{2}$ **d** $\frac{13\pi}{12}$

6 a $t = \sqrt{2} - 1$

b $\sec\frac{\pi}{8} = \sqrt{4 - 2\sqrt{2}}, \sin\frac{\pi}{8} = \frac{\sqrt{2-\sqrt{2}}}{2}, \cos\frac{\pi}{8} = \frac{\sqrt{2+\sqrt{2}}}{2}$

7 $\frac{1 + \sin x - \cos x}{\sin x + \cos x - 1} \equiv \frac{1 + t^2 + 2t - 1 + t^2}{2t + 1 - t^2 - 1 - t^2} \equiv \frac{1+t}{1-t}$

and $\frac{1 + \sin x}{\cos x} \equiv \frac{1 + t^2 + 2t}{1 - t^2} \equiv \frac{1+t}{1-t}$

8 $\tan^2\theta - \sin^2\theta \equiv \frac{4t^2((1+t^2)^2 - (1-t^2)^2)}{(1-t^2)^2(1+t^2)^2}$

$\equiv \frac{4t^2 4t^2}{(1-t^2)^2(1+t^2)^2} \equiv \tan^2\theta\sin^2\theta$

9 $\sin\theta\cos\theta\tan\theta \equiv \frac{4t^2}{(1+t^2)^2}$

$1 - \cos^2\theta \equiv 1 - \frac{(1-t^2)^2}{(1+t^2)^2} \equiv \frac{4t^2}{(1+t^2)^2}$

10 $\frac{1+\sin\theta}{1-\sin\theta} - \frac{1-\sin\theta}{1+\sin\theta} \equiv \frac{(1+t)^2}{(1-t)^2} - \frac{(1-t)^2}{(1+t)^2} \equiv \frac{8t+8t^3}{(1-t^2)^2}$

$4\tan\theta\sec\theta \equiv \frac{8t+8t^3}{(1-t^2)^2}$

11 $\frac{1+\tan^2 x}{1-\tan^2 x} \equiv \frac{(1-t^2)^2 + 4t^2}{(1-t^2)^2 - 4t^2} \equiv \frac{(1+t^2)^2}{(1-t^2)^2 - 4t^2}$

$\frac{1}{\cos^2 x - \sin^2 x} \equiv \frac{(1+t^2)^2}{(1-t^2)^2 - 4t^2}$

12 $\frac{1}{1-\sin\theta} - \frac{1}{1+\sin\theta} \equiv \frac{1+t^2}{1+t^2-2t} - \frac{1+t^2}{1+t^2+2t}$

$\equiv \frac{(1+t^2)4t}{(1-t^2)^2} \equiv 2\tan\theta\sec\theta$

13 $\tan\theta + \frac{\cos\theta}{1+\sin\theta} \equiv \frac{2t}{1-t^2} + \frac{1-t^2}{1+t^2+2t} \equiv \frac{1+t^2}{1-t^2} \equiv \sec\theta$

14 $(\sin\theta + \cos\theta)(\tan\theta + \cot\theta) = \left(\frac{2t + 1 - t^2}{1+t^2}\right)\left(\frac{4t^2 + (1-t^2)^2}{2t(1-t^2)}\right)$

$= \frac{1 + 2t + 2t^3 - t^4}{2t(1-t^2)}$

$\sec\theta + \operatorname{cosec}\theta = \frac{1+t^2}{1-t^2} + \frac{1+t^2}{2t} = \frac{1 + 2t + 2t^3 - t^4}{2t(1-t^2)}$

15 a $3\cos x - \sin x = -1$

$\Rightarrow 3\left(\frac{1-t^2}{1+t^2}\right) - \frac{2t}{1+t^2} = -1 \Rightarrow t^2 + t - 2 = 0$

b $\theta = 1.57, 4.07$ (2 d.p.)

16 a $\sin\theta + \cos\theta = -\frac{1}{5}$

$\Rightarrow \frac{2t}{1+t^2} + \frac{1-t^2}{1+t^2} = -\frac{1}{5} \Rightarrow 2t^2 - 5t - 3 = 0$

b $\theta = 2.50, 5.36$ (2 d.p.)

17 a $6\tan\theta + 12\sin\theta + \cos\theta = 1$

$\Rightarrow \frac{12t}{1-t^2} + \frac{24t}{1+t^2} + \frac{1-t^2}{1+t^2} = 1 \Rightarrow t^4 - 6t^3 - t^2 + 18t = 0$

$\Rightarrow t(t-2)(t^2 - 4t - 9) = 0$

b $\theta = 0, 2.21, 2.79, 4.26, 6.28$

18 a $5\cot x + 4\operatorname{cosec} x = \frac{9}{4}$

$\Rightarrow \frac{5 - 5t^2}{2t} + \frac{4 + 4t^2}{2t} = \frac{9}{4} \Rightarrow 2t^2 + 9t - 18 = 0$

b $x = 1.97, 3.47$

19 a $\frac{dp}{dx} = 10\left(4\cos 5x - 8\sin 5x - 4\cos 10x - \frac{16}{3}\sin 10x\right)$

$= \frac{10(4 - 4t^4 - 16t - 16t^3 - 4(1 - 6t^2 + t^4) - \frac{64}{3}(t - t^3))}{(1+t^2)^2}$

$= \frac{-80t(3t^3 - 2t^2 - 9t + 14)}{3(1+t^2)^2} = \frac{-80t(t+2)(3t^2 - 8t + 7)}{3(1+t^2)^2}$

b The maxima and minima do not change, whereas we might expect blood pressure to vary with each heartbeat.
Also this model has a fixed period, whereas heart rates are not constant, and will vary with, for example, physical activity. This model doesn't capture changing heart rates.

c At a pressure low-point (local minimum) we have $\frac{dp}{dx} = 0$. From part **a**, this happens when $t = 0, -2$.
We can see from the figure that the solution $t = 0$ corresponds to the maximum at $x = 0$.
Thus at the first minimum we have $t = \tan\frac{5x}{2} = -2$, and hence $x = \frac{2}{5}(\pi - \arctan(-2)) = 0.814$ (3 d.p)

Challenge

a Writing $t = \frac{1}{4}$ then $\tan\theta = \frac{2t}{1-t^2} = \frac{\frac{1}{2}}{\frac{15}{16}} = \frac{8}{15}$,

$\sin\theta = \frac{2t}{1+t^2} = \frac{\frac{1}{2}}{\frac{17}{16}} = \frac{8}{17}$, $\cos\theta = \frac{1-t^2}{1+t^2} + \frac{\frac{15}{16}}{\frac{17}{16}} = \frac{15}{17}$

b

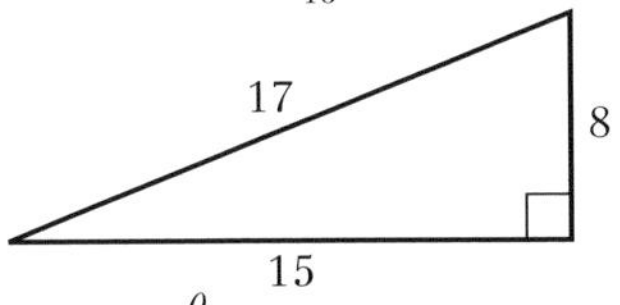

c If $t = \tan\frac{\theta}{2}$ is rational, then so are $\sin\theta = \frac{2t}{1+t^2}$ and $\cos\theta = \frac{1-t^2}{1+t^2}$. So we can construct the triangle

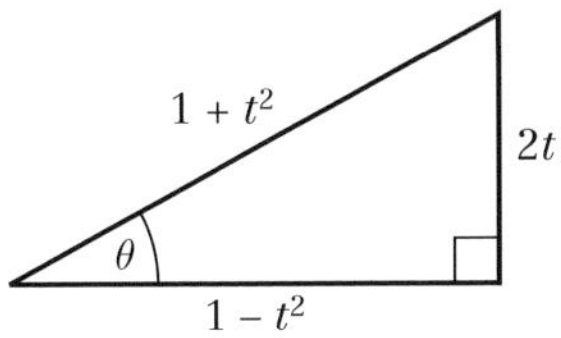

where all sides are of rational length. Write the length of each side as a fraction in lowest possible terms, then scale the triangle by the lowest common multiple of each of the denominators. The resulting triangle is similar, so in particular is right-angled with angle θ, but each side is integer length and the sides have no common multiples.

d Using the above construction, every rational value of $\tan\frac{\theta}{2}$ between 0 and 1 gives rise to a primitive Pythagorean triple. Note that the same triple is generated by triangles with acute angles θ and $90 - \theta$, so we get a unique triple for every $0 \leqslant \theta \leqslant \frac{\pi}{4}$.
But there are infinitely many values of $0 \leqslant \theta \leqslant \frac{\pi}{4}$ such that $t = \tan\frac{\theta}{2}$ is rational.

CHAPTER 6

Prior knowledge check

1 a $-3x^2\sin(1+x^3)$ **b** $\frac{1}{(1+x^2)\arctan(x)}$

c $\frac{-(\sin x + \cos x)}{e^x \sin^2 x}$

2 Auxiliary equation $\lambda^2 + 2\lambda + 2 = 0$ has solution $\lambda = -1 \pm i$, so general solution is $y(x) = Ae^{-x}\sin x + Be^{-x}\cos x$

3 a $e^x = 1 + x + \frac{x^2}{2!} + \frac{x^3}{3!} + \ldots$

b $\sin x = x - \frac{x^3}{3!} + \frac{x^5}{5!} - \frac{x^7}{7!} + \ldots$

c $\ln(1+x) = x - \frac{x^2}{2} + \frac{x^3}{3} - \frac{x^4}{4} + \ldots$

Exercise 6A

1 a $1 + \frac{1}{2}(x-1) - \frac{1}{8}(x-1)^2 + \frac{1}{16}(x-1)^3 - \frac{5}{128}(x-1)^4 + \ldots$

b 1.095 (3 d.p.)

2 a $1 + \frac{x-e}{e} - \frac{(x-e)^2}{2e^2} + \ldots$

b $\sqrt{3} + 4\left(x - \frac{\pi}{3}\right) + 4\sqrt{3}\left(x - \frac{\pi}{3}\right)^2 + \frac{40}{3}\left(x - \frac{\pi}{3}\right)^3 + \ldots$

c $\cos 1 - \sin 1(x-1) - \frac{\cos 1}{2}(x-1)^2 + \frac{\sin 1}{6}(x-1)^3 + \frac{\cos 1}{24}(x-1)^4 + \ldots$

3 a i $\frac{\sqrt{2}}{2}\left(1 - x - \frac{1}{2}x^2 + \frac{1}{6}x^3 + \frac{1}{24}x^4 - \ldots\right)$

ii $\ln 5 + \frac{1}{5}x - \frac{1}{50}x^2 + \frac{1}{375}x^3 - \frac{1}{2500}x^4 + \ldots$

iii $\frac{1}{2}\left(-\sqrt{3} + x + \frac{\sqrt{3}}{2!}x^2 - \frac{1}{3!}x^3 - \frac{\sqrt{3}}{4!}x^4 + \ldots\right)$

b 1.649 (4 s.f.)

4 b $e^{-1}\left(-1 + \frac{1}{2}(x+1)^2 + \frac{1}{3}(x+1)^3 + \frac{1}{8}(x+1)^4 + \ldots\right)$

5 a $(x-1) + \frac{5}{2}(x-1)^2 + \frac{11}{6}(x-1)^3 + \frac{1}{4}(x-1)^4 + \ldots$

b 0.4059 (4 d.p.)

6 $-\frac{3}{4} + \frac{25}{16}x - \frac{75}{64}x^2 + \ldots$

7 $\frac{\sqrt{3}}{2} + 1\left(x - \frac{\pi}{6}\right) - \sqrt{3}\left(x - \frac{\pi}{6}\right)^2 - \frac{2}{3}\left(x - \frac{\pi}{6}\right)^3 + \frac{\sqrt{3}}{3}\left(x - \frac{\pi}{6}\right)^4 + \ldots$

8 a $\left.\frac{dy}{dx}\right|_3 = -\frac{1}{16}$

$\left.\frac{d^2y}{dx^2}\right|_3 = \frac{3}{128}$

b $y = \frac{1}{\sqrt{(1+x)}} = \frac{1}{2} - \frac{1}{16}(x-3) + \frac{3}{256}(x-3)^2 + \ldots$

9 $\frac{13}{5} + \frac{12}{5}(x - \ln 5) + \frac{13}{10}(x - \ln 5)^2 + \frac{2}{5}(x - \ln 5)^3 + \frac{13}{120}(x - \ln 5)^4 + \ldots$

10 a Let $f(x) = \sinh(ax)$, then $f'(x) = a\cosh(ax)$ so
$f(x) = f(\ln 2) + f'(\ln 2)(x - \ln 2) + \ldots$
$= \ldots + \frac{a(2^a + 2^{-a})}{2}(x - \ln 2) + \ldots$
If $a = 2$ then $\frac{a(2^a + 2^{-a})}{2} = 4 + \frac{1}{4} = \frac{17}{4}$

b $f''(x) = 4\sinh(2x), f'''(x) = 8\cosh(2x)$
Hence $f(x) = \frac{15}{8} + \frac{17}{4}(x - \ln 2) + \frac{15}{4}(x - \ln 2)^2 + \frac{17}{6}(x - \ln 2)^3$

11 $f(x) = \ln x$, $f'(x) = \frac{1}{x}$, $f''(x) = -\frac{1}{x^2}$, $f'''(x) = \frac{2}{x^3}$
$f^k(x) = (-1)^{k-1}\frac{(k-1)!}{x^k} \Rightarrow f^k(2) = (-1)^{k-1}\frac{(k-1)!}{2^k}$
Substituting into the Taylor series expansion gives
$f(x) = \ln 2 + \sum_{n=1}^{\infty}\frac{1}{n!}(-1)^{n-1}\frac{(n-1)!}{2^n}(x-2)^n$
$= \ln 2 + \sum_{n=1}^{\infty}(-1)^{n-1}\frac{(x-2)^n}{n2^n}$

Challenge

a $\ln(\cos 2x) = -2(x-\pi)^2 - \frac{4}{3}(x-\pi)^4 - \ldots$

b -0.1433 (4 d.p.)

Exercise 6B

1 a $\frac{7}{5}$ b $\frac{3}{2}$ c -2 d 2

2 a 4 b $-\frac{1}{2}$ c $\frac{1}{3}$ d $\frac{1}{4}$

3 a -1 b 4

4 a 1 b 2 c $\frac{2}{3}$ d $\frac{5}{6}$

5 a $\sin x = x - \frac{x^3}{3!} + \frac{x^5}{5!} - \ldots$, $e^{-x} = 1 - x + \frac{x^2}{2!} - \frac{x^3}{3!} + \ldots$
b $-\frac{1}{2}$

6 a $\ln x = (x-1) - \frac{1}{2}(x-1)^2 + \frac{1}{3}(x-1)^3 - \ldots$
$\sqrt{x} = 1 + \frac{1}{2}(x-1) - \frac{1}{8}(x-1)^2 + \frac{1}{16}(x-1)^3 + \ldots$
b 2

7 a $\sinh x = x + \frac{x^3}{3!} + \frac{x^5}{5!} + \ldots$ b $\frac{2}{3}$

8 a $\sqrt{1+4x} = 3 + \frac{2}{3}(x-2) - \frac{2}{27}(x-2)^2 + \ldots$ b $\frac{1}{36}$

Challenge

a $\sqrt{1+5y} = 1 + \frac{5}{2}y - \frac{25}{8}y^2 + \frac{125}{16}y^3 - \ldots$ b $\frac{5}{2}$

Exercise 6C

1 $y = 1 + \frac{x}{2} + x^2 + \frac{x^3}{3} + \frac{x^4}{6} + \ldots$

2 $y = x - \frac{x^3}{6} + \ldots$

3 $y = 2 - x + x^2 - \frac{x^3}{6} \ldots$

4 $y = 1 + 2x - \frac{1}{2}x^2 - \frac{2}{3}x^3 + \frac{1}{8}x^4 + \ldots$

5 $y = 1 - (x-1) + \frac{5}{2}(x-1)^2 - \frac{5}{3}(x-1)^3 + \ldots$

6 $y = 1 + x - x^2 + \frac{1}{2}x^4 + \ldots$

7 a Differentiating $(1+2x)\frac{dy}{dx} = x + 2y^2$ with respect to x

$$(1+2x)\frac{d^2y}{dx^2} + 2\frac{dy}{dx} = 1 + 4y\frac{dy}{dx} \quad ①$$

Differentiating ① gives

$$(1+2x)\frac{d^3y}{dx^3} + 2\frac{d^2y}{dx^2} + 2\frac{d^2y}{dx^2} = 4y\frac{d^2y}{dx^2} + 4\left(\frac{dy}{dx}\right)^2$$

$$\Rightarrow (1+2x)\frac{d^3y}{dx^3} + 4(1-y)\frac{d^2y}{dx^2} = 4\left(\frac{dy}{dx}\right)^2 \quad ②$$

b $y = 1 + 2x + \frac{5}{2}x^2 + \frac{8}{3}x^3 \ldots$

8 $y = \sqrt{2} + \sqrt{2}\left(x - \frac{\pi}{4}\right) + \frac{3\sqrt{2}}{2}\left(x - \frac{\pi}{4}\right)^2 + \ldots$

9 a i Differentiating $\frac{dy}{dx} - x^2 - y^2 = 0$ with respect to x,

gives $\frac{d^2y}{dx^2} - 2y\frac{dy}{dx} - 2x = 0 \quad ①$

ii Differentiating ① gives

$$\frac{d^3y}{dx^3} - 2y\frac{d^2y}{dx^2} - 2\left(\frac{dy}{dx}\right)^2 - 2 = 0$$

So $\frac{d^3y}{dx^3} - 2y\frac{d^2y}{dx^2} - 2\left(\frac{dy}{dx}\right)^2 = 2 \quad ②$

b $\frac{d^4y}{dx^4} - 2y\frac{d^3y}{dx^3} - 6\left(\frac{dy}{dx}\right)\frac{d^2y}{dx^2} = 0$

c $y = 1 + x + x^2 + \frac{4}{3}x^3 + \frac{7}{6}x^4 + \ldots$

10 Differentiating $\cos x\frac{dy}{dx} + y\sin x + 2y^3 = 0$, ① with respect to x, gives

$$\cos x\frac{d^2y}{dx^2} - \sin x\frac{dy}{dx} + y\cos x + \sin x\frac{dy}{dx} + 6y^2\frac{dy}{dx} = 0, \quad ②$$

Differentiating again

$$\cos x\frac{d^3y}{dx^3} - \sin x\frac{d^2y}{dx^2} - y\sin x + \cos x\frac{dy}{dx} + 6y^2\frac{d^2y}{dx^2} + 12y\left(\frac{dy}{dx}\right)^2 = 0, \quad ③$$

Substituting $x_0 = 0$, $y_0 = 1$ into ① gives $\left.\frac{dy}{dx}\right|_0 + 2(1) = 0$, so $\left.\frac{dy}{dx}\right|_0 = -2$

Substituting $x_0 = 0$, $y_0 = 1$, $\left.\frac{dy}{dx}\right|_0 = -2$ into ② gives

$\left.\frac{d^2y}{dx^2}\right|_0 + 1 + 6(1)(-2) = 0$, so $\left.\frac{d^2y}{dx^2}\right|_0 = 11$

Substituting $x_0 = 0$, $y_0 = 1$, $\left.\frac{dy}{dx}\right|_0 = -2$, $\left.\frac{d^2y}{dx^2}\right|_0 = 11$ into ③ gives

$\left.\frac{d^3y}{dx^3}\right|_0 + (1)(-2) + 6(1)(11) + 12(1)(-2)^2$,

so $\left.\frac{d^3y}{dx^2}\right|_0 = -112$

Substituting these values into the Taylor series, gives

$y = 1 + (-2)x + \frac{11}{2!}x^2 + \frac{(-112)}{3!}x^3 + \ldots$

$y = 1 - 2x + \frac{11}{2}x^2 - \frac{56}{3}x^3 + \ldots$

Ignoring terms in x^4 and higher powers,

$y \approx 1 - 2x + \frac{11}{2}x^2 - \frac{56}{3}x^3$.

11 a Repeated differentiation gives:

$$\frac{d^3y}{dx^3} = 4\frac{dy}{dx} + 4x\frac{d^2y}{dx^2} - 2\frac{dy}{dx} = 2\frac{dy}{dx} + 4x\frac{d^2y}{dx^2}$$

$$\frac{d^4y}{dx^4} = 2\frac{d^2y}{dx^2} + 4\frac{d^2y}{dx^2} + 4x\frac{d^3y}{dx^3} = 6\frac{d^2y}{dx^2} + 4x\frac{d^3y}{dx^3}$$

$$\frac{d^5y}{dx^5} = 6\frac{d^3y}{dx^3} + 4\frac{d^3y}{dx^3} + 4x\frac{d^4y}{dx^4} = 4x\frac{d^4y}{dx^4} + 10\frac{d^3y}{dx^3}$$

$p = 4$ and $q = 10$

b $y = 2 + 2(x-1) + 2(x-1)^2 + \frac{10}{3}(x-1)^3 + \frac{13}{3}(x-1)^4 + \frac{77}{15}(x-1)^5 + \ldots$

Mixed exercise 6

1 Let $f(x) = \left(x - \frac{\pi}{4}\right)\cot x$ and $a = \frac{\pi}{4} \Rightarrow f(a) = 0$

$f'(x) = \left(x - \frac{\pi}{4}\right)(-\operatorname{cosec}^2 x) + \cot x \Rightarrow f'(a) = 1$

$f''(x) = \left(x - \frac{\pi}{4}\right)2\cot x\operatorname{cosec}^2 x + (-2\operatorname{cosec}^2 x) \Rightarrow f''(a) = -4$

$f'''(x) = \left(x - \frac{\pi}{4}\right)(-2\operatorname{cosec}^4 x - 4\cot^2 x\operatorname{cosec}^2 x) + 6\cot x\operatorname{cosec}^2 x \Rightarrow f'''(a) = 12$

Substituting into the Taylor series expansion gives

$f(x) = 0 + 1\left(x - \frac{\pi}{4}\right) + \frac{-4}{2!}\left(x - \frac{\pi}{4}\right)^2 + \frac{12}{3!}\left(x - \frac{\pi}{4}\right)^3 + \ldots$

$= \left(x - \frac{\pi}{4}\right) - 2\left(x - \frac{\pi}{4}\right)^2 + 2\left(x - \frac{\pi}{4}\right)^3 + \ldots$ as required

2 a $f'(0) = \frac{1}{2}$, $f''(0) = \frac{1}{4}$

b $f'''(x) = \frac{e^x(1-e^x)}{(e^x+1)^3} \Rightarrow f'''(0) = \frac{1(1-1)}{(1+1)^3} = 0$

c $\ln 2 + \frac{x}{2} + \frac{x^2}{8} + \ldots$
$x < 0$

3 a $1 - 8x^2 + \frac{32}{3}x^4 - \frac{256}{45}x^6 + \ldots$

4 $e^{\cos x} = e(e^{\cos x - 1})$

$= e\left(1 + \left(-\frac{x^2}{2} + \frac{x^4}{24} + \ldots\right) + \frac{\left(-\frac{x^2}{2} + \ldots\right)^2}{2} + \ldots\right)$

$= e\left(1 - \frac{x^2}{2} + \frac{x^4}{24} + \frac{x^4}{8} + \ldots\right) \approx e\left(1 - \frac{x^2}{2} + \frac{x^4}{6}\right)$

5 a $y = 2x + \frac{3}{2}x^2 + \frac{1}{2}x^3 + \dots$ **b** 0.2155

6 $-3x^2 - 2x^3 - \dots$

7 $y = 2 + 4x + x^2 - \frac{2}{3}x^3 + \dots$

8 a $y = x + \dfrac{x^3}{6} + \dots$ **b** $\frac{2}{3}$

9 If $f(x) = \cosh(x)$ then $f^{(n)}(x) = \sinh(x)$ if n is odd, and $f^{(n)}(x) = \cosh(x)$ if n is even.

Also, $\sinh(\ln 2) = \frac{3}{4}$, $\cosh(\ln 2) = \frac{5}{4}$.

Hence a general expression for the nth term is

a $\dfrac{5}{4n!}(x - \ln 2)^n$ when n is even

b $\dfrac{3}{4n!}(x - \ln 2)^n$ when n is odd

10 2

11 2

12 a $\dfrac{d}{dx}(e^x) = \dfrac{d}{dx}\left(1 + x + \dfrac{x^2}{2!} + \dfrac{x^3}{3!} + \dfrac{x^4}{4!} + \dots + \dfrac{x^r}{r!} + \dfrac{x^{r+1}}{(r+1)!} + \dots\right)$

$= 1 + \dfrac{2x}{2!} + \dfrac{3x^2}{3!} + \dfrac{4x^3}{4!} + \dots + \dfrac{(r+1)x^r}{(r+1)!} + \dots$

$= 1 + x + \dfrac{x^2}{2!} + \dfrac{x^3}{3!} + \dots + \dfrac{x^r}{r!} + \dots$

$= e^x$

b $\dfrac{d}{dx}(\sin x) = \dfrac{d}{dx}\left(x - \dfrac{x^3}{3!} + \dfrac{x^5}{5!} - \dots + (-1)^r\dfrac{x^{2r+1}}{(2r+1)!} + \dots\right)$

$= 1 - \dfrac{3x^2}{3!} + \dfrac{5x^4}{5!} - \dots + (-1)^r\dfrac{(2r+1)x^{2r}}{(2r+1)!} + \dots$

$= 1 - \dfrac{x^2}{2!} + \dfrac{x^4}{4!} - \dfrac{x^6}{6!} + \dots + (-1)^r\dfrac{x^{2r}}{(2r)!} + \dots = \cos x$

c $\dfrac{d}{dx}(\cos x) = \dfrac{d}{dx}\left(1 - \dfrac{x^2}{2!} + \dfrac{x^4}{4!} - \dfrac{x^6}{6!} + \dots + (-1)^r\dfrac{x^{2r}}{(2r)!} + (-1)^{r+1}\dfrac{x^{2r+2}}{(2r+2)!} + \dots\right)$

$= -\dfrac{2x}{2!} + \dfrac{4x^3}{4!} - \dfrac{6x^5}{6!} + \dots + (-1)^r\dfrac{2rx^{2r-1}}{(2r)!} + (-1)^{r+1}\dfrac{(2r+2)x^{2r+1}}{(2r+2)!} + \dots$

$= -x + \dfrac{x^3}{3!} - \dfrac{x^5}{5!} + \dots + (-1)^{r+1}\dfrac{x^{2r+1}}{(2r+1)!} + \dots$

$= -\left(x - \dfrac{x^3}{3!} + \dfrac{x^5}{5!} - \dots + (-1)^r\dfrac{x^{2r+1}}{(2r+1)!} + \dots\right)$

$= -\sin x$

13 $y = 2(x-1) + \frac{1}{2}(x-1)^2 - \frac{1}{2}(x-1)^3 + \dots$

14 a You can write $\cos x = 1 - \left(\dfrac{x^2}{2} - \dfrac{x^4}{24} + \dots\right)$; it is not necessary to have higher powers

$\sec x = \dfrac{1}{\cos x} = \dfrac{1}{1 - \left(\frac{x^2}{2} - \frac{x^4}{24} + \dots\right)}$

$= \left(1 - \left(\dfrac{x^2}{2} - \dfrac{x^4}{24} + \dots\right)\right)^{-1}$

Using the binomial expansion but only requring powers up to x^4

$\sec x = 1 + (-1)\left(-\left(\dfrac{x^2}{2} - \dfrac{x^4}{24}\right)\right) + \dfrac{(-1)(-2)}{2!}\left(-\left(\dfrac{x^2}{2} - \dfrac{x^4}{24}\right)\right)^2 + \dots$

$= 1 + \left(\dfrac{x^2}{2} - \dfrac{x^4}{24}\right) + \dfrac{x^4}{4} + \text{higher powers of } x$

$= 1 + \dfrac{x^2}{2} + \frac{5}{24}x^4 + \dots$

b $x + \dfrac{x^3}{3} + \frac{2}{15}x^5 + \dots$ **c** $\frac{1}{2}$

15 a $1 + x - 4x^2 - \frac{13}{3}x^3 + \dots$ **b** $-\frac{7}{4}$

16 a $y = 2 + x - x^2 - \dfrac{x^3}{6} + \dots$

b Differentiating with respect to x gives

$\dfrac{d^4y}{dx^4} + 2x\dfrac{d^2y}{dx^2} + 2\dfrac{dy}{dx} + x^2\dfrac{d^3y}{dx^3} + 2x\dfrac{d^2y}{dx^2} + \dfrac{d^2y}{dx^2} = 0$ ①

Substituting $x = 0$, $\left.\dfrac{dy}{dx}\right|_0 = 1$, $\left.\dfrac{d^2y}{dx^2}\right|_0 = -2$ and $\left.\dfrac{d^3y}{dx^3}\right|_0 = -1$ into ① gives,

at $x = 0$, $\dfrac{d^4y}{dx^4} + 2(1) + (-2) = 0$, so $\dfrac{d^4y}{dx^4} = 0$

17 a $f'(x) = (1+x)^2\dfrac{1}{1+x} + 2(1+x)\ln(1+x)$

$= (1+x)(1 + 2\ln(1+x))$

$f''(x) = (1+x)\left(\dfrac{2}{1+x}\right) + (1 + 2\ln(1+x))$

$= 3 + 2\ln(1+x)$

$f'''(x) = \dfrac{2}{1+x}$

b $x + \frac{3}{2}x^2 + \frac{1}{3}x^3 + \dots$

18 a $x - \dfrac{x^2}{2} + \dfrac{x^3}{6} - \dfrac{x^4}{12} + \dots$ **b** 0.116 (3 d.p.)

19 a $f(x) = e^{\tan x} = e^{x + \frac{x^3}{3} + \dots} = e^x \times e^{\frac{x^3}{3}}$

(As only terms up to x^3 are required, only first two terms of $\tan x$ are needed.)

$= \left(1 + x + \dfrac{x^2}{2!} + \dfrac{x^3}{3!} + \dots\right)\left(1 + \dfrac{x^3}{3} + \dots\right)$

no other terms required.

$= \left(1 + \dfrac{x^3}{3} + x + \dfrac{x^2}{2!} + \dfrac{x^3}{3!} + \dots\right)$

$= 1 + x + \dfrac{x^2}{2} + \dfrac{x^3}{2} + \dots$

b $1 - x + \dfrac{x^2}{2} - \dfrac{x^3}{2} + \dots$

c -2

20 $\frac{1}{3}$

21 a $\dfrac{d^3y}{dx^3} = -\dfrac{1}{y}\left(\dfrac{dy}{dx}\left(3\dfrac{d^2y}{dx^2} + 1\right)\right)$

b $y = 1 + x - x^2 + \dfrac{5x^3}{6} + \dots$

c The approximation is best for small values of x (close to 0). $x = 0.2$, therefore, would be acceptable, but not $x = 50$.

22 a

$f(x) = \ln\cos x$	$f(0) = 0$
$f'(x) = \dfrac{-\sin x}{\cos x} = -\tan x$	$f'(0) = 0$
$f''(x) = -\sec^2 x$	$f''(0) = -1$
$f'''(x) = -2\sec^2 x\tan x$	$f'''(0) = 0$
$f''''(x) = -2\sec^4 x - 4\sec^2 x\tan^2 x$	$f''''(0) = -2$

Substituting into Maclaurin:

$\ln\cos x = (-1)\dfrac{x^2}{2!} + (-2)\dfrac{x^4}{4!} + \dots = -\dfrac{x^2}{2} - \dfrac{x^4}{12} - \dots$

b Using $1 + \cos x \equiv 2\cos^2\left(\dfrac{x}{2}\right)$,

$\ln(1 + \cos x) = \ln\left(2\cos^2\left(\dfrac{x}{2}\right)\right) = \ln 2 + 2\ln\cos\left(\dfrac{x}{2}\right)$

so $\ln(1 + \cos x) = \ln 2 + 2\left(-\frac{1}{2}\left(\dfrac{x}{2}\right)^2 - \frac{1}{12}\left(\dfrac{x}{2}\right)^4 - \dots\right)$

$= \ln 2 - \dfrac{x^2}{4} - \dfrac{x^4}{96} - \dots$

c $\frac{1}{2}$

23 a Let $y = 3^x$, then $\ln y = \ln 3^x = x\ln 3 \Rightarrow y = e^{x\ln 3}$

so $3^x = e^{x\ln 3}$

$1 + x\ln 3 + \dfrac{x^2(\ln 3)^2}{2} + \dfrac{x^3(\ln 3)^3}{6} + \dots$

b 1.73 (3 s.f.)

24 a $f(x) = \operatorname{cosec} x$

$f'(x) = -\operatorname{cosec} x \cot x$

i $f''(x) = -\operatorname{cosec}x(-\operatorname{cosec}^2x) + \cot x(\operatorname{cosec}x\cot x)$
$= \operatorname{cosec}x(\operatorname{cosec}^2x + \cot^2x)$
$= \operatorname{cosec}x(\operatorname{cosec}^2x + (\operatorname{cosec}^2x - 1))$
$= \operatorname{cosec}x(2\operatorname{cosec}^2x - 1)$

ii $f'''(x) = \operatorname{cosec}x(-4\operatorname{cosec}^2x\cot x) - \operatorname{cosec}x\cot x(2\operatorname{cosec}^2x - 1)$
$= -\operatorname{cosec}x\cot x(6\operatorname{cosec}^2x - 1)$

b $\sqrt{2} - \sqrt{2}\left(x - \frac{\pi}{4}\right) + \frac{3\sqrt{2}}{2}\left(x - \frac{\pi}{4}\right)^2 - \frac{11\sqrt{2}}{6}\left(x - \frac{\pi}{4}\right)^3 + \ldots$

25 a $f'(x) = \dfrac{-\pi\sin\left(\frac{\pi x}{2}\right)}{1 + 2\cos\left(\frac{\pi x}{2}\right)}$

$f''(x) = -\dfrac{\pi^2\cos\left(\frac{\pi x}{2}\right)}{2\left(1 + 2\cos\left(\frac{\pi x}{2}\right)\right)} - \dfrac{\pi^2\sin^2\left(\frac{\pi x}{2}\right)}{\left(1 + 2\cos\left(\frac{\pi x}{2}\right)\right)^2}$

b $f(1) = 0$, $f'(1) = -\pi$ and $f''(1) = -\pi^2$, so
$f(x) = -\pi(x - 1) - \frac{\pi^2}{2}(x - 1)^2 + \ldots$

c $\ln(2 - x) = -(x - 1) - \frac{1}{2}(x - 1)^2 - \frac{1}{3}(x - 1)^3 - \ldots$

Hence $\lim\limits_{x\to1}\dfrac{\ln\left(1 + 2\cos\left(\frac{\pi x}{2}\right)\right)}{3\ln(2 - x)} = \dfrac{\pi}{3}$

Challenge

a Base case: $n = 1$ we have $\frac{d}{dx}\ln x = \frac{1}{x}$

Suppose that $\frac{d^n}{dx^n}\ln x = (-1)^{n+1}\frac{(n - 1)!}{x^n}$, then

$\frac{d^{n+1}}{dx^{n+1}}\ln x = \frac{d}{dx}(-1)^{n+1}\frac{(n - 1)!}{x^n} = (-1)^{n+2}\frac{n!}{x^{n+1}}$

b $\ln x = \ln a + \sum\limits_{n=1}^{\infty}(-1)^{n+1}\frac{(x - a)^n}{na^n}$

c We have $a_n = (-1)^{n+1}\frac{(x - a)^n}{na^n}$ so

$\left|\frac{a_{n+1}}{a_n}\right| = \left|\frac{(x - a)^{n+1}na^n}{(x - a)^n(n + 1)a^{n+1}}\right| = \left|\frac{x - a}{a}\right|\frac{n}{n + 1}$ hence

$\lim\limits_{n\to\infty}\left|\frac{a_{n+1}}{a_n}\right| = \left|\frac{x - a}{a}\right|$, and $\left|\frac{x - a}{a}\right| < 1$ is satisfied if $0 < x < 2a$.

d At $x = 2a$ the series takes the form

$\ln a + \sum\limits_{n=1}^{\infty}\frac{(-1)^{n+1}}{n} = \ln a + \sum\limits_{n=1}^{\infty}(-1)^{n+1}b_n$

where $b_n = \frac{1}{n}$. We can easily verify the three conditions of the alternating series test

$\frac{1}{n} \geqslant 0$ for all n

$\frac{1}{n} \geqslant \frac{1}{n + 1}$ for all n

$\lim\limits_{n\to\infty}\frac{1}{n} = 0$

Hence the alternating series test implies that the series converges at $x = 2a$, so we have convergence for any $0 < x \leqslant 2a$.

CHAPTER 7

Prior knowledge check

1 a $3e^{3x}\cos x - e^{3x}\sin x$ **b** $\frac{1}{2\sqrt{x}}(\ln x + 2)$

2 Let $t = \tan\frac{x}{2}$, then

$\operatorname{cosec}x - \cot x\cos x = \frac{1 + t^2}{2t} - \frac{(1 - t^2)^2}{2t(1 + t^2)} = \frac{4t^2}{2t(1 + t^2)}$
$= \frac{2t}{1 + t^2} = \sin x$

3 $3t^2 + 2t - 1 = 0$

Exercise 7A

1 a i $\frac{dy}{dx} = 5e^{5x}$, $\frac{d^2y}{dx^2} = 25e^{5x}$, $\frac{d^3y}{dx^3} = 125e^{5x}$

ii $\frac{d^ny}{dx^n} = 5^ne^{5x}$

b i $\frac{dy}{dx} = -e^{-x}$, $\frac{d^2y}{dx^2} = e^{-x}$, $\frac{d^3y}{dx^3} = -e^{-x}$

ii $\frac{d^ny}{dx^n} = (-1)^ne^{-x}$

c i $\frac{dy}{dx} = mx^{m-1}$, $\frac{d^2y}{dx^2} = m(m - 1)x^{m-2}$,
$\frac{d^3y}{dx^3} = m(m - 1)(m - 2)x^{m-3}$

ii $\frac{d^ny}{dx^n} = \frac{m!}{(m - n)!}x^{m-n}$, provided $m \geqslant n$

d i $\frac{dy}{dx} = (1 - x)e^{-x}$, $\frac{d^2y}{dx^2} = (x - 2)e^{-x}$, $\frac{d^3y}{dx^3} = (3 - x)e^{-x}$

ii $\frac{d^ny}{dx^n} = (-1)^n(x - n)e^{-x}$

2 a $96x^2 - 12x + 10$

b $\frac{2\cos x}{x} - \sin x\left(\ln x + \frac{1}{x^2}\right)$

c $e^{3x}(5\cos 2x - 12\sin 2x)$

d $6x\ln(2x + 1) + \frac{12x^2}{2x + 1} - \frac{4x^3}{(2x + 1)^2}$

e $12(5x^2 - 2x + 1)$

f $9\left(3\sqrt{2x} - \frac{1}{(2x)^{\frac{3}{2}}}\right)\cosh 3x + 3\left(\frac{9}{\sqrt{2x}} + \frac{1}{(2x)^{\frac{5}{2}}}\right)\sinh 3x$

g $16(x^2 - x + 3)\cosh 2x + 32(2x - 1)\sinh 2x$

h $-4\cos x\sinh x$

3 a $\frac{8 - (\ln x)^2}{4x^{\frac{3}{2}}(\ln x)^3}$

b $\frac{11x^3 - 6x^3\ln x + 54x^2 + 81x + 54}{x^3(x + 3)^4}$

c $-\frac{2e^x(e^{2x} + 4e^x + 1)}{(e^x - 1)^4}$

d $\frac{30\sin x}{x^6} - \frac{24\cos x}{x^5} - \frac{9\sin x}{x^4} + \frac{2\cos x}{x^3} + \frac{\sin x}{4x^2}$

4 $\frac{dy}{dx} = (\cos x - \sin x)e^x$

By Leibnitz's theorem:

$\frac{d^6y}{dx^6} = e^x(\cos x - 6\sin x - 15\cos x + 20\sin x + 15\cos x - 6\sin x - \cos x) = 8e^x\sin x$

$\frac{d^6y}{dx^6} + 8\frac{dy}{dx} - 8y = 8e^x\sin x + 8e^x(\cos x - \sin x) - 8e^x\cos x$
$= 0$

5 Let $u = 2x^3$ and $v = e^{2x}$

$\frac{du}{dx} = 6x^2$, $\frac{d^2u}{dx^2} = 12x$, $\frac{d^3u}{dx^3} = 12$, $\frac{d^ku}{dx^k} = 0$ for $k > 3$

$\frac{d^kv}{dx^k} = 2^ke^{2x}$

$\frac{d^ny}{dx^n} = \binom{n}{0}u\frac{d^nv}{dx^n} + \binom{n}{1}\frac{du}{dx}\frac{d^{n-1}v}{dx^{n-1}} + \binom{n}{2}\frac{d^2u}{dx^2}\frac{d^{n-2}v}{dx^{n-2}} + \binom{n}{3}\frac{d^3u}{dx^3}\frac{d^{n-3}v}{dx^{n-3}}$

$= 2x^3(2^ne^{2x}) + n(6x^2)(2^{n-1}e^{2x}) + \frac{n(n - 1)}{2}(12x)(2^{n-2}e^{2x}) + \frac{n(n - 1)(n - 2)}{6}(12)(2^{n-3}e^{2x})$

$= 2^{n-2}e^{2x}(8x^3 + 12nx^2 + 6n(n - 1)x + n(n - 1)(n - 2))$

6 a **Base case**: if $n = 1$ then $\frac{dy}{dx} = -\frac{1}{x^2} = (-1)^1\frac{1!}{x^2}$

Inductive step: Suppose that the claim is true for n.

Then $\frac{d^{n+1}y}{dx^{n+1}} = \frac{d}{dx}\left(\frac{d^n y}{dx^n}\right) = \frac{d}{dx}\left((-1)^n\frac{n!}{x^{n+1}}\right)$

$= -(-1)^n\frac{n!(n+1)}{x^{n+2}}$

$= (-1)^{n+1}\frac{(n+1)!}{x^{n+2}}$

b From part **a**,

$\frac{d^n}{dx^n}(\ln x) = \frac{d^{n-1}}{dx^{n-1}}\left(\frac{1}{x}\right) = (-1)^{n-1}\frac{(n-1)!}{x^n}$ if $n \geqslant 1$.

Hence by Leibniz's theorem

$\frac{d^n}{dx^n}(x^3\ln x) = \sum_{k=0}^{n}\binom{n}{k}\frac{d^k}{dx^k}(x^3)\frac{d^{n-k}}{dx^{n-k}}(\ln x)$

$= (-1)^{n-1}\frac{(n-1)!}{x^{n-3}} + 3n(-1)^{n-2}\frac{(n-2)!}{x^{n-3}}$

$+ \frac{n(n-1)}{2}6(-1)^{n-3}\frac{(n-3)!}{x^{n-3}}$

$+ \frac{n(n-1)(n-2)}{6}6(-1)^{n-4}\frac{(n-4)!}{x^{n-3}}$

$= \frac{(-1)^n(n-4)!}{x^{n-3}}(-(n-1)(n-2)(n-3) + 3n(n-2)(n-3)$

$-3n(n-1)(n-3) + n(n-1)(n-2))$

$= \frac{6(-1)^n(n-4)!}{x^{n-3}}$

7 For m even we have $\frac{d^m}{dx^m}(\sinh kx) = k^m\sinh kx$, and for m odd we have $\frac{d^m}{dx^m}(\sinh kx) = k^m\cosh kx$

Let $f(x) = x^2$ and $g(x) = \sinh kx$. Then $f^{(m)}(x) = 0$ for all $m \geqslant 3$. So by Leibniz's theorem

$(fg)^{(n)}(x) = f(x)g^{(n)}(x) + nf'(x)g^{(n-1)}(x) + \frac{n(n-1)}{2}f''(x)g^{(n-2)}(x)$

And so if n is even we can write this as

$(fg)^{(n)}(x) = k^{n-2}\sinh kx(k^2x^2 + n(n-1)) + 2nk^{n-1}x\cosh kx$

whereas if n is odd we can write this as

$(fg)^{(n)}(x) = k^{n-2}\cosh kx(k^2x^2 + n(n-1)) + 2nk^{n-1}x\sinh kx$

Challenge

a When $n = 1$,

$F'(x) = \sum_{k=0}^{1}\binom{1}{1}f^{(k)}(x)g^{(1-k)}(x)$

$= f^{(0)}(x)g^{(1)}(x) + f^{(1)}(x)g^{(0)}(x)$

$= f(x)g'(x) + f'(x)g(x)$

b By part **a**, Leibniz's theorem holds for $n = 1$.

Suppose that the theorem is correct for some n. Then

$F^{(n+1)}(x) = \frac{d}{dx}\sum_{k=0}^{n}\binom{n}{k}f^{(k)}(x)g^{(n-k)}(x)$

$= \sum_{k=0}^{n}\binom{n}{k}(f^{(k+1)}(x)g^{(n-k)}(x) + f^{(k)}(x)g^{(n+1-k)}(x))$

$= \sum_{k=1}^{n+1}\binom{n}{k-1}f^{(k)}(x)g^{(n+1-k)}(x)$

$+ \sum_{k=0}^{n}\binom{n}{k}f^{(k)}(x)g^{(n+1-k)}(x)$

$= f(x)g^{(n+1)}(x) + f^{(n+1)}(x)g(x)$

$+ \sum_{k=1}^{n}\left(\binom{n}{k-1} + \binom{n}{k}\right)f^{(k)}(x)g^{(n+1-k)}(x)$

$= \sum_{k=0}^{n+1}\binom{n+1}{k}f^{(k)}(x)g^{(n+1-k)}(x)$

So the theorem holds for all n by induction.

Exercise 7B

1 a $\frac{2}{5}$ **b** 4 **c** 0
d $\frac{1}{\pi}$ **e** 0 **f** $\frac{4}{5}$

2 a 0 **b** no limit **c** 0

3 a 1 **b** 1 **c** $\frac{1}{e}$

4 a $\frac{2x^2 + x - 1}{3x^2 - 2x - 1} = \frac{2x^2 + x - 1}{(3x+1)(x-1)}$

If $\frac{2x^2 + x - 1}{3x^2 - 2x - 1} \equiv A + \frac{B}{3x+1} + \frac{C}{x-1}$, then

$2x^2 + x - 1 \equiv A(3x^2 - 2x - 1) + B(x-1) + C(3x+1)$

$x = 1{:}\ 2 = A(0) + B(0) + C(4) \Rightarrow C = \frac{1}{2}$

$x = -\frac{1}{3}{:}\ -\frac{10}{9} = -\frac{4}{3}B \Rightarrow B = \frac{5}{6}$

$x = 0{:}\ -1 = -A - B + C \Rightarrow -1 = -A - \frac{5}{6} + \frac{1}{2} \Rightarrow A = \frac{2}{3}$

So $\frac{2x^2 + x - 1}{3x^2 - 2x - 1} \equiv \frac{2}{3} + \frac{\frac{5}{6}}{3x+1} + \frac{\frac{1}{2}}{x-1}$

b $\frac{2}{3}$

c $\lim_{x\to\infty}\left(\frac{2x^2 + x - 1}{3x^2 - 2x - 1}\right) = \lim_{x\to\infty}\left(\frac{4x+1}{6x-2}\right) = \lim_{x\to\infty}\left(\frac{4}{6}\right) = \frac{2}{3}$

5 a $\lim_{x\to 3}\left(\frac{x^2 - 5x + 6}{4x}\right) = \frac{0}{4}$. The limit is not in an indeterminate form, so L'Hospital's rule cannot be applied.

b 0

6 a The limit of the numerator is 1 and hence the limit of the fraction is not an indeterminate form.

b $\frac{1}{4}$

7 1 **8** 1 **9** 0

10 $\lim_{x\to k}\sqrt{x} - \sqrt{k} = 0$ and $\lim_{x\to k}\sqrt[3]{x} - \sqrt[3]{k} = 0$ so we can apply L'Hospital's rule. Differentiating we have

$\frac{d}{dx}(\sqrt{x} - \sqrt{k}) = \frac{1}{2\sqrt{x}}$ and $\frac{d}{dx}(\sqrt[3]{x} - \sqrt[3]{k}) = \frac{1}{3x^{\frac{2}{3}}}$

Hence $\lim_{x\to k}\frac{\sqrt{x} - \sqrt{k}}{\sqrt[3]{x} - \sqrt[3]{k}} = \lim_{x\to k}\frac{3x^{\frac{2}{3}}}{2x^{\frac{1}{2}}} = \frac{3\sqrt[6]{k}}{2}$

11 1

12 $\frac{d}{dh}(\sin(x+h) - \sin(x)) = \cos(x+h)$ and $\frac{d}{dh}(h) = 1$

hence by L'Hospital's rule

$\lim_{h\to 0}\frac{\sin(x+h) - \sin x}{h} = \lim_{h\to 0}\cos(x+h) = \cos x$

13 a Total after 5 years $= 1000 \times 1.05^5 \approx 1276.28$

b Nominal interest is 10%, so $\frac{10}{12}$% is paid each month. Hence total after 12 months increases by a factor $\left(1 + \frac{0.1}{12}\right)^{12} \approx 1.1047$, implying an effective rate of 10.47%.

c $A_n(r) = A\left(1 + \frac{r}{n}\right)^n$

d Write $A_n(r) = Ae^{n\ln(1+\frac{r}{n})}$. By L'Hospital's rule,

$\lim_{n\to\infty}n\ln\left(1 + \frac{r}{n}\right) = \lim_{n\to\infty}\frac{\ln(1 + \frac{r}{n})}{\frac{1}{n}} = \lim_{n\to\infty}\frac{r}{(1 + \frac{r}{n})} = r$ so

$A_\infty(r) = \lim_{n\to\infty}A_n(r) = Ae^r$

Exercise 7C

1 a $\frac{1}{2\sqrt{2}}\ln\left|\frac{\sqrt{2} + \tan\frac{x}{2}}{\sqrt{2} - \tan\frac{x}{2}}\right| + c$ **b** $\ln\left|\frac{1 + \tan\frac{x}{2}}{1 - \tan\frac{x}{2}}\right| + c$

c $\frac{\ln|\tan\frac{x}{2}|}{2} - \frac{\tan^2\frac{x}{2}}{4} + c$ **d** $\frac{4}{1 - \tan\frac{x}{2}} + c$

2 **a** 1.2465 (4 d.p.) **b** 0.2218 (4 d.p.)
c 0.4636 (4 d.p.) **d** −0.0693 (4 d.p.)

3 **a** $\int \frac{1}{12 - 13\sin x}\,dx = \int \frac{1}{12 - \frac{26t}{1+t^2}} \times \frac{2}{1+t^2}\,dt$

$= \int \frac{1}{6t^2 - 13t + 6}\,dt$

b 0.3048 (4 d.p.)

4 Evaluating the integral using the Weierstrass substitution:

$$\int_0^{\frac{\pi}{2}} \frac{1}{a + \cos 2x}\,dx = \int_0^1 \frac{2}{(a+1) + (a-1)t^2}\,dt$$
$$= \frac{2}{a-1}\frac{\sqrt{a-1}}{\sqrt{a+1}}\arctan\left(\frac{\sqrt{a-1}}{\sqrt{a+1}}\right)$$

Substituting $a = 2$: $\frac{2}{1} \times \frac{1}{\sqrt{3}}\arctan\left(\frac{1}{\sqrt{3}}\right) = \frac{\pi}{3\sqrt{3}}$

5 Using the substitution $x = \tan\frac{t}{2}$

$$\frac{dx}{dt} = \frac{\sec^2\frac{t}{2}}{2} = \frac{1 + \tan^2\frac{t}{2}}{2} \Rightarrow dx = \frac{1 + \tan^2\frac{t}{2}}{2}\,dt$$

Transforming the limits: $x = 0 \Rightarrow t = 0$, $x = 1 \Rightarrow t = \frac{\pi}{2}$

$$\cos t = \frac{1 - x^2}{1 + x^2} \Rightarrow t = \arccos\left(\frac{1 - x^2}{1 + x^2}\right)$$

$$\int_0^1 \frac{\arccos\left(\frac{1-x^2}{1+x^2}\right)}{1 + x^2}\,dx = \int_0^{\frac{\pi}{2}} \frac{t}{1 + \tan^2\frac{t}{2}} \times \frac{1 + \tan^2\frac{t}{2}}{2}\,dt$$

$$= \int_0^{\frac{\pi}{2}} \frac{t}{2}\,dt = \left[\frac{t^2}{4}\right]_0^{\frac{\pi}{2}} = \frac{\pi^2}{16} - 0 = \frac{\pi^2}{16}$$

Challenge

$(\sqrt{2} - 1)\pi$

Mixed exercise 7

1 **a** $60x^3 - 24x^2 + 36x - 44$

b $8e^{4x}(\sec^2 2x(\tan 2x + 2) + 2\tan 2x)$

c $\frac{-96x^{\frac{7}{2}} - 88x^{\frac{3}{2}}}{(1 + 4x^2)^3} + \frac{9x^{-\frac{1}{2}}}{2(1 + 4x^2)} - \frac{3x^{-\frac{3}{2}}}{8}\arctan 2x$

2 $2\tan x\sec^2 x$

3 **a** By Leibnitz's theorem

$(f(gh))''(x) = f''(x)(gh)(x) + 2f'(x)(gh)'(x) + f(x)(gh)''(x)$
$= f''(x)g(x)h(x) + 2f'(x)(g'(x)h(x) + g(x)h'(x))$
$+ f(x)(g''(x)h(x) + 2g'(x)h'(x) + g(x)h''(x))$

b $2e^x(2\cos 2x\cos 3x - 3\sin 2x\sin 3x - 6\sin 2x\cos 3x - 6\sin 3x\cos 2x)$

4 $\frac{d^3}{dx^3}\left(\frac{\sqrt{3x+2}}{\cos x}\right) = \sqrt{3x+2}\,(5\tan x\sec^3 x + \tan^3 x\sec x)$

$+ \frac{9(\sec^3 x + \tan^2 x\sec x)}{2\sqrt{3x+2}} - \frac{27\tan x\sec x}{4(3x+2)^{\frac{3}{2}}} + \frac{81\sec x}{8(3x+2)^{\frac{5}{2}}}$

5 **a** First check the base case $n = 1$. We have

$\frac{dy}{dx} = \cos x = \sin\left(\frac{\pi}{2} + x\right)$

Now suppose the claim holds for some n, then

$$\frac{d^{n+1}y}{dx^{n+1}} = \frac{d}{dx}\left(\frac{d^n y}{dx^n}\right) = \frac{d}{dx}\sin\left(\frac{n\pi}{2} + x\right)$$
$$= \cos\left(\frac{n\pi}{2} + x\right) = \sin\left(\frac{(n+1)\pi}{2} + x\right)$$

So the claim holds for all n by induction.

b Applying Leibnitz's theorem we get

$$\frac{d^n y}{dx^n} = \sum_{k=0}^{n}\binom{n}{k}\frac{d^k}{dx^k}(x^2)\frac{d^{n-k}}{dx^{n-k}}(\sin x)$$
$$= x^2\sin\left(\frac{n\pi}{2} + x\right) + 2nx\sin\left(\frac{(n-1)\pi}{2} + x\right)$$
$$+ n(n-1)\sin\left(\frac{(n-2)\pi}{2} + x\right)$$
$$= \sin\left(\frac{n\pi}{2} + x\right)(x^2 + n - n^2) - 2nx\cos\left(\frac{n\pi}{2} + x\right)$$

6 **a** $\frac{2}{3}$ **b** $\frac{1}{3}$ **c** 3 **d** $-\frac{\pi}{9}$

7 $-\frac{1}{2}$

8 $\frac{1}{n2^{n-1}}$

9 1

10 **a** $\frac{\sqrt{3}}{2}\ln\left|\frac{\sqrt{3} + \tan\frac{x}{2}}{\sqrt{3} - \tan\frac{x}{2}}\right| + c$ **b** $\ln\left|\frac{\tan^2\frac{x}{2} - \tan\frac{x}{2} - 1}{\tan^2\frac{x}{2} - 1}\right| + c$

c $\sqrt{2}\ln\left|\frac{\tan\frac{x}{2} + \sqrt{2} - 1}{\tan\frac{x}{2} - \sqrt{2} - 1}\right| + c$

11 **a** 0.2407 (4 s.f.) **b** $\frac{\sqrt{3}}{6}$ or 0.2887 (4 s.f.)

12 **a** $\int \frac{1}{4\cos x - 3\sin x}\,dx = \int \frac{1}{\frac{4(1-t^2)}{1+t^2} - \frac{6t}{1+t^2}} \times \frac{2}{1+t^2}\,dt$

$= \int \frac{2}{4 - 4t^2 - 6t}\,dt = \int \frac{-1}{2t^2 + 3t - 2}\,dt$

b −0.3429 (4 s.f.)

13 $\int_{\frac{\pi}{3}}^{\frac{\pi}{2}} \frac{1 - \operatorname{cosec} x}{\sin x}\,dx = \int_{\frac{1}{\sqrt{3}}}^{1} \frac{1 - \frac{1+t^2}{2t}}{\frac{2t}{1+t^2}} \times \frac{2}{1+t^2}\,dt$

$= \int_{\frac{1}{\sqrt{3}}}^{1} \frac{4t - 2 - 2t^2}{4t^2}\,dt = \left[\ln t + \frac{1}{2t} - \frac{t}{2}\right]_{\frac{1}{\sqrt{3}}}^{1}$

$= -\ln\left(\frac{1}{\sqrt{3}}\right) - \frac{\sqrt{3}}{2} + \frac{1}{2\sqrt{3}}$

$= \ln\sqrt{3} - \frac{1}{\sqrt{3}}$

Challenge

We check the base case, $n = 1$. By L'Hospital's rule

$\lim_{n\to\infty}\frac{x}{e^x} = \lim_{n\to\infty}\frac{1}{e^x} = 0$

Now we suppose that the claim holds for some n. Then by L'Hospital's rule once again we have

$\lim_{n\to\infty}\frac{x^{n+1}}{e^x} = \lim_{n\to\infty}\frac{(n+1)x^n}{e^x} = (n+1)\lim_{n\to\infty}\frac{x^n}{e^x} = 0$

CHAPTER 8

Prior knowledge check

1 $P(x_0 - h, x_0^2 - 2hx_0 + h^2 + bx_0 - bh)$ and $Q(x_0 + h, x_0^2 + 2hx_0 + h^2 + bx_0 + bh)$

Gradient

$= \frac{(x_0^2 + 2hx_0 + h^2 + bx_0 + bh) - (x_0^2 - 2hx_0 + h^2 + bx_0 - bh)}{(x_0 + h) - (x_0 - h)}$

$= \frac{4hx_0 + 2bh}{2h} = 2x_0 + b$

2 $y = 2e^x(1 - e^{-x}(\sin x + \cos x))$

3 2317

Exercise 8A

1 87.3 (3 s.f.)
2 2.24 (3 s.f.)
3 **a** 0.21 (2 d.p.) **b** 2.854, 3.363 (3 d.p.)
4 £8400
5 0.885 (3 s.f.)

Online Full worked solutions are available in SolutionBank.

Exercise 8B

1 a 3   b 3.195 (3 d.p.)

2 4.464 (3 d.p.)

3 a $\left(\frac{dy}{dx}\right)_0 = \sin 2 = 0.909297...$

$\frac{y_1 - 2}{0.2} = 0.909297... \Rightarrow y_1 = 2.1819$ (5 s.f.)

b 1.999 (4 s.f.)

4 810

5 10.8 (3 s.f.)

6 a $\left(\frac{dy}{dx}\right)_0 = 1^2 - 1 + 1 - 2 = -1$

$\frac{y_1 - 1}{0.1} = -1 ... \Rightarrow y_1 = 0.9$

b 0.862 (3 d.p.)

c $y = \frac{1}{2}e^{2-2x} + \frac{1}{2}x^2 + 1 - x$ (o.e.)

d 0.85516..., % error = 0.80% (2 s.f.)

Challenge

-5; $\frac{dy}{dx}$ is undefined at $x = 1$, general solution curve has vertical asymptote.

Exercise 8C

1 a 4.1, 4.252, 4.45852   b 1.4, 1.936, 2.700324

c 1.1, 1.2441, 1.437463   d 2.1, 2.195304, 2.286855

2 a 2.0625   b 3.114647   c 1.14   d 1.66

3 a 0.9   b 0.8052, 0.7212

4 1.12, 1.326844, 1.584322

5 a 2   b 0.31

6 −3.02455

Exercise 8D

1 0.7206 (4 s.f.)

2 14.41 (4 s.f.)

3 a 1.202 (4 s.f.)

b Increase the number of intervals.

4 a Simpson's rule can only be used with an even number of intervals.

b 0.9223 (4 s.f.)

5 a 0.4471 (4 s.f.)   b 0.44648   c 0.14%

6 a 19.84 (4 s.f.)

b $\int_1^3 x\sinh x\,dx = [x\cosh x]_0^3 - \int_1^3 \cosh x\,dx$

$= [x\cosh x - \sinh x]_1^3$

$= \left(\frac{3e^3 + 3e^{-3}}{2} - \frac{e^3 - e^{-3}}{2}\right) - \left(\frac{e^1 + e^{-1}}{2} - \frac{e^1 - e^{-1}}{2}\right)$

$= e^3 + 2e^{-3} - e^{-1}$

c 0.0115%

7 a $x = 0 \Rightarrow t = 0$, $x = 2 \Rightarrow t = 1$

Area $= \pi\int_0^1((t - t^2)^2(1 + 2t))\,dt$

$= \pi\int_0^1(t^2 - 3t^4 + 2t^5)\,dt$

b 0.2127 (4 s.f.)

c Exact area $= \frac{\pi}{15}$, so percentage error

$= \left(\frac{0.2127 - \frac{\pi}{15}}{\frac{\pi}{15}}\right) \times 100\% = 1.56...\% < 1.6\%$

d Use more intervals.

Mixed exercise 8

1 2124.098 (3 d.p.)

2 a 0.05   b 4.581, 25.775

3 £9000

4 7.52 (3 s.f.)

5 a $\left(\frac{dv}{dt}\right)_0 = -4$, $\frac{v_1 - 2}{0.1} = -4 \Rightarrow v_1 = 1.6$

b 1.56   c $v = 5t - \frac{5}{2} + \frac{9}{2}e^{-2t}$ d $-\frac{3}{2} + \frac{9}{2}e^{-0.4}$, 2.87%

6 2.1, 1.979, 1.681

7 a $-\frac{16}{9}$ b 3.191 (4 s.f.)

8 a 2.830 (4 s.f.)   b Use more intervals.

9 a 0.70668 (5 d.p.)   b 0.70659 (5 d.p.)   c 0.013%

Challenge

a Assume parabola is $y = ax^2 + bx + c$, and let $x_0 = -h$, $x_1 = 0$ and $x_2 = h$, so $y_0 = ah^2 - bh + c$, $y_1 = c$ and $y^2 = ah^2 + bh + c$. Then the area under the curve is given by

$\int_{-h}^{h}(ax^2 + bx + c)\,dx = \left[a\frac{x^3}{3} + b\frac{x^2}{2} + cx\right]_{-h}^{h} = \frac{2ah^3}{3} + 2ch$

$= \frac{1}{3}h(2ah^2 + 6h) = \frac{1}{3}h(y_0 + 4y_1 + y_2)$

b Divide $[x_0, x_n]$ into an even number n of subintervals of equal length h, then $h = \frac{x_n - x_0}{n}$

There are a total of $n + 1$ points with the x-coordinates $x_0, x_0 + h, x_0 + 2h, ..., x_0 + nh = x_n$, and the corresponding y-coordinates are $y_0, y_1, y_2, ..., y_n$.

Area under curve $\approx \frac{h}{3}(y_0 + 4y_1 + y_2) + \frac{h}{3}(y_2 + 4y_3 + y_4) + ...$

$+ \frac{h}{3}(y_{n-2} + 4y_{n-1} + y_n)$

$= \frac{h}{3}(y_0 + 4y_1 + 2y_2 + 4y_3 + ... + 4y_{n-1} + y)$

$= \frac{h}{3}(y_0 + 4(y_1 + y_3 + ...) + 2(y_2 + y_4 + ...) + y_n)$

CHAPTER 9

Prior knowledge check

1 a $y = Ax^2 + 1$ b $y = \frac{\frac{2}{3}x^3 + c}{x}$ c $y = Ae^{3x} + Be^x$

2 a $y = \frac{2}{9}e^{-x}$, $\frac{dy}{dx} = -\frac{2}{9}e^{-x}$, $\frac{d^2y}{dx^2} = \frac{2}{9}e^{-x}$

$\frac{d^2y}{dx^2} - 4\frac{dy}{dx} + 4y = \frac{2}{9}e^{-x} + \frac{8}{9}e^{-x} + \frac{8}{9}e^{-x} = 2e^{-x}$

b $y = e^{2x}(A + Bx) + \frac{2}{9}e^{-x}$

Exercise 9A

1 a $y^2 = 2x^2(\ln x + c)$ b $y^3 = 3x^3(\ln x + c)$

c $y = \frac{-x}{\ln x + c}$ d $y^3 = x^3(Ax - 1)$

2 a Given $z = y^{-2}$, $y = z^{-\frac{1}{2}}$ and $\frac{dy}{dx} = -\frac{1}{2}z^{-\frac{3}{2}}\frac{dz}{dx}$

So $\frac{dy}{dx} + (\frac{1}{2}\tan x)y = -(2\sec x)y^3$

$\Rightarrow -\frac{1}{2}z^{-\frac{3}{2}}\frac{dz}{dx} + (\frac{1}{2}\tan x)z^{-\frac{1}{2}} = -2\sec x\, z^{-\frac{3}{2}}$

$\therefore \frac{dz}{dx} - z\tan x = 4\sec x$

b $y = \sqrt{\frac{\cos x}{4x + c}}$

3 a Given that $z = x^{\frac{1}{2}}$, $x = z^2$ and $\frac{dx}{dt} = 2z\frac{dz}{dt}$

So the equation $\frac{dx}{dt} + t^2x = t^2x^{\frac{1}{2}}$ becomes

$2z\frac{dz}{dt} + t^2z^2 = t^2z$

Divide through by $2z$: $\frac{dz}{dt} + \frac{1}{2}t^2z = \frac{1}{2}t^2$

b $x = (1 + ce^{-\frac{1}{6}t^3})^2$

4 a Let $z = y^{-1}$, then $y = z^{-1}$ and $\frac{dy}{dx} = -z^{-2}\frac{dz}{dx}$

So $\frac{dy}{dx} - \frac{1}{x}y = \frac{(x+1)^3}{x}y^2$ becomes

$-z^{-2}\frac{dz}{dx} - \frac{1}{x}z^{-1} = \frac{(x+1)^3}{x}z^{-2}$

Multiply through by $-z^2$: $\frac{dz}{dx} + \frac{1}{x}z = -\frac{(x+1)^3}{x}$

b $y = \frac{4x}{4c - (x+1)^4}$

5 a $(1 + x^2)\frac{dz}{dx} + 2xz = 1$ **b** $y = \sqrt{\frac{x + c}{1 + x^2}}$

c $y = \sqrt{\frac{x + 4}{1 + x^2}}$

6 $\frac{dy}{dx} = \frac{dy}{dz} \times \frac{dz}{dx} = \frac{1}{-(n-1)y^{-n}} \times \frac{dz}{dx}$

So differential equation becomes

$-\frac{y^n}{n-1} \times \frac{dz}{dx} + Py = Qy^n$

$\Rightarrow \frac{dz}{dx} - (n-1)Py^{-(n-1)} = -Q(n-1)$

and then $\frac{dz}{dx} - (n-1)Pz = -Q(n-1)$

7 a Differential equation becomes $\frac{du}{dx} = \frac{1}{1+u}$

b This solves to give $u + \frac{1}{2}u^2 = x + c$.

$2(y + 2x) + (y + 2x)^2 - 2x = k \ (k = 2c)$

$\Rightarrow 4x^2 + 4xy + y^2 + 2y + 2x = k$

Challenge

Substitute $y = \frac{1}{v}$, $\frac{dy}{dx} = -\frac{1}{v^2}\frac{dv}{dx}$

Differential equation becomes

$x^2\left(-\frac{1}{v^2}\frac{dv}{dx}\right) - \frac{x}{v} = \frac{1}{v^2}$

$\Rightarrow x\frac{dv}{dx} + v = -\frac{1}{x}$

Integrate both sides to get $xv = -\ln x + C$

Substitute $v = \frac{1}{y}$ to get $y = \frac{-x}{\ln x + C}$

Exercise 9B

1 a $y = \frac{A}{x^4} + \frac{B}{x}$ **b** $y = (A + B\ln x) \times \frac{1}{x^2}$

c $y = \frac{A}{x^2} + \frac{B}{x^3}$ **d** $y = \frac{A}{x^7} + Bx^4$

e $y = Ax^7 + \frac{B}{x^2}$ **f** $y = \frac{1}{x}(A\cos\ln x + B\sin\ln x)$

2 a $y = \frac{z}{x} \Rightarrow xy = z$ and $x\frac{dy}{dx} + y = \frac{dz}{dx}$

Also $x\frac{d^2y}{dx^2} + \frac{dy}{dx} + \frac{dy}{dx} = \frac{d^2z}{dx^2}$

So the equation $x\frac{d^2y}{dx^2} + (2 - 4x)\frac{dy}{dx} - 4y = 0$

becomes $\frac{d^2z}{dx^2} - 4\left(\frac{dz}{dx} - y\right) - 4y = 0$

which rearranges to give $\frac{d^2z}{dx^2} - 4\frac{dz}{dx} = 0$

b $z = A + Be^{4x}$ **c** $y = \frac{A}{x} + \frac{B}{x}e^{4x}$

3 a $y = \frac{z}{x^2} \Rightarrow x^2y = z$

So $x^2\frac{dy}{dx} + 2xy = \frac{dz}{dx}$ (1)

and $x^2\frac{d^2y}{dx^2} + 2x\frac{dy}{dx} + 2x\frac{dy}{dx} + 2y = \frac{d^2z}{dx^2}$ (2)

The differential equation becomes

$\left(x^2\frac{d^2y}{dx^2} + 4x\frac{dy}{dx} + 2y\right) + \left(2x^2\frac{dy}{dx} + 4xy\right) + 2x^2y = e^{-x}$

Using results (1) and (2),

$\frac{d^2z}{dx^2} + 2\frac{dz}{dx} + 2z = e^{-x}$

b $z = e^{-x}(A\cos x + B\sin x + 1)$

c $y = \frac{e^{-x}}{x^2}(A\cos x + B\sin x + 1)$

4 a $z = \sin x \Rightarrow \frac{dz}{dx} = \cos x$

So $\frac{dy}{dx} = \frac{dy}{dz} \times \cos x$

and $\frac{d^2y}{dx^2} = \frac{d^2y}{dz^2}\cos^2 x - \frac{dy}{dz}\sin x$

The equation becomes

$\cos^3 x\frac{d^2y}{dz^2} - \cos x\sin x\frac{dy}{dz} + \cos x\sin x\frac{dy}{dz} - 2y\cos^3 x = 2\cos^5 x$

Dividing by $\cos^3 x$ gives

$\frac{d^2y}{dz^2} - 2y = 2\cos^2 x = 2(1 - z^2)$

b $y = Ae^{\sqrt{2}\sin x} + Be^{-\sqrt{2}\sin x} + \sin^2 x$

5 a $x = ut$, $\frac{dx}{dt} = u + t\frac{du}{dt}$, $\frac{d^2x}{dt^2} = 2\frac{du}{dt} + t\frac{d^2u}{dt^2}$

So differential equation becomes

$t^2\left(2\frac{du}{dt} + t\frac{d^2u}{dt^2}\right) - 2t\left(u + t\frac{du}{dt}\right) = -2(1 - 2t^2)ut$

which rearranges to give $t^3\left(\frac{d^2u}{dt^2} - 4u\right) = 0$

$\Rightarrow \frac{d^2u}{dt^2} - 4u = 0$

b $x = t(Ae^{2t} + Be^{-2t})$ **c** $x = t\left(\frac{3}{4e^2}e^{2t} + \frac{5}{4e^{-2}}e^{-2t}\right)$

Challenge

$y = A\ln x + B + 3x^2$

Exercise 9C

1 a $\frac{dx}{dt} = \frac{du}{dt}t + u$

So $t(ut)\left(\frac{du}{dt}t + u\right) - u^2t^2 = 3t^4$

which rearranges to $u\frac{du}{dt} = 3t$.

b Solve the differential equation in u and t to get $\frac{1}{2}u^2 = \frac{3}{2}t^2 + c$, and then use $u = \frac{x}{t} = 3$ to find $c = 3$.

So $u^2 = 3t^2 + 6 \Rightarrow x^2 = 3t^4 + 6t^2 \Rightarrow x = \sqrt{3t^4 + 6t^2}$

The particular solution is $x = t\sqrt{3t^2 + 6}$.

c The function increases without limit so the displacement gets very large.

2 a $\frac{dv}{dt} = \frac{dz}{dt}t + z$

So $3z^2t^3\left(\frac{dz}{dt}t + z\right) = z^3t^3 + t^3$, which rearranges to

$3z^2t\frac{dz}{dt} = 1 - 2z^3$

b Differential equation in z and t solves to give $|1 - 2z^3| = \frac{A}{t^2}$

If $v = 2$ for $t = 1$, then $z = 2$, and $A = |-15| \times 1 = 15$.

Then $t^2(2z^3 - 1) = 15 \Rightarrow 2v^3 - t^3 = 15t$.

The particular solution is $v = \sqrt[3]{\frac{t^3 + 15t}{2}}$

c 2.668; 0.632

3 a $s = \frac{v}{t}, \frac{ds}{dt} = \frac{1}{t}\frac{dv}{dt} - \frac{v}{t^2}, \frac{d^2s}{dt^2} = \frac{1}{t}\frac{d^2v}{dt^2} - \frac{2}{t^2}\frac{dv}{dt} + \frac{2v}{t^3}$

So equation becomes

$t\left(\frac{1}{t}\frac{d^2v}{dt^2} - \frac{2}{t^2}\frac{dv}{dt} + \frac{2v}{t^3}\right) + (2-t)\left(\frac{1}{t}\frac{dv}{dt} - \frac{v}{t^2}\right) - (1+2t)\frac{v}{t} = e^{2t}$

Rearranging terms gives

$\frac{d^2v}{dt^2} + \left(-\frac{2}{t} + \frac{2-t}{t}\right)\frac{dv}{dt} + \left(\frac{2v}{t^2} - \frac{(2-t)v}{t^2} - \frac{(1+2t)v}{t}\right) = e^{2t}$

which simplifies to $\frac{d^2v}{dt^2} - \frac{dv}{dt} - 2v = e^{2t}$.

b Auxiliary equation has roots 2 and −1, so the complementary function is $v = Ae^{2t} + Be^{-t}$. To find the particular integral, try $v = \lambda te^{2t}$.

Then $\frac{dv}{dt} = \lambda e^{2t} + 2\lambda te^{2t}$ and $\frac{d^2v}{dt^2} = 4\lambda e^{2t} + 4\lambda te^{2t}$

So $\frac{d^2v}{dt^2} - \frac{dv}{dt} - 2v = 4\lambda e^{2t} + 4\lambda te^{2t} - (\lambda e^{2t} + 2\lambda te^{2t}) - 2\lambda te^{2t} = 3\lambda e^{2t}$

Letting $\lambda = \frac{1}{3}$ gives a particular integral of $v = \frac{1}{3}te^{2t}$.

Therefore the general solution is

$v = Ae^{2t} + Be^{-t} + \frac{1}{3}te^{2t}$

c $s = \frac{Ae^{2t} + Be^{-t}}{t} + \frac{1}{3}e^{2t}$; $t \neq 0$

4 a $\frac{dx}{dt} = u + t\frac{du}{dt}, \frac{d^2x}{dt^2} = 2\frac{du}{dt} + t\frac{d^2u}{dt^2}$

So differential equation becomes

$t\left(2\frac{du}{dt} + t\frac{d^2u}{dt^2}\right) - 2\left(u + t\frac{du}{dt}\right) + \left(\frac{2+t^2}{t}\right)ut = t^4$

which rearranges to the required equation.

b $x = t(A\cos t + B\sin t + t^2 - 2)$

c As t gets large, x gets large; the spring will reach its elastic limit and/or break.

Mixed exercise 9

1 a Given that $z = y^{-1}$, then $y = z^{-1}$ so $\frac{dy}{dx} = -z^{-2}\frac{dz}{dx}$

The equation $x\frac{dy}{dx} + y = y^2\ln x$ becomes

$-xz^{-2}\frac{dz}{dx} + z^{-1} = z^{-2}\ln x$

Dividing through by $-xz^{-2}$ gives $\frac{dz}{dx} - \frac{z}{x} = -\frac{\ln x}{x}$

b $y = \frac{1}{1 + cx + \ln x}$, where c is a constant.

2 a Given that $z = y^2$, $y = z^{\frac{1}{2}}$ and $\frac{dy}{dx} = \frac{1}{2}z^{-\frac{1}{2}}\frac{dz}{dx}$, the differential equation becomes

$\cos x\, z^{-\frac{1}{2}}\frac{dz}{dx} - z^{\frac{1}{2}}\sin x + z^{-\frac{1}{2}} = 0$

Divide through by $z^{-\frac{1}{2}}$: $\cos x\frac{dz}{dx} - z\sin x = -1$

b $z = c\sec x - x\sec x$

c $y^2 = c\sec x - x\sec x$, where c is a constant

3 a Given that $z = \frac{y}{x}$, $y = zx$ so $\frac{dy}{dx} = z + x\frac{dz}{dx}$

The equation $(x^2 - y^2)\frac{dy}{dx} - xy = 0$ becomes

$(x^2 - z^2x^2)\left(z + x\frac{dz}{dx}\right) - xzx = 0$

$\Rightarrow (1 - z^2)z + (1 - z^2)x\frac{dz}{dx} - z = 0$

$\Rightarrow x\frac{dz}{dx} = \frac{z}{1 - z^2} - z$

$\Rightarrow x\frac{dz}{dx} = \frac{z^3}{1 - z^2}$

b $2y^2(\ln y + c) + x^2 = 0$, where c is a constant

4 a $z = \frac{y}{x} \Rightarrow y = xz$ and $\frac{dy}{dx} = z + x\frac{dz}{dx}$

So $\frac{dy}{dx} = \frac{y(x + y)}{x(y - x)}$ becomes $z + x\frac{dz}{dx} = \frac{xz(x + xz)}{x(xz - x)}$

$\Rightarrow z + x\frac{dz}{dx} = \frac{z(1 + z)}{(z - 1)}$

So $x\frac{dz}{dx} = \frac{z(1 + z)}{z - 1} - z = \frac{2z}{z - 1}$

b $\frac{y}{2x} - \frac{1}{2}\ln y = \frac{1}{2}\ln x + c$, where c is a constant.

5 a Given that $z = \frac{y}{x}$, $y = zx$ and $\frac{dy}{dx} = z + x\frac{dz}{dx}$

The equation $\frac{dy}{dx} = \frac{-3xy}{y^2 - 3x^2}$ becomes

$z + x\frac{dz}{dx} = \frac{-3x^2z}{z^2x^2 - 3x^2}$

So $x\frac{dz}{dx} = \frac{-3z}{z^2 - 3} - z = \frac{-z^3}{z^2 - 3}$

b $\ln y + \frac{3x^2}{2y^2} = c$, where c is a constant.

6 a Let $u = x + y$, then $\frac{du}{dx} = 1 + \frac{dy}{dx}$ and so

$\frac{dy}{dx} = (x + y + 1)(x + y - 1)$ becomes

$\frac{du}{dx} - 1 = (u + 1)(u - 1) = u^2 - 1$

$\Rightarrow \frac{du}{dx} = u^2$

b $y = \frac{-1}{x + c} - x$, where c is a constant

7 a Given that $u = y - x - 2$, $\frac{du}{dx} = \frac{dy}{dx} - 1$

So $\frac{dy}{dx} = (y - x - 2)^2$ becomes $\frac{du}{dx} + 1 = u^2$

$\Rightarrow \frac{du}{dx} = u^2 - 1$

b $y = x + 2 + \frac{1 + Ae^{2x}}{1 - Ae^{2x}}$, where A is a positive constant.

8 a $v = u^{-\frac{1}{2}}, \frac{dv}{dt} = -\frac{1}{2}u^{-\frac{3}{2}}\frac{du}{dt}$

Equation becomes $-\frac{1}{2}u^{-\frac{3}{2}}\frac{du}{dt} \times t + u^{-\frac{1}{2}} = 2t^3u^{-\frac{3}{2}}$ which rearranges to $\frac{du}{dt} - \frac{2u}{t} = -4t^2$.

b Using integrating factor $e^{-2\int\frac{1}{t}dt} = e^{-2\ln t} = t^{-2}$, get

$\frac{d}{dt}(ut^{-2}) = -4 \Rightarrow ut^{-2} = -4t + c$, and $u = -4t^3 + ct^2$.

Then the general solution for the original equation is $v = \frac{1}{\sqrt{t^2(c - 4t)}}$

Given that $v = \frac{1}{2}$ when $t = 1$, $\frac{1}{\sqrt{c - 4}} = \frac{1}{2}$, so $c = 8$ and the particular solution is $v = \frac{1}{\sqrt{t^2(8 - 4t)}}$

9 a $y = \frac{A}{x} + \frac{B}{x^2} + \frac{1}{2}\ln x - \frac{3}{4}$ **b** $y = \frac{4}{x} - \frac{9}{4x^2} + \frac{1}{2}\ln x - \frac{3}{4}$

10 $y = \frac{1}{2}\cos(\sin x) + \frac{5}{2}\sin(\sin x) + \frac{1}{2}e^{\sin x}$

11 a $t = e^u$, $u = \ln t$, $\frac{du}{dt} = \frac{1}{t}, \frac{d^2u}{dt^2} = -\frac{1}{t^2}$

$\frac{dx}{dt} = \frac{dx}{du} \times \frac{du}{dt}, \frac{d^2x}{dt^2} = \frac{d^2x}{du^2} \times \frac{1}{t^2} - \frac{dx}{du} \times \frac{1}{t^2}$

So equation becomes

$t^2\left(\frac{d^2x}{du^2} \times \frac{1}{t^2} - \frac{dx}{du} \times \frac{1}{t^2}\right) - 2t\left(\frac{dx}{du} \times \frac{du}{dt}\right) + 2x = 4\ln(e^u)$

which rearranges to $\frac{d^2x}{du^2} - 3\frac{dx}{du} + 2x = 4u$

b $x = At^2 + Bt + 2\ln t + 3$

c As t gets very large, the distance of the particle from its original position becomes very large.

12 a $\frac{dx}{dt} = v + t\frac{dv}{dt}, \frac{d^2x}{dt^2} = 2\frac{dv}{dt} + t\frac{d^2v}{dt^2}$

Equation becomes

$2t^2\left(2\frac{dv}{dt} + t\frac{d^2v}{dt^2}\right) - 4t\left(v + t\frac{dv}{dt}\right) + (4 - 2t^2)tv = t^4$

which rearranges to $2\frac{d^2v}{dt^2} - 2v = t$

b $x = Ate^t + Bte^{-t} - \frac{1}{2}t^2$

13 a $u = v^{-1} \Rightarrow \frac{dv}{du} = -\frac{1}{u^2}\frac{du}{dt}$

$1000\frac{dv}{dt} - 500v + tv^2 = 0$ becomes

$-1000u^{-2}\frac{du}{dt} - 500u^{-1} + tu^{-2} = 0$

$\Rightarrow \frac{du}{dt} + 0.5u - 0.001t = 0$

$\Rightarrow \frac{du}{dt} + 0.5u = 0.001t$

b $v = \frac{500e^{0.5t}}{e^{0.5t}(t - 2) + A}$ **c** $v = \frac{500e^{0.5t}}{e^{0.5t}(t - 2) + 252}$

d $v \to 0$ as $t \to \infty$ so not valid for large values of t.

Challenge

Let $u = \frac{dy}{dx}$, so equation becomes $\frac{du}{dx} = u^2$

$\Rightarrow \int \frac{1}{u^2}du = \int dx \Rightarrow -\frac{1}{u} = x + B$

$\Rightarrow \frac{dy}{dx} = -\frac{1}{x + B} \Rightarrow y = A - \ln(x + b)$

Review exercise 2

1 $\cos\frac{x}{2} = \sqrt{1 - \left(\frac{12}{13}\right)^2} = \frac{5}{13}$

$t = \tan\frac{x}{2} = \frac{\sin\frac{x}{2}}{\cos\frac{x}{2}} = \frac{12}{13} \div \frac{5}{13} = \frac{12}{5} \Rightarrow \cot x = \frac{1 - t^2}{2t} = -\frac{119}{120}$

2 a $\sin^2\theta = \frac{2 + \sqrt{3}}{4}, \cos^2\theta = 1 - \left(\frac{2 + \sqrt{3}}{4}\right) = \frac{2 - \sqrt{3}}{4}$

$\Rightarrow \tan\theta = -\sqrt{\frac{\sin^2\theta}{\cos^2\theta}} = -\sqrt{\frac{2 + \sqrt{3}}{2 - \sqrt{3}}} = -2 - \sqrt{3}$

b $t = \tan\theta = -2 - \sqrt{3}$

$\Rightarrow \sin 2\theta = \frac{2t}{1 + t^2} = -\frac{1}{2}$ and $\cos 2\theta = \frac{1 - t^2}{1 + t^2} = -\frac{\sqrt{3}}{2}$

c $\theta = \frac{7\pi}{12}$

3 a $\sec x + \tan x = \frac{1 + t^2}{1 - t^2} + \frac{2t}{1 - t^2} = \frac{(1 + t)^2}{(1 + t)(1 - t)} = \frac{1 + t}{1 - t}$

b $\tan\left(\frac{\pi}{4} + \frac{x}{2}\right) = \frac{1 + \tan\frac{x}{2}}{1 - (1 \times \tan\frac{x}{2})} = \frac{1 + t}{1 - t} = \sec x + \tan x$

4 $2\cos^2\frac{\theta}{2} - 1 = 2\left(\frac{1}{\sqrt{1 + t^2}}\right)^2 - 1 = \frac{2 - (1 + t^2)}{1 + t^2} = \frac{1 - t^2}{1 + t^2} = \cos\theta$

5 a $3\left(\frac{1 - t^2}{1 + t^2}\right) - 4\left(\frac{2t}{1 + t^2}\right) - 4 = 0$

$\Rightarrow \frac{3 - 3t^2 - 8t - 4 - 4t^2}{1 + t^2} = 0 \Rightarrow 7t^2 + 8t + 1 = 0$

b $x = 4.71, 6.00$ (2 d.p.)

6 a $2\left(\frac{2t}{1 + t^2}\right) + \left(\frac{1 - t^2}{1 + t^2}\right) - 1 = 0$

$\Rightarrow \frac{4t + 1 - t^2 - 1 - t^2}{1 + t^2} = 0 \Rightarrow t^2 - 2t = 0$

b $0, 2\pi, 2.21$ (2 d.p.)

7 a $v = \frac{ds}{dx} = 2\cos 4x \times 4 + 4\cos 2x \times 2 = 8(\cos 4x + \cos 2x)$

$t = \tan x \Rightarrow \sin 2x = \frac{2t}{1 + t^2}, \cos 2x = \frac{1 - t^2}{1 + t^2}$

$\cos 4x = \cos^2 2x - \sin^2 2x$

$\Rightarrow v = 8\left(\left(\frac{1 - t^2}{1 + t^2}\right)^2 - \left(\frac{2t}{1 + t^2}\right)^2 + \left(\frac{1 - t^2}{1 + t^2}\right)\right)$

$= \frac{16}{(1 + t^2)^2}(1 - 3t^2)$

b Least value of s occurs at $x = \frac{5\pi}{6}$ and is -4.196 m.

It is a minimum because $\left.\frac{ds}{dx}\right|_{\frac{5\pi}{6}} = 0$ and $\left.\frac{d^2s}{dx^2}\right|_{\frac{5\pi}{6}} > 0$.

8 a $t = \tan\frac{x}{8}$, so $\sin\frac{x}{4} = \frac{2t}{1 + t^2}$ and $\cos\frac{x}{4} = \frac{1 - t^2}{1 + t^2}$

$f'(x) = 5\cos\frac{x}{2} + \frac{11}{4}\cos\frac{x}{4} - 5\sin\frac{x}{4}$

$= 5\cos^2\left(\frac{x}{4}\right) - 5\sin^2\left(\frac{x}{4}\right) + \frac{11}{4}\cos\frac{x}{4} - 5\sin\frac{x}{4}$

$= 5\left(\frac{1 - t^2}{1 + t^2}\right)^2 - 5\left(\frac{2t}{1 + t^2}\right)^2 + \frac{11}{4}\left(\frac{1 - t^2}{1 + t^2}\right) - 5\left(\frac{2t}{1 + t^2}\right)$

$= \frac{(9t^4 - 40t^3 - 120t^2 - 40t + 31)}{4(1 + t^2)^2}$

$= \frac{(t + 1)(9t^3 - 49t^2 - 71t + 31)}{4(1 + t^2)^2}$

b 6π

c Accept values in the range [4.8, 5].

d It is the second-lowest trough on the left. Accept values in the range [91.2, 95].

9 a $-2\left(x - \frac{\pi}{4}\right) + \frac{4}{3}\left(x - \frac{\pi}{4}\right)^3 - \frac{4}{15}\left(x - \frac{\pi}{4}\right)^5 + \ldots$

b -0.416147 (6 d.p.)

10 a $-\ln 2 + \sqrt{3}\left(x - \frac{\pi}{6}\right) - 2\left(x - \frac{\pi}{6}\right)^2 + \frac{4\sqrt{3}}{3}\left(x - \frac{\pi}{6}\right)^3 + \ldots$

b -0.735166 (6 d.p.)

11 a $\frac{dy}{dx} = \sec^2 x$

$\frac{d^2y}{dx^2} = 2\sec^2 x\tan x$

$\frac{d^3y}{dx^3} = 4\sec^2 x\tan^2 x + 2\sec^4 x$

b $1 + 2\left(x - \frac{\pi}{4}\right) + 2\left(x - \frac{\pi}{4}\right)^2 + \frac{8}{3}\left(x - \frac{\pi}{4}\right)^3 + \ldots$

c Let $x = \frac{3\pi}{10} \Rightarrow x - \frac{\pi}{4} = \frac{\pi}{20}$

$\tan\frac{3\pi}{10} = 1 + 2\left(\frac{\pi}{20}\right) + 2\left(\frac{\pi}{20}\right)^2 + \frac{8}{3}\left(\frac{\pi}{20}\right)^3$

$= 1 + \frac{\pi}{10} + \frac{\pi^2}{200} + \frac{\pi^3}{3000}$

12 a $(x - 1) - \frac{1}{2}(x - 1)^2 + \frac{1}{3}(x - 1)^3 + \ldots$ **b** -2

13 a $\sinh x = x + \frac{1}{6}x^3 + \frac{1}{120}x^5 + \ldots$ **b** $\frac{1}{2}$

14 a $\frac{d^3y}{dx^3} = 1$ **b** $2 - x - 2x^2 + \frac{1}{6}x^3 + \ldots$

15 a Differentiate the equation with respect to x:

$$2\frac{dy}{dx} + (1 + 2x)\frac{d^2y}{dx^2} = 1 + 8y\frac{dy}{dx}$$

$$(1 + 2x)\frac{d^2y}{dx^2} = 1 + 8y\frac{dy}{dx} - 2\frac{dy}{dx} = 1 + 2(4y - 1)\frac{dy}{dx}$$

b $2\frac{d^2y}{dx^2} + (1 + 2x)\frac{d^3y}{dx^3} = 8\left(\frac{dy}{dx}\right)^2 + 2(4y - 1)\frac{d^2y}{dx^2} \ldots$

c $\frac{1}{2} + x + \frac{3}{2}x^2 + \frac{4}{3}x^3 + \ldots$

16 **a** $1 + x + 2x^2 + 2x^3 + \ldots$ **b** 1.12 (2 d.p.)

17 **a** $1.5 + 0.8x - 0.208x^2 + 0.131\,982x^3 + \ldots$

b 1.578 (3 d.p.)

18 **a** $-\frac{1}{y}\frac{dy}{dx}\left(3\frac{d^2y}{dx^2} + 1\right)$ **b** $1 + x - x^2 + \frac{5}{6}x^3 + \ldots$

c The series expansion up to and including the term in x^3 can be used to estimate y if x is small. So it would be sensible to use it at $x = 0.2$ but not at $x = 50$.

19 **a** $1 + \frac{3}{2}x^2 + 2x^3 + \frac{5}{4}x^4 + \ldots$ **b** 1.08 (2 d.p.)

20 Let $u = x^3$ and $v = e^{3x}$

$$\frac{du}{dx} = 3x^2, \frac{d^2u}{dx^2} = 6x, \frac{d^3u}{dx^3} = 6, \frac{d^ku}{dx^k} = 0 \text{ for } k > 3$$

$$\frac{dv}{dx} = 3e^{3x}, \frac{d^2v}{dx^2} = 9e^{3x} \Rightarrow \frac{d^kv}{dx^k} = 3^ke^{3x}$$

$$\frac{d^ny}{dx^n} = u\frac{d^nv}{dx^n} + \binom{n}{1}\frac{du}{dx}\frac{d^{n-1}v}{dx^{n-1}} + \binom{n}{2}\frac{d^2u}{dx^2}\frac{d^{n-2}v}{dx^{n-2}} + \binom{n}{3}\frac{d^3u}{dx^3}\frac{d^{n-3}v}{dx^{n-3}}$$

$$= x^3(3^ne^{3x}) + n(3x^2)(3^{n-1}e^{3x}) + \frac{n(n-1)}{2}(6x)(3^{n-2}e^{3x})$$

$$+ \frac{n(n-1)(n-2)}{6}(6)(3^{n-3}e^{3x})$$

$$= 3^{n-3}e^{3x}(27x^3 + 27nx^2 + 9n(n-1)x + n(n-1)(n-2))$$

21 Let $u = e^x$ and $v = \sin x$

$$\frac{d^ku}{dx^k} = e^x$$

$$\frac{dv}{dx} = \cos x, \frac{d^2v}{dx^2} = -\sin x, \frac{d^3v}{dx^3} = -\cos x, \frac{d^4v}{dx^4} = \sin x,$$

$$\frac{d^5v}{dx^5} = \cos x, \frac{d^6v}{dx^6} = -\sin x$$

$$\frac{dy}{dx} = e^x(\sin x + \cos x)$$

$$\frac{d^6y}{dx^6} = e^x(\sin x + 6\cos x - 15\sin x - 20\cos x + 15\sin x + 6\cos x - \sin x)$$

$$= -8e^x\cos x$$

$$\frac{d^6y}{dx^6} + 8\frac{dy}{dx} + 8y = -8e^x(\cos x) + 8e^x(\sin x + \cos x) - 8e^x(\sin x) = 0$$

22 $\frac{1}{2}$

23 $\lim_{x\to 0}\frac{\ln x}{\frac{1}{x}} = \lim_{x\to 0}\frac{\frac{1}{x}}{\frac{-1}{x^2}} = \lim_{x\to 0}(-x) = 0$

24 $\frac{1}{2}$

25 $\lim_{x\to 0}\frac{e^x - \cos x}{x} = \lim_{x\to 0}\frac{e^x + \sin x}{1} = 1$

26 **a** $t = \tan\frac{x}{2} \Rightarrow dx = \frac{2}{1 + t^2}dt,$

$$\frac{1}{1 - \sin x + \cos x} = \frac{1}{1 - \frac{2t}{1 + t^2} + \frac{1 - t^2}{1 + t^2}} = \frac{1 + t^2}{2(1 - t)}$$

$$\Rightarrow \int\frac{1}{1 - \sin x + \cos x}dx = \int\frac{1 + t^2}{2(1 - t)} \times \frac{2}{1 + t^2}dt$$

$$= \int\frac{1}{1 - t}dt$$

b $\int_0^{\frac{\pi}{4}}\frac{1}{1 - \sin x + \cos x}dx = 0.535$ (3 d.p.)

27 $t = \tan\frac{x}{2} \Rightarrow dx = \frac{2}{1 + t^2}dt$

$$\frac{1}{3\sin x - 4\cos x} = \frac{1}{3\left(\frac{2t}{1 + t^2}\right) - 4\left(\frac{1 - t^2}{1 + t^2}\right)} = \frac{1 + t^2}{2(2t - 1)(t + 2)}$$

$$\int\frac{1}{3\sin x - 4\cos x}dx = \int\frac{1 + t^2}{2(2t - 1)(t + 2)} \times \frac{2}{1 + t^2}dt$$

$$= \int\frac{2}{5(2t - 1)} - \frac{1}{5(t + 2)}dt$$

$$\Rightarrow \int_{\frac{\pi}{2}}^{\frac{7\pi}{6}}\frac{1}{3\sin x - 4\cos x}dx$$

$$= \frac{1}{5}\left[\ln\left|2\tan\frac{x}{2} - 1\right| - \ln\left|\tan\frac{x}{2} + 2\right|\right]_{\frac{\pi}{2}}^{\frac{7\pi}{6}} = \frac{1}{5}\ln(6 + 5\sqrt{3})$$

$a = 6, b = 5$

28 0.734 (3 s.f.)

29 **a** 0.2 **b** 2.065, 2.406

30 £8063

31 1.537 (3 d.p.)

32 **a** $y_1 = 1 + 0.2\cos(1) = 1.108$ (3 d.p.) **b** 0.964 (3 d.p.)

33 10 660

34 **a** 0.8 **b** 0.4413 (3 d.p.)

35 2.1, 2.202 (3 d.p.), 2.298 (3 d.p.)

36 2.94

37 **a** 1.09 (3 s.f.) **b** Increase the number of intervals.

38 **a** 3.8637 (4 d.p.)

b $u = x \Rightarrow \frac{du}{dx} = 1$ and $v' = \cosh x \Rightarrow v = \sinh x$

LHS $= x\sinh x - \cosh x = [x\sinh x - \cosh x]_1^2$

$= -\frac{3}{2}e^{-2} + e^{-1} + \frac{1}{2}e^2 =$ RHS

c 0.11%

39 **a** $y = \frac{Ce^{2x} - 2x - 1}{4}$ **b** $y = \frac{9e^{2x} - 2x - 1}{4}$

40 **a** $y = vx, \frac{dy}{dx} = x\frac{dv}{dx} + v$

$$x\frac{dv}{dx} + v = \frac{(4x + vx)(x + vx)}{x^2} = 4 + 5v + v^2$$

$$\Rightarrow x\frac{dv}{dx} = 4 + 4v + v^2 = (2 + v)^2$$

b $v = -2 - \frac{1}{\ln x + c}$

41 **a** $y = vx, \frac{dy}{dx} = x\frac{dv}{dx} + v$

$$x\frac{dv}{dx} + v = \frac{3x - 4vx}{4x + 3vx} = \frac{3 - 4v}{4 + 3v}$$

$$x\frac{dv}{dx} = \frac{3 - 4v}{4 + 3v} - v = -\frac{3v^2 + 8v - 3}{3v + 4}$$

b $3v^2 + 8v - 3 = \frac{C}{x^2}$

c $y = xv \Rightarrow v = \frac{y}{x} \Rightarrow \frac{3y^2}{x^2} + \frac{8y}{x} - 3 = \frac{C}{x^2}$

$\Rightarrow 3y^2 + 8yx - 3x^2 = C$

$y = 7$ at $x = 1 \Rightarrow C = 200$

Factorising the LHS, $(3y - x)(y + 3x) = 200$

42 **a** $\frac{d\mu}{dx} = -2y^{-3}\frac{dy}{dx} \Rightarrow \frac{dy}{dx} = -\frac{y^3}{2}\frac{d\mu}{dx}$

So $-\frac{1}{2}\frac{d\mu}{dx} + 2x\mu = xe^{-x^2}$

$$\Rightarrow \frac{d\mu}{dx} - 4x\mu = -2xe^{-x^2}$$

b $\mu = \frac{1}{3}e^{-x^2} + Ce^{2x^2}$ **c** $\frac{1}{y^2} = \frac{1}{3}e^{-x^2} + \frac{2}{3}e^{2x^2}$

43 a $\frac{dy}{dx} = v + x\frac{dv}{dx}, \frac{d^2y}{dx^2} = 2\frac{dv}{dx} + x\frac{d^2v}{dx^2}$

So $x^2\left(x\frac{d^2v}{dx^2} + 2\frac{dv}{dx}\right) - 2x\left(v + x\frac{dv}{dx}\right) + (2 + 9x^2)vx = x^5$

$\Rightarrow x^3\frac{d^2v}{dx^2} + 9x^3v = x^5 \Rightarrow \frac{d^2v}{dx^2} + 9v = x^2$

b $v = A\cos 3x + B\sin 3x + \frac{1}{9}x^2 - \frac{2}{81}$

c $y = Ax\cos 3x + Bx\sin 3x + \frac{1}{9}x^3 - \frac{2}{81}x$

44 a $2t^{\frac{1}{2}}\frac{dy}{dt}$

b $4t\frac{d^2y}{dt^2} + 2\frac{dy}{dt} + \left(6t^{\frac{1}{2}} - \frac{1}{t^{\frac{1}{2}}}\right)2t^{\frac{1}{2}}\frac{dy}{dt} - 16ty = 4te^{2t}$

$\Rightarrow 4t\frac{d^2y}{dt^2} + 12t\frac{dy}{dt} - 16ty = 4te^{2t}$

c $y = Ae^{x^2} + Be^{-4x^2} + \frac{1}{6}e^{2x^2}$

$\Rightarrow \frac{d^2y}{dt^2} + 3\frac{dy}{dt} - 4y = e^{2t}$

45 a $t\frac{dy}{dt}$

b $\frac{d^2y}{dx^2} = \frac{dt}{dx} \times \frac{d}{dt}\left(\frac{dy}{dx}\right) = t\frac{d}{dt}\left(t\frac{dy}{dt}\right)$

$= t\left(\frac{dy}{dt} + t\frac{d^2y}{dt^2}\right) = t^2\frac{d^2y}{dt^2} + t\frac{dy}{dt}$

c $\left(t^2\frac{d^2y}{dt^2} + t\frac{dy}{dt}\right) - (1 - 6t)t\frac{dy}{dt} + 10yt^2 = 5t^2\sin 2t$

$\Rightarrow \frac{d^2y}{dt^2} + 6\frac{dy}{dt} + 10y = 5\sin 2t$

d $y = e^{-3e^x}(A\cos(e^x) + B\sin(e^x)) + \frac{1}{6}\sin(2e^x) - \frac{1}{3}\cos(2e^x)$

46 a $\frac{dy}{dt} = -2x^{-3}\frac{dx}{dt}, \frac{d^2y}{dt^2} = 6x^{-4}\left(\frac{dx}{dt}\right)^2 - 2x^{-3}\frac{d^2x}{dt^2}$

Divide the differential equation by $-x^4$:

$-2x^{-3}\frac{d^2y}{dt^2} + 6x^{-4}\left(\frac{dx}{dt}\right)^2 = -x^{-2} + 3$

$\Rightarrow \frac{d^2y}{dt^2} = -y + 3 \Rightarrow \frac{d^2y}{dt^2} + y = 3$

b $y = A\cos t + B\sin t + 3$ **c** $x = \frac{1}{\sqrt{\cos t + 3}}$ **d** $\frac{1}{\sqrt{2}}$

Challenge

1 $\frac{\tan x + \tan y}{\cot x + \cot y} \equiv \frac{\left(\frac{2t}{1-t^2}\right) + \left(\frac{2s}{1-s^2}\right)}{\left(\frac{1-t^2}{2t}\right) + \left(\frac{1-s^2}{2s}\right)} \equiv \frac{4st}{(1 - s^2)(1 - t^2)}$

$\equiv \left(\frac{2t}{1 - t^2}\right)\left(\frac{2s}{1 - s^2}\right) \equiv \tan x\tan y$

2 $\frac{d^n}{dx^n}(x^3e^x\cosh x) = \sum_{k=0}^{n}\binom{n}{k}\frac{d^k}{dx^k}(x^3)\frac{d^{n-k}}{dx^{n-k}}(e^x\cosh x)$

$= \sum_{k=0}^{3}\binom{n}{k}\frac{d^k}{dx^k}(x^3)\,2^{n-k-1}e^{2x}$

$= e^{2x}2^{n-4}\left(6\binom{n}{3} + 12x\binom{n}{2} + 12x^2\binom{n}{1} + 8x^3\right)$

$= 2^{n-4}e^{2x}(8x^3 + 12nx^2 + 6n(n - 1)x + n(n - 1)(n - 2))$

3 a $\frac{du}{dx} = u^3 \Rightarrow \frac{1}{u^2} = B - 2x \Rightarrow u = (B - 2x)^{-\frac{1}{2}}$

Then integrate both sides with respect to x.

b $A = \frac{3}{2}, B = \frac{9}{4}$

Exam-style practice: AS level

1 $x > \sqrt{3}, -\sqrt{3} < x < -1, x < -3$

2 a If $t = \tan\frac{x}{2}$, then $\sin x = \frac{2t}{1 + t^2}$ and $\cos x = \frac{1 - t^2}{1 + t^2}$

Hence, $2\sin x - 5\cos x = 2\frac{2t}{1 + t^2} - 5\frac{1 - t^2}{1 + t^2} = 2$

$4t - 5(1 - t^2) = 2(1 + t^2) \Rightarrow 4t - 5 + 5t^2 = 2(1 + t^2)$

$\Rightarrow 3t^2 + 4t - 7 = 0$

b $x = 3.95, \frac{\pi}{2}$

3 1.195

4 a $k = 4$ **b** $x = -4$ **c** 33.7

5 a $-14\mathbf{i} - 5\mathbf{j} - 6\mathbf{k}$ **b** 8.02

c 38 dice **d** Plastic wastage

Exam-style practice: A level

1 a $7x + 2y + 4z = 7$ **b** $\frac{104}{3}$ **c** 0.930 radians

2 a 0.687 78 **b** 0.687 95 **c** 0.02% error

3 a $xy\frac{dy}{dx} + 3x^2 + y^2 \Rightarrow \frac{dy}{dx} + \frac{3x}{y} + \frac{y}{x}$ (1)

$y = zx \Rightarrow \frac{dy}{dx} = z + x\frac{dz}{dx}$ (2)

Substituting (2) into (1) gives: $z + x\frac{dz}{dx} + \frac{3}{z} + z = 0$

$\Rightarrow x\frac{dz}{dx} + \frac{3}{z} + 2z = 0 \Rightarrow x\frac{dz}{dx} + \frac{3 + 2z^2}{z} = 0$

b $3x^4 + 2x^2y^2 = 53$

c $x = 2.050 = 205$ metres

d Velocity of jumper tends to infinity as distance from top of the cliff tends to 0. Hence the model is unsuitable for very small values of x.

4 a L'Hospital's rule is only applicable for the limits of functions which tend to $\frac{\pm\infty}{\pm\infty}$ or $\frac{0}{0}$

The function given tends to $\frac{1}{0}$, hence L'Hospital's rule is cannot be used.

b $-\frac{14}{29}$

5 a $\frac{x^2}{a^2} - \frac{y^2}{b^2} = 1$

Substitute in $y = mx + c$: $\frac{x^2}{a^2} - \frac{(mx + c)^2}{b^2} = 1$

$\Rightarrow b^2x^2 - a^2(mx+c)^2 = a^2b^2$

$\Rightarrow b^2x^2 - a^2(m^2x^2 + 2mc + c^2) = a^2b^2$

$\Rightarrow (b^2 - a^2m^2)x^2 - 2mca^2x - a^2(c^2 + b^2) = 0$

This is in the form of a quadratic equation.

For $y = mx + c$ to be a tangent, discriminant = 0:

$4m^2c^2a^4 = -4a^2(b^2 - a^2m^2)(c^2 + b^2)$

$\Rightarrow m^2c^2a^2 = -b^4 - b^2c^2 + a^2m^2c^2 + a^2m^2b^2$

$\Rightarrow b^2 + c^2 = a^2m^2$

b $y = x + 1, y = -\frac{17}{11}x + \frac{67}{11}$

6 $y = 1 + x - \frac{3x^2}{2} + \frac{2x^2}{3}$

7 $\{x: x < -1 - \sqrt{7}\} \cup \{x: -\sqrt{6} < x < \sqrt{7} - 1\}$

8 For $y = e^x\sin x$, let $u = e^x$.

Hence $\frac{d^ku}{dx^k} = e^x$ for all values of k

Let $v = \sin x$, hence $\frac{dv}{dx} = \cos x, \frac{d^2v}{dx^2} = -\sin x,$

$\frac{d^3v}{dx^3} = -\cos x, \frac{d^4v}{dx^4} = \sin x, \frac{d^5v}{dx^5} = \cos x, \frac{d^6v}{dx^6} = -\sin x$

Apply Leibnitz's theorem:

$e^x\sin x + 6e^x\cos x - 15e^x\sin x - 20e^x\cos x + 15e^x\sin x$

$+ 6e^x\cos x - e^x\sin x = -8e^x\cos x = \frac{d^6y}{dx^6}$

$8\frac{dy}{dx} = 8e^x(\cos x + \sin x)$

Hence $\frac{d^6y}{dx^6} + 8\frac{dy}{dx} = -8e^x\cos x + 8e^x\cos x + 8e^x\sin x$

$= 8e^x\sin x = 8y$

Index